Joel D. Monnett

MOLECULAR MOTION IN POLYMERS BY ESR

MMI PRESS SYMPOSIUM SERIES

A series of monographs based on special symposia held at the Michigan Molecular Institute

Editor: Hans-Georg Elias

Volume 1 MOLECULAR MOTION IN POLYMERS BY ESR
Edited by Raymond F. Boyer and Steven E. Keinath

Volume 2 BLOCK COPOLYMERS: SCIENCE AND TECHNOLOGY
Edited by Dale J. Meier

Additional volumes in preparation

ISSN 0195-3966

The publisher will accept continuation orders for this series, which may be cancelled at any time and which provide for automatic billing and shipping of each title in the series upon publication. Please write for details.

MMI PRESS SYMPOSIUM SERIES continues the MIDLAND MACRO–MOLECULAR MONOGRAPHS Series previously published by Gordon and Breach, Science Publishers, Inc. The following titles from that series are available from Gordon and Breach, One Park Avenue, New York, New York 10016 U.S.A.:

Volume 1 TRENDS IN MACROMOLECULAR SCIENCE
Edited by Hans-Georg Elias

Volume 2 ORDER IN POLYMER SOLUTIONS
Edited by Karel Šolc

Volume 3 POLYMERIZATION OF ORGANIZED SYSTEMS
Edited by Hans-Georg Elias

Volume 4 MOLECULAR BASIS OF TRANSITIONS AND RELAXATIONS
Edited by Dale J. Meier

Volume 5 POLYMERIC DELIVERY SYSTEMS
Edited by Robert J. Kostelnik

Volume 6 FLOW-INDUCED CRYSTALLIZATION IN POLYMER SYSTEMS
Edited by Robert L. Miller

Volume 7 SILYLATED SURFACES
Edited by Donald E. Leyden and Ward T. Collins

ISSN 0141-0342

Papers presented at the Eighth
Midland Macromolecular Meeting
held August 21-25, 1978
in Midland, Michigan

MOLECULAR MOTION IN POLYMERS BY ESR

Edited by
Raymond F. Boyer
and
Steven E. Keinath

Michigan Molecular Institute
(formerly Midland Macromolecular Institute)

Published for MMI Press by
harwood academic publishers
chur london new york

Harwood Academic Publishers GmbH
Poststrasse 22
7000 Chur, Schweiz

Editorial Office for the United Kingdom:
Chansitor House
37/38, Chancery Lane
London WC2A 7EL

Editorial Office for the United States of America:
Post Office Box 786
Cooper Station
New York, New York 10003

Library of Congress Cataloging in Publication Data

Midland Macromolecular Meeting, 8th, 1978.
Molecular motion in polymers by ESR.

(MMI Press symposium series ; v. 1 ISSN 0195-3966)
Bibliography: p.
Includes index.
1. Polymers and polymerization—Congresses.
2. Electron paramagnetic resonance spectroscopy—Congresses. I. Boyer, Raymond F. II. Keinath, Steven E., 1954- III. Michigan Molecular Institute.
IV. Title. V. Series.
QD380.M53 1978 547'.84 79-67768
ISBN 3-7186-0012-9

ISBN 3-7186 0012-9. ISSN 0195-3966. Library of Congress catalog card number 79-67768.

Printed in the United States of America

This volume is dedicated to those individuals who have pioneered the applications of spin probe and spin label research to the study of molecular motion and structure in synthetic and biological macromolecules.

Preface

This book presents the collected proceedings of the eighth Midland Macromolecular Symposium, held at MMI on August 21–25, 1978. The contents reflect the research efforts of eleven invited speakers and four contributing speakers. Prof. Buchachenko was unable to attend the meeting but has submitted two coauthored papers on his research activities in this field. Two general discussion sessions were held during the symposium, the proceedings of which are presented at the end of this volume. The discussions give a unique presentation of the current state of thought and development in the field of polymer research using electron spin resonance techniques. Brief question and answer periods followed each talk; these discussions are appended to the respective papers and also reflect currently accepted beliefs and sometimes new ideas and uncertain hunches of the speakers.

The main emphasis of the symposium concerned the use of spin label and spin probe techniques to characterize motion, reactivity, and structure in polymer solutions and melts. While most of the contributions deal with work on synthetic polymers, a few authors have addressed the subject of biological polymers. Extensive coverage was not attempted in the areas of biological macromolecules, radiation damage, fracture mechanics, or free radical polymerization kinetics.

There have been a number of ESR symposia in recent years on various topics, in many cases placing emphasis on transient free radicals. This symposium appears especially timely in its coverage of the research on polymer transitions and molecular dynamics utilizing stable free radical species. This field is generating current widespread interest and is in a particularly dynamic state of growth. Many of the investigations reported herein had their beginnings well within the past decade.

The book is conveniently divided into six sections. The papers can be read individually, although the reader may find the introductory chapters helpful in providing a useful background to ESR techniques, instrument capabilities, and spin labeling methods. Three sections are devoted to spin label and spin probe studies of chain dynamics and polymer transitions in the solid state and in solution. The more adventurous reader may enjoy the section on new techniques. The general discussion sessions provide new ideas and convey the spirit and dynamism of the symposium.

We would like to express our sincere thanks to all of the speakers and participants at the symposium, and to all those individuals who have aided our effort in publishing this volume.

Midland, Michigan
April 1979

RAYMOND F. BOYER
STEVEN E. KEINATH

LIST OF PARTICIPANTS

Ms. Susan Barnum, Midland Macromolecular Institute, Midland, Michigan 48640

Prof. Lawrence J. Berliner, Ohio State University, Columbus, Ohio 43210

Prof. Albert M. Bobst, University of Cincinnati, Cincinnati, Ohio 45221

Dr. Raymond F. Boyer, Midland Macromolecular Institute, Midland, Michigan 48640

Dr. Ian M. Brown, McDonnell Douglas Research Laboratories, St. Louis, Missouri 63166

Prof. A. L. Buchachenko, Institute of Chemical Physics, Academy of Sciences of the USSR, Moscow, USSR

Prof. Anthony T. Bullock, University of Aberdeen, Old Aberdeen, Scotland

Prof. G. Gordon Cameron, University of Aberdeen, Old Aberdeen, Scotland

Dr. Timothy Chen, Midland Macromolecular Institute, Midland, Michigan 48640

Dr. Robert B. Clarkson, Varian Associates, Inc., Palo Alto, California 94303

Ms. Lisa R. Denny, Midland Macromolecular Institute, Midland, Michigan 48640

Prof. George W. Eastland, Saginaw Valley State College, University Center, Michigan 48710

Dr. Hans-G. Elias, Midland Macromolecular Institute, Midland, Michigan 48640

Dr. Mohamed El-Sabbah, Midland Macromolecular Institute, Midland, Michigan 48640

Dr. Bonnie J. Epperson, Becton, Dickinson & Company, Research Triangle Park, North Carolina 27709

Dr. Amitava Gupta, Jet Propulsion Laboratory, California Institute of Technology, Pasadena, California 91103

Dr. John R. Harbour, Xerox Research Center of Canada, Ltd., Mississauga, Ontario, Canada

Dr. Arthur Heiss, Bruker Instruments, Inc., Billerica, Massachusetts 08121

Dr. Peter Hellman, Midland Macromolecular Institute, Midland, Michigan 48640

Dr. James S. Hyde, Milwaukee County Medical Complex, Milwaukee, Wisconsin 53226

Mr. Ira W. Hutchison, Midland Macromolecular Institute and Dow Corning Corporation, Midland, Michigan 48640

Dr. Leszek Jarecki, Midland Macromolecular Institute, Midland, Michigan 48640
Mr. Steven E. Keinath, Midland Macromolecular Institute, Midland, Michigan 48640
Prof. Philip L. Kumler, State University of New York, Fredonia, New York 14063
Prof. Naoshi Kusumoto, Kumamoto University, Kumamoto, Japan
Mr. William Loris, Midland Macromolecular Institute, Midland, Michigan 48640
Dr. D. B. Losee, Philip Morris Research Center, Richmond, Virginia 23261
Dr. Dale J. Meier, Midland Macromolecular Institute, Midland, Michigan 48640
Dr. Robert L. Miller, Midland Macromolecular Institute, Midland, Michigan 48640
Prof. Wilmer G. Miller, University of Minnesota, Minneapolis, Minnesota 55455
Ms. Cindy Peck, Midland Macromolecular Institute, Midland, Michigan 48640
Prof. Bengt Rånby, Royal Institute of Technology, Stockholm, Sweden
Dr. Sabz Ali, Midland Macromolecular Institute, Midland, Michigan 48640
Dr. Peter M. Smith, American Cyanamid Company, Stamford, Connecticut 06904
Prof. Junkichi Sohma, Hokkaido University, Sapporo, Japan
Dr. Karel Šolc, Midland Macromolecular Institute, Midland, Michigan 48640
Prof. Pertti Törmälä, Tampere University of Technology, Tampere, Finland
Dr. K. Varadarajan, Kent State University, Kent, Ohio 44242
Dr. Richard W. Von Korff, Midland Macromolecular Institute, Midland, Michigan 48640
Dr. Donald Ware, Bruker Instruments, Inc., Billerica, Massachusetts 08121
Dr. Robert J. Warner, Midland Macromolecular Institute, Midland, Michigan 48640
Prof. Arthur Yelon, Ecole Polytechnique, Montreal, Quebec, Canada

Contents

INTRODUCTION AND HISTORICAL OVERVIEW

Applications of ESR to Polymer Research—An Overview of Important Developments

Bengt Rånby

Department of Polymer Technology, The Royal Institute of Technology, S-100 44 Stockholm, Sweden

The development of ESR spectroscopy from the first observations by Zavoisky in 1945 and the first applications to polymer problems in 1959–60 is briefly reviewed. The introduction of flow systems, the improvements in stability and sensitivity of the ESR instruments, the application of on-line computers to ESR instruments for recording and simulation of spectra, and the development of stable free radicals as spin probes, spin traps, and spin labels in polymer systems has made ESR spectroscopy a new method of great potential in polymer research.

1. INTRODUCTION

Applications of ESR methods to polymer research was treated at a Nobel Symposium in Lidingö, Stockholm, in 1972[1]. The main emphasis at that meeting was on transient radicals during polymerization and during polymer degradation induced by high and low energy irradiation. More stable and long-lived radicals, e.g., nitroxides, were known at that time but applied to polymers only to a limited extent. Two papers at the symposium described such applications. An extensive review of ESR spectroscopy in polymer research was published as a book in 1977[2]. The two largest chapters of this book—together 184 pp. of totally 347 pp.—are still on the short-lived transient radicals formed during initiation and propagation of polymerization and during polymer degradation processes initiated by irradiation. The book also presents more recent topics like polymer reactions with reactive gases, polymer radicals formed during mechanical fracture, grafting and crosslinking, nitroxide radicals used as spin probes, spin traps and spin labels and other stable polymer radicals. The present ESR symposium is highly specialized by emphasizing spin probes and spin labels to study motion of polymers in solution and in bulk. This is a timely subject of large importance in polymer research.

2. EARLY APPLICATIONS

The first successful experimental observation of the ESR phenomenon as energy absorption was reported by Zavoisky in 1945[3]. The magnetic polarity of the electron was well-known in the 1930's and defined as a spin quantum number. The ESR observation required a homogeneous and constant magnetic field (H) of reasonable strength and an electromagnetic wave of matching frequency for resonance (ν) according to the basic equation:

$$\Delta E = h\,\nu = g_e\,\beta_e\,H \qquad (1)$$

where ΔE is the energy difference between the two spin states of the electron in the magnetic field (H), β_e the magnetic moment of the electron and g_e the free spin factor (the "Landé g-factor") of a value close to 2. For a frequency in the usual microwave region for radar (9500 MHz, about 3 cm wave-length) a magnetic field of about 3400 G is required for resonance absorption of energy by free electrons. Zavoisky used a much longer wave-length (25 m) and a weaker magnetic field (from 10 to about 300 G). The absorption lines he obtained were rather broad bands of electromagnetic loss vs. H (Fig. 1). These experiments were based on theories of paramagnetic susceptibility and relaxation by Gorter[4] and Frenkel[5]. Zavoisky's ESR experiments are part of the history of science.

An early note by Bamford in 1955 reports an unresolved ESR signal from acrylonitrile polymerization in the gel phase.

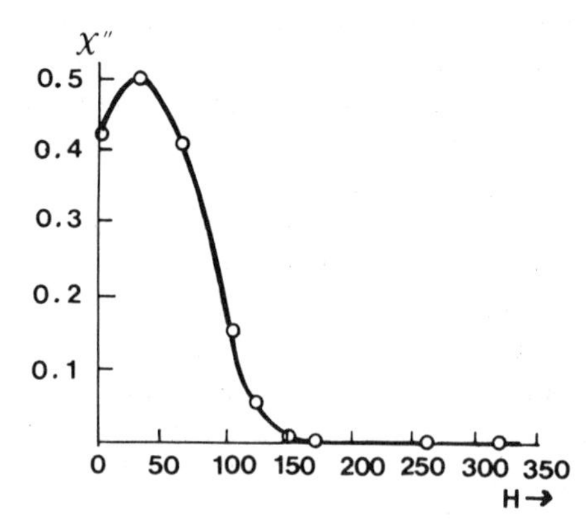

FIGURE 1 The first absorption maximum in an oscillating field of an unpaired electron spin, reported by Zavoisky[3] for Mn^{2+} ions in methanol at room temperature. The ESR spectrum for Mn^{2+} is a sextet, due to its nuclear spin ($\pm 5/2, \pm 3/2, \pm 1/2$). χ'' is paramagnetic absorption vs. magnetic field (H). (Reprinted from *J. Physics (USSR)*, **9**, 211 (1945) with permission of Akademii Nauk, CCCR).

The first successful applications of ESR measurements to polymer research were reported in 1959–60 independently by Bresler *et al.* in the USSR and Bamford and Ingram in the UK. Bresler *et al.* studied radical reactions during polymerization of common monomers like methyl methacrylate and vinyl acetate in the precipitated phase[6] (Fig. 2). In other ESR studies Bresler *et al.*[7] demonstrated the formation of radicals during mechanical disintegration of plastics, correctly interpreted as due to mechanical degradation of polymer chains[8] (Fig. 3). Bamford *et al.* made ESR studies of free radical polymerization by irradiating monomer crystals[9] and by forming "pop-corn" during rapid chain growth[10]. Under these conditions sufficiently high radical concentrations were obtained for the recording of rather well resolved ESR spectra. Some spectra were at first difficult to interpret, e.g., the 5- and 9-line spectra from polymerizing methyl methacrylate and poly(methyl methacrylate) degraded by ionizing irradiation. The radical structure is interpreted to be

$$-CH_2-\overset{\overset{\displaystyle CH_3}{|}}{\underset{\underset{\displaystyle C(=O)-O-CH_3}{|}}{C}}\cdot$$

in both cases[11]. The hydrogens of the CH_2 group are nonequivalent due to steric hindrance.

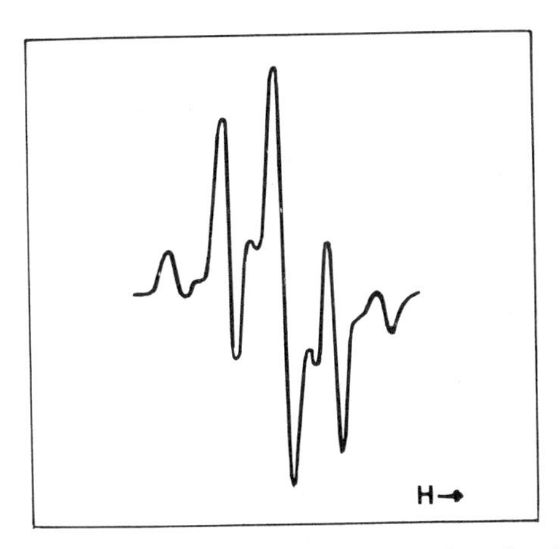

FIGURE 2 ESR spectrum of free radicals of methyl methacrylate, photoinitiated to polymerize and precipitated to a gel[6]. (Reprinted from *Vysokomol. Soyedin.*, **1**, 132 (1959) with permission of Akademii Nauk, CCCR).

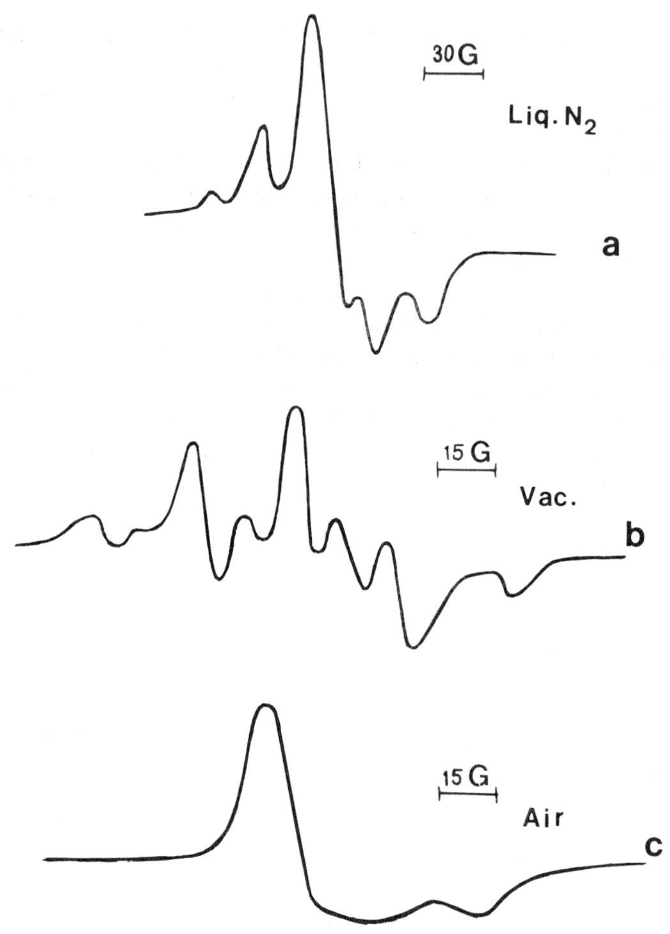

FIGURE 3 ESR spectra of free radicals formed in poly(methyl methacrylate) during mechanical disintegration: (a) in liquid N_2, (b) in vacuum, and (c) after access to air[7]. (Reprinted from *Z. tech.fiziki.*, **29**, 358 (1959) with permission of Akademii Nauk, CCCR).

3. DEVELOPMENT OF FLOW SYSTEMS

A serious problem in ESR studies of free radical polymerization in normal systems is the low concentration of radical species during the reaction. The transient radicals in solution are usually very reactive and short-lived. This gives low steady state concentrations of radicals also with high rates of initiation. The problem was recognized during the 1950's by Melville and others. ESR measurements in solution required radical concentrations of at least 10^{-6} M at that time while radical polymerizations normally occur with transient radical concentrations of 10^{-8} to 10^{-9} M. These difficulties were overcome in the early 1960's by development of flow systems, originally by Dixon and Norman[12] for ESR studies of oxidation reactions in solution. A flow system (Fig. 4) is particularly valuable for studies of redox initiated radical polymerizations and widely used[13, 14]. The reducing and oxidizing solutions, one or both solutions containing monomer, are mixed immediately before entering the cell which is located in the ESR cavity. High transient radical concentrations of growing chains (10^{-5} to 10^{-4} M) are easily reached. Several radical spectra in the same ESR recording, e.g., from certain monomers and during copolymerization, can be identified, interpreted and measured when a monomer blend is initiated in a flow cell. As an illustration Fig. 5 shows ESR spectra of radicals from allyl alcohol initiated in a flow cell with HO· radicals[15]. Three radicals are identified (two alkyl and one allyl radical). Even

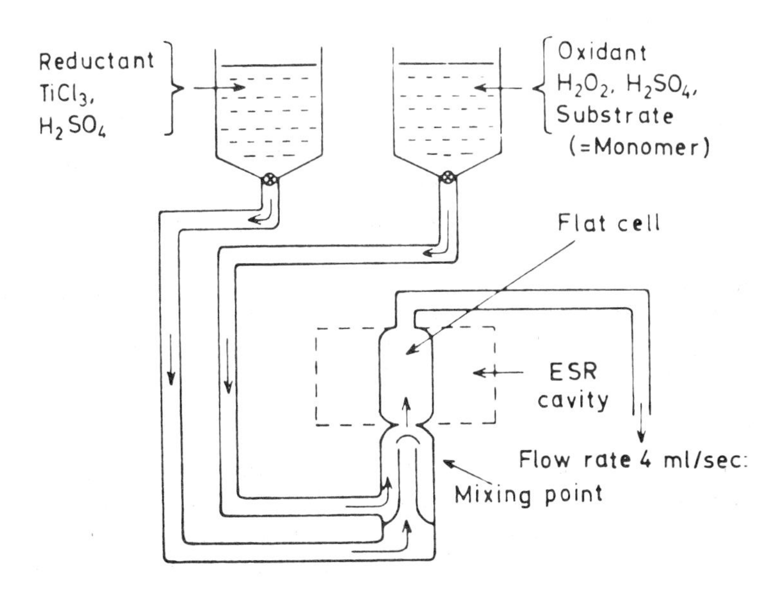

CHAIN INITIATION : 1) $H_2O_2 + Ti(III) \rightarrow Ti(IV) + OH^- + HO\cdot$

2) $HO\cdot + CH_2 = \underset{X}{CH} \rightarrow HO - CH_2 - \underset{X}{\dot{C}H}$

CONDITIONS : e.g. $[H_2O_2]$ = 0.15 MOLE/L, $[TiCl_3]$ = 0.007 MOLE/L,

$[H_2SO_4]$ = 0.022 MOLE/L

[MONOMER] = $10^{-2} \sim 10^{-1}$ MOLE/L, TEMP. 22 °C., AIR.

FIGURE 4 Flow system for production of high concentrations of free radicals[12]. (Reprinted from *J. Chem. Soc.*, 3119 (1963) with permission of The Chemical Society).

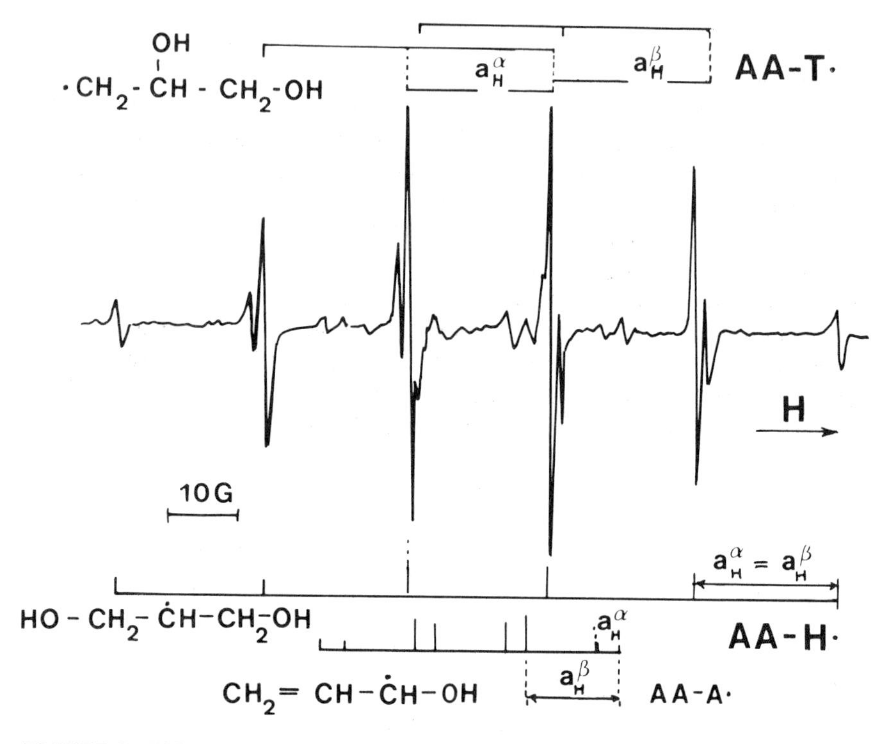

FIGURE 5 ESR spectrum of radicals from allyl alcohol initiated with HO· radicals in a flow system[15]. AA–T· refers to a "tail" radical of allyl alcohol, AA–H· to a "head" radical, and AA–A· to an allylic type radical of allyl alcohol (structures as given). (Reprinted from *Appl. Polym. Symp.*, **26**, 327 (1975) with permission of John Wiley and Sons, Inc.).

more complex spectra can be resolved, e.g., in copolymerization studies[16, 17] where five or more radicals may appear simultaneously and still be resolved and identified in the ESR recordings and their relative concentrations measured (Fig. 6). More complex flow methods are developed for consecutive additions of reactants, spin trapping, etc.

4. THE DEVELOPMENT OF ESR INSTRUMENTS

ESR instruments have been developed to a high degree of technical sophistication in recent years. Since the start of ESR research in Stockholm in 1963, we have operated three generations of instruments.

Our *first generation ESR instrument*, built by JEOL Co. in 1963 (Model JES-3B), was one of the first ESR instruments commercially available. We used it until 1972. It was only partially transistorized, had limited resolution

(~ 0.1 G) and low sensitivity (~ 10^{12} spin/G). Organic free radicals could be clearly detected at concentrations of ~10^{-5} M and their spectra recorded and analyzed at 10^{-4} to 10^{-3} M. The ESR spectra in Figs. 5 and 6 are recorded with the JES-3B instrument.

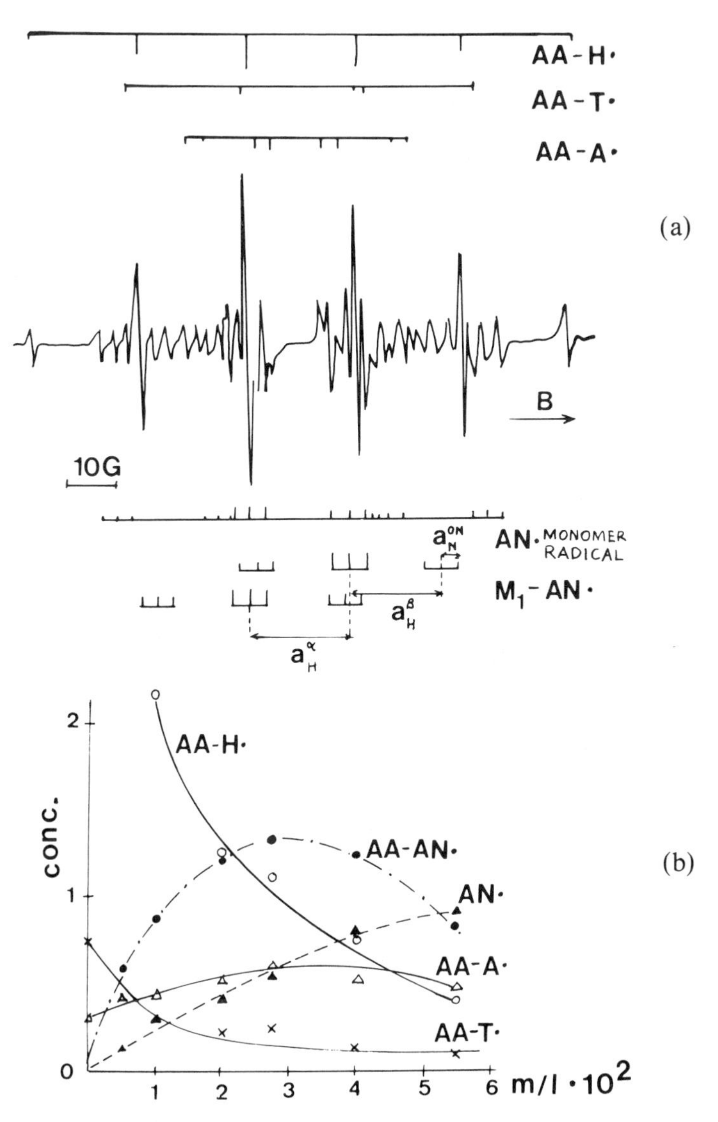

FIGURE 6 (a) ESR spectrum of radicals and (b) relative radical concentrations during copolymerization of allyl alcohol (AA) with acrylonitrile (AN) initiated with HO· radicals in a flow system[16, 17]. AA–H·, AA–T·, and AA–A· refer to "head", "tail", and allylic type radicals of allyl alcohol (compare caption to Fig. 5). (Reprinted from "Nobel Symposium 22", p. 53, 1973 with permission of Almqvist & Wiksell).

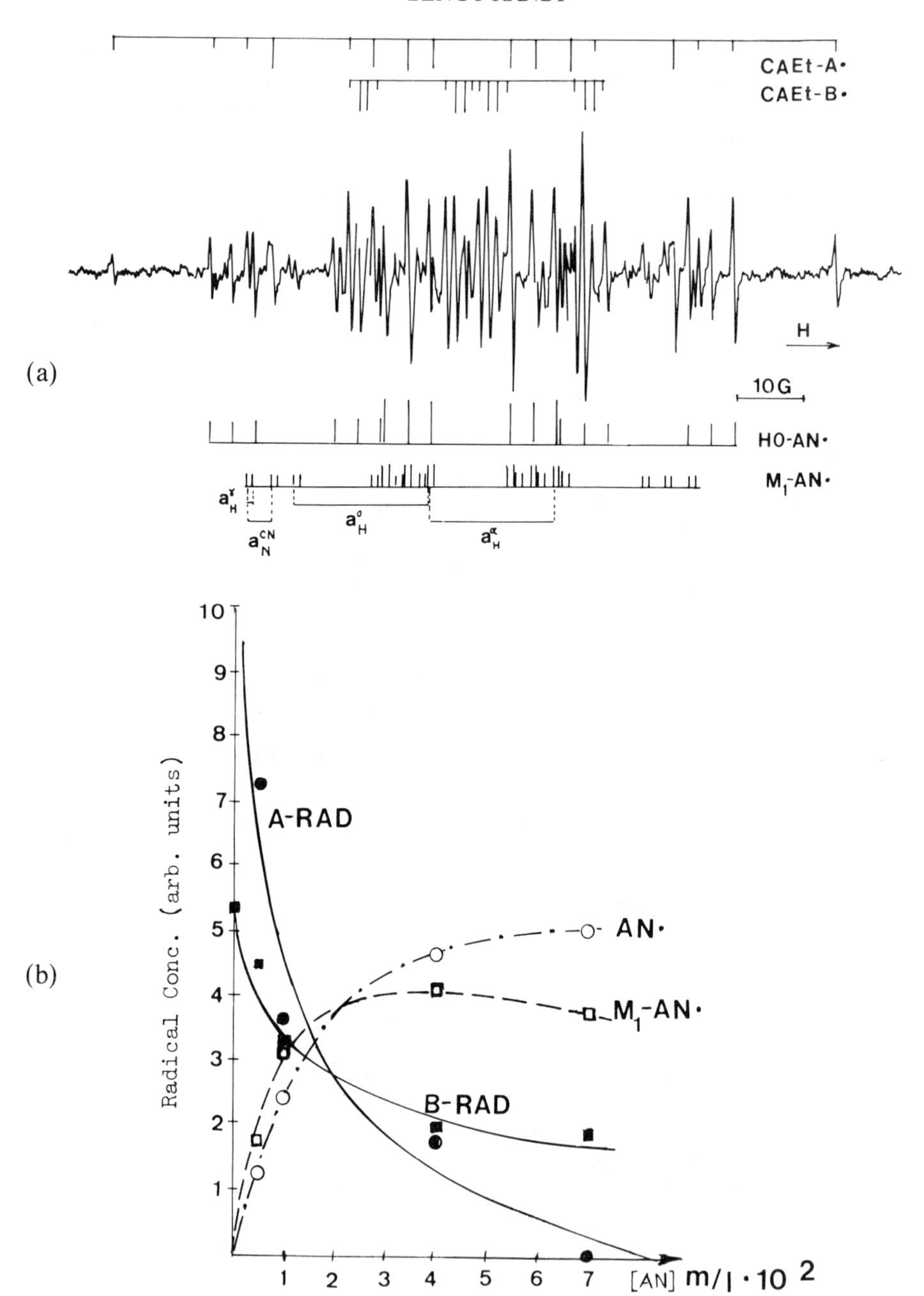

FIGURE 7 (a) ESR spectrum of radicals and (b) relative radical concentrations during copolymerization of ethylcrotonate (CAEt) and acrylonitrile (AN) initiated with HO· radicals in a flow system. [CAEt] = 0.055 mol/L, [AN] = 0.020 mol/L in (a) and [AN] varied in (b)[18]. (Reprinted from *Macromolecules,* **10**, 797 (1977) with permission of American Chemical Society).

Our *second generation ESR instrument* built in 1968 was also from JEOL Co. (Model JES-ME-1X). It is still in use. Its solid state electronics gives increased resolution (~0.03 G) and higher sensitivity (~5×10^{10} spin/G) at 100 KHz modulation with cylindrical cavity. It allows a variation of microwave output attenuation of 30 dB. Its maximum microwave output is 200 mW with 0.1 mW effect on the sample at maximum damping. The frequency stabilization is given as ~1 part in 10^6 in the X-band (~9400 MHz). The ESR spectrum in Fig. 7 is recorded with the JES-ME-1X instrument.

Our *third generation ESR instrument* was built by Bruker-Physik AG, Karlsruhe, West-Germany in 1976 (Type B-ER 420). It has a sensitivity of 2×10^{10} spin/G at 100 KHz modulation with a rectangular cavity. This means about ten times higher sensitivity than our second generation instrument. The resolution is better than 0.02 G. The stabilization of the magnetic field is better than 10^{-6} which means a fluctuation of ~0.003 G. This is again one power of ten better stability than our second generation instrument. The variation of microwave output attenuation is 60 dB as standard which means a minimum

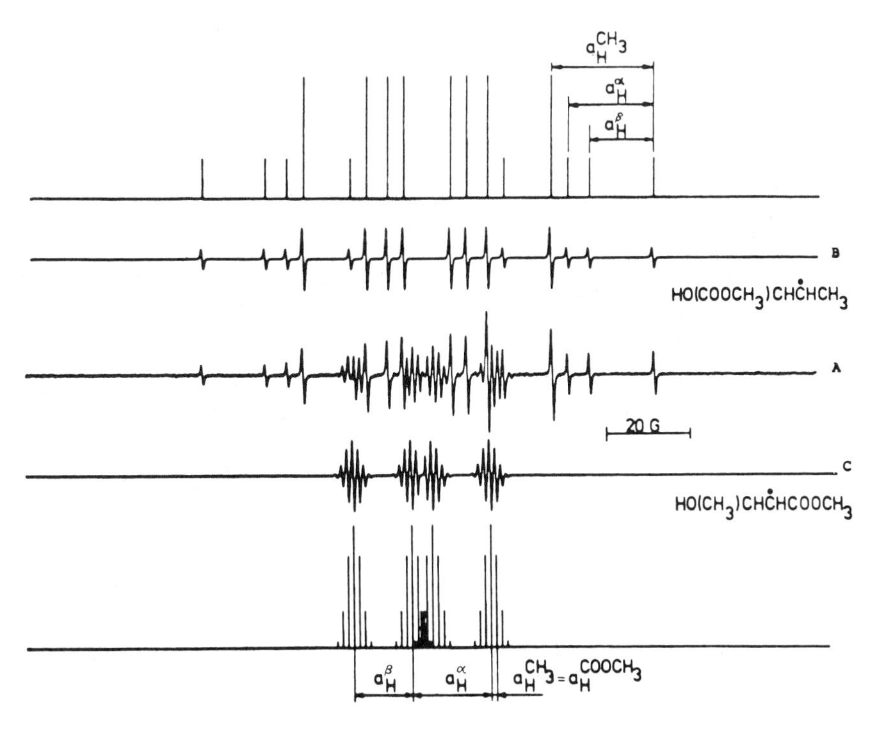

FIGURE 8 ESR spectrum of radicals from methylcrotonate (CAMe) initiated with HO· radicals in a flow system. A is a recording of an experimental spectrum, B and C are computer simulated spectra of the radicals given. [CAMe] = 0.055 mol/L[19].

microwave energy at the sample of 0.2 μW. A most important development is the automatic frequency control (AFC) which gives a lock stability for the standard cavity ($Q \sim 6000$) of 2 parts in 10^7 for 0 to 30 dB attenuation and 1 part in 10^5 for 50 dB. A composite ESR spectrum recorded with the B-ER 420 instrument is given in Fig. 8 together with simulated spectra for the components radicals.

5. COMPUTER APPLICATIONS TO ESR SPECTROSCOPY

To the Bruker ESR instrument (Type B-ER 420) a computer is attached on line. It is of the type NIC-1180 from the Nicolet Instrument Corp. The computer is a programmable laboratory data system (a minicomputer from 1975/76) with a 16 K semiconductor memory and 20-bit word length. It is complete and directly connected to the ESR instrument. The computer also has a cartridge disc memory pack (DIABLO) of the capacity 1.2 million 20-bit words, running at a speed of 50 K words per sec. The memory capacity corresponds to about 300 ESR spectra with 4000 recorded points per spectrum. The teletype attached is of type 33. The NIC-1180 computer is used for the following operations:

1. Recording of ESR spectra with 256 to 8000 points/spectrum and write-out of spectra (first derivative).

2. Derivation and integration of recorded spectra.

3. Addition and subtraction of recorded and simulated spectra.

4. Simulation of first order ESR spectra, i.e., stick spectra (position and intensity of lines). With line shape and line width fed in (Gaussian or Lorentzian), 1st derivative spectra can also be simulated as shown in Fig. 8.

5. Simulation of higher order ESR spectra, e.g., nitroxide spin probe, spin trap, and spin label spectra for different splittings (due to the nuclear spin of N) and different relaxation and correlation times.

A *new generation computer* for the Bruker B-ER 420 ESR instrument is available (Bruker Aspect 2000) this year. They have larger word length (24-bit words) which gives a large dynamic range (1 to 1.6×10^7). Aspect 2000 has the same disc memory pack (DIABLO) as NIC-1180 which means a capacity of about 250 ESR spectra with 4000 recorded points. The new computer (Aspect 2000) is easier to operate also for timesharing with other instruments. The advantages in comparison with the NIC-1180 are:

1. Easier data accumulation.

2. Fast display and extensive manipulation of data, i.e., for integration, spectrum expansion, and difference spectra.

3. Simulation of Gaussian or Lorentzian lines and line spectra.

4. *g*-Factor calibration of recorded and simulated spectra.

5. Resolution enhancement with Fourier transform.

6. Dual analog-to-digital converter with a high sampling rate (250 KHz or 100 KHz).

The computer applications to ESR spectroscopy opens new possibilities in polymer research, especially for the recording and interpretation of complex spectra of poor resolution.

6. THE APPLICATION OF SPIN PROBE, SPIN TRAP, AND SPIN LABEL TECHNIQUES

Nitroxides as a class of stable free radicals were discovered in 1956[20, 21]. They were used as spin probes in polymers in 1969[22, 23] and as spin traps from 1970[24,25,26,27]. Spin-labeling of macromolecules was first applied to biopolymers, as reviewed by Berliner[28].

The use of *spin traps* is a most useful method to accumulate and "preserve" short-lived transient radicals. Sufficiently high concentrations are easily reached. The method also has disadvantages. When a nitroxide group is coupled to a radical, the unpaired electron is moved one or two atoms away from the original location. This decreases the hyperfine coupling from the original radical structure and makes it more difficult to identify. In addition the reactivity of the different radicals to nitroxide addition usually varies which may be misleading when concentrations of the original radicals are estimated.

The use of *spin probes* is a very simple and convenient method to study various types of molecular motion in a polymer system. It requires nitroxide radicals of sufficient solubility in the polymer matrix and stability for introduction into the polymer system, e.g., in a polymer melt or a polymer solution.

The use of *spin labels* is more difficult experimentally as it involves a chemical attachment of the nitroxide groups to the polymer chains. The results obtained regarding chain motion refer more directly to the polymer chains than spin probe measurements.

The field of nitroxide radicals in polymer applications until 1976 was reviewed in our ESR book[29]. Spin probe and spin label techniques are extensively described by others at this meeting and are not further treated here.

7. CONCLUSIONS

In the ESR applications to polymer research, initiated some 20 years ago, important progress has occured in four main areas.

1. Flow systems have increased radical concentrations in solution and made it possible to study initiation and propagation of radical polymerization using ESR.

2. Improvements in the sensitivity and resolving power of the ESR instruments have made it possible to identify and study free radicals in solution at concentrations of 10^{-6} mole/L with a spectral resolution better than 0.02 G.

3. With introduction of on-line computers to the ESR instruments, long series of ESR spectra can be recorded by rapid scanning, derivation and integration of spectra can be made, and ESR spectra of first order and higher orders can be simulated and recorded. Simulated and recorded spectra can be added and subtracted for easy identification of free radicals.

4. Applications of stable free radicals as spin probes, spin traps, and spin labels have opened new areas of polymer science for ESR research: accumulation of very short-lived transient radicals using spin traps for identification and new methods to study molecular motions in solution and bulk phase using spin probes and spin labels.

REFERENCES

1. "ESR Applications to Polymer Research", (eds. P.O. Kinell, B. Rånby, and V. Runnström-Reio), Nobel Symposium 22, Sept. 1972, Almqvist & Wiksell, Stockholm, 1973.
2. B. Rånby and J.F. Rabek, "ESR Spectroscopy in Polymer Research", Springer-Verlag, Berlin-Heidelberg-New York, 1977.
3. E. Zavoisky, *J. Physics (USSR)*, **9**, 211 (1945).
4. C. Gorter, *Phys. Zeitschr.*, **34**, 462 (1933), and C. Gorter and R. de L. Kronig, *Physica*, **3**, 1009 (1936).
5. J. Frenkel, *J. Physics (USSR)*, **9**, (1945).
6. S.E. Bresler, E.N. Kazbekov, E.M. Saminskii, *Vysokomol. Soedin.*, **1**, 132, 1374 (1959).
7. S.E. Bresler, S.N. Zhurkov, E.N. Kazbekov, E.M. Saminskii, and E.E. Tomashevskii, *Zh. Tekh. Fiz. (J. Tech. Phys. USSR)*, **29**, 358 (1959).
8. S.E. Bresler, et al., *Rubber Chem. Technol.*, **33**, 462, 469 (1960) (translation of ref. 6).
9. C.H. Bamford, A.D. Jenkins, M.C.R. Symons, and M.G. Townsend, *J. Polymer Sci.*, **34**, 181 (1959). Cf. a note on polyacrylonitrile radicals, *Nature*, **175**, 894 (1955).
10. C.H. Bamford, A.D. Jenkins, and J.C. Ward, *Nature*, **186**, 713 (1960).
11. D.J.E. Ingram, M.C.R. Symons, and M.G. Townsend, *Trans. Faraday Soc.*, **54**, 409 (1958).
12. W.T. Dixon and R.O.C. Norman, *Nature*, **196**, 891 (1962), and *J. Chem. Soc.*, 3119 (1963).
13. H. Fischer, *Adv. Polym. Sci.*, **5**, 463 (1968).
14. K. Takakura and B. Rånby, *Adv. Chem. Ser.* **No. 91**, 125 (1969).
15. B. Rånby, *Appl. Polymer Symp.* **No. 26**, 327 (1975).
16. Z. Izumi and B. Rånby, *Pure and Appl. Chem., Macromol. Chem.*, **8**, 107 (1973).
17. B. Rånby, "Nobel Symp. 22", (ref. 1), 53 (1973).
18. G. Canbäck and B. Rånby, *Macromolecules*, **10**, 797 (1977).
19. G. Canbäck and B. Rånby, unpublished data.
20. D.R. Johnson, M. Rogers, and G. Trappe, *J. Chem. Soc.*, 1093 (1956).
21. M. Rogers, *J. Chem. Soc.*, 2101 (1956).

22. V.B. Stryukov, Yu.S. Karimov, and E.G. Rozantsev, *Vysokomol. Soedin.*, **B 9**, 493 (1967) and further in *ibid.*, **A 10**, 626 (1968).
23. G.P. Rabold, *J. Polymer Sci., Al*, **7**, 1187, 1203 (1969).
24. M.J. Perkins, P. Ward, and A. Hortsfield, *J. Chem. Soc.*, **B**, 395 (1970), and further in M.J. Perkins, in "Essays on Free Radical Chemistry", London, Chem. Soc., 1970, p. 97.
25. E.G. Janzen, *Acc. Chem. Res.*, **4**, 31 (1971).
26. H. Fischer, *Acc. Chem. Res.*, **4**, 110 (1971).
27. C. Lagercrantz, *J. Phys. Chem.*, **75**, 3466 (1971).
28. L.J. Berliner, "Spin Labeling: Theory and Applications", Academic Press, New York, 1976.
29. B. Rånby and J.F. Rabek, ref. 2, chapter 11 (pp. 294–311).

DISCUSSION

R. F. Boyer (Midland Macromolecular Institute, Midland, Michigan): Will you please given an example of spin trapping?

B. Rånby: Several examples of spin trapping of initiating and propagating radicals are given in our book [B. Rånby and J. F. Rabek, "ESR Applications to Polymer Research", Springer-Verlag, Berlin, 1977, p. 301–305] I should mention that the following have been used

$C_6H_5{-}N{=}O \qquad (CH_3)_3{-}C{-}N{=}O \qquad \text{2,4,6-tri-}t\text{-Bu-}C_6H_2{-}N{=}O$

successfully in polymerization studies of methyl methacrylate, styrene, and vinyl acetate.

A. M. Bobst (University of Cincinnati, Cincinnati, Ohio): Will spin traps only trap free radicals or can they also interact with nucleophiles?

B. Rånby: Spin traps can react with other compounds but they usually only react with free radicals because of kinetic factors. Strong nucleophiles may compete with the spin traps.

G. G. Cameron (University of Aberdeen, Old Aberdeen, Scotland): Many spin traps (nitroso compounds), e.g., 2-methyl-2-nitroso propane will also react with good nucleophiles such as carbanions. This can be exploited for labeling polymers in reactions such as

$$\sim\!R^{\ominus}\,Li^{\oplus} + t\,BuNO \longrightarrow \sim\!R{-}\underset{t\,Bu}{\underset{|}{N}}{-}O^{\ominus}\,Li^{\oplus} \xrightarrow[\text{oxidation}]{\text{hydrolysis}} \sim\!R{-}\underset{t\,Bu}{\underset{|}{N}}{-}O\cdot$$

(nitroxide)

J. S. Hyde (Milwaukee County Medical Complex, Milwaukee, Wisconsin): Could you give an overview of the utility of microwave power saturation measurements to obtain structural, chemical, or motional information of free radicals in organic polymers?

B. Rånby: We consider it desirable to get ESR measurements using a very low level of microwave power on the sample (our new instrument: ~0.2 μW). Free radicals saturate at many different microwave power levels, e.g., certain radicals in polyethylene saturate in the mW range. We are planning a systematic study of the saturation effect using increasing microwave power on the samples, to try to identify different radicals present.

A. Gupta (Jet Propulsion Laboratory, California Institute of Technology, Pasadena, California): What is the minimum radical lifetime you can measure with the Nicolet 1180 computer, in a stopped-flow experiment, for example?

B. Rånby: The NIC-1180 computer can be programmed to sweep and record at speeds from 0.2 μsec to 16.7 sec per sampling point. If you use the setting, 1000 points per spectrum, you can record one ESR spectrum in 0.2 msec. This means that you can measure radicals with a minimum lifetime of about 1 msec. The new computer, ASPECT-2000, which we have on order from Bruker has about the same sweep speed but a more flexible sweep control, e.g., non-linear sweeps.

Spin Labels: Historical Overview and Some Future Prospects in Macromolecular Research

Lawrence J. Berliner

Department of Chemistry, The Ohio State University
140 W. 18th Avenue, Columbus, Ohio 43210

This chapter intends to quite briefly review the past history of spin labeling, to survey both current and proposed aspects of spin label and spin probe synthesis and finally to describe a few new problems and approaches, some of which may be covered in detail by the other contributors.

SPIN LABELING

Since the nitroxide moiety,

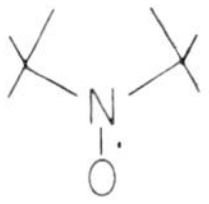

has been the most versatile, stable functionality found for ESR labeling studies we find that reference to a covalently bound spin *label* or noncovalently bound spin *probe* usually refers to nitroxide compounds. In fact, however, the first biological application incorporating an organic paramagnetic spin probe into a biological macromolecule utilized instead a phenothiazine based radical, chlorpromazine radical cation, which was shown by Ohnishi and McConnell[1] to intercalate DNA perpendicular to the fiber axis. Since this radical was relatively short lived, certainly in protic (solvent) environments, the need for a more versatile molecule was obvious.

The great wealth of information and development in the spin labeling field came with the utilization of stable di-*t*-alkyl nitroxides developed first and chronologically foremost by the groups of Hoffman and Henderson,[2] Rozantsev,[3] and Rassat.[4] The development in the applications, theory and chemistry has been phenomenal over the intervening 14 years since McConnell and coworkers published their first biochemical-biopolymer application with nitroxides.[5] The field as it applies to physical, organic, biological, polymer and biomedical chemistry has been recently compended in a two volume text by Berliner.[6,7]

PHYSICAL INFORMATION OBTAINABLE FROM NITROXIDE SPECTRA

While many of the parameters obtainable from ESR analysis of nitroxide spin labels and spin probes will manifest themselves in the specialized chapters of many of the other contributors, it is worthwhile to briefly outline them below.

A. Motion

While the magnetic resonance theory allows calculation of molecular tumbling rates, and in many cases the anisotropy of this tumbling, there are valuable conclusions to be drawn from the qualitative aspects of the spectra as well.

i) Kinetics—the change in dynamic state of a nitroxide group incorporated into a macromolecule (e.g., melting) may be followed by the growth and/or disappearance of the spectral lines that represent the intact and melted macromolecule, respectively.

ii) Site topography—the geometry (e.g., length and width) or hydrophobicity of a specific site incorporated nitroxide may be systematically varied in length or width (i.e., inclusion of one or more methylene groups in the 'arm' between nitroxide and matrix backbone). Changes in motion with variations in spin label structure represents the movement of the 'molecular dipstick' from the backbone to the 'surface' of the macromolecular matrix.

iii) Conformational changes (segmental motion)—cooperative or non-cooperative local or global structural changes in a macromolecule are monitored by the change in nitroxide tumbling rate with changes in its immediate molecular environment.

iv) Rotational correlation times—the theory is well based for calculating the rotational tumbling time of a nitroxide moiety.[8] There are at least two possible origins of this tumbling time to consider in order to properly ascribe its significance to the macromolecular system in which it resides.

a) The nitroxide tumbles faster than the macromolecule—since one always observes the nitroxide moiety, not the macromolecule, it is always wise to reason whether the calculated nitroxide rotational correlation time is at all close to that expected for the macromolecule. Of much significance to polymer structural studies is the aspect of *segmental motion,* where local oscillatory motions may be quite fast compared to an essentially immobile macromolecular matrix.

b) The nitroxide is as rigid as the macromolecular backbone—the correlation time equals that of the polymer. This time range will likely approach what we term the very slow tumbling region where the experimental techniques of Hyde must be employed.[9, 10]

B. Polarity

The nitroxide bond is polarizable by both solvent and e.g., extrinsic Lewis acids towards its ionic resonance form:

$$\ddot{N}-O\cdot \rightleftharpoons \dot{N}^{\oplus}-\ddot{O}:^{\ominus}$$

resulting in an increased electron-nuclear coupling constant. Thus as Table I exemplifies, the hyperfine coupling constant decreases going from very polar (e.g., water) towards apolar (hexane) environments.

Thus under ideal experimental conditions both the polarity of some internal polymer site may be probed with copolymerized nitroxide moieties as well as changes therein associated with mechanical or other distortions of the macromolecule.

C. Intra(macro)molecular distances

Magnetic resonance theory accurately describes the interactions between magnetic dipoles of known separation. These interactions may be electron spin-nuclear spin, electron spin-electron spin, and even electron spin-fluorophore.

Interaction	Sensitivity	Formulation	References
Nitroxide—Metal ion	10–15 Å	$1/r^6$	12,13
Nitroxide–Nitroxide	≤ 10 Å	$1/r^3$	14,15
Nitroxide–Nuclear spin	10–25 Å	$1/r^6$	16
Nitroxide–Fluorophore	4–6 Å	triplet quenching	17
		singlet quenching	18

D. Orientation

As Figure 1 displays, the hyperfine and g-tensor are anisotropic yielding angular dependent line splittings which change with the magnetic field

TABLE I
Isotropic ESR parameters for di-*t*-butyl nitroxide[a,b]

No.	Solvent	A_0	g_0
1	Hexane	15.10	2.0061
2	Heptane–pentane (1:1)[c]	15.13	2.0061
3	2-Hexene	15.17	2.0061
4	1,5-Hexadiene	15.30	2.0061
5	Di-*n*-propylamine	15.32	2.0061
6	Piperidine	15.40	2.0061
7	*n*-Butylamine	15.41	2.0060
8	Methyl propionate	15.45	2.0061
9	Ethyl acetate	15.45	2.0061
10	Isopropylamine	15.45	2.0060
11	2-Butanone	15.49	2.0060
12	Acetone	15.52	2.0061
13	Ethyl acetate saturated with water	15.59	2.0060
14	*N,N*-Dimethylformamide	15.63	2.0060
15	EPA[d](5:5:2)[c]	15.63	2.0060
16	Acetonitrile	15.68	2.0060
17	Dimethylsulfoxide	15.74	2.0059
18	*N*-Methylpropionamide	15.76	2.0059
19	2-Methyl-2-butanol	15.78	2.0059
20	EPA[d] (5:5:10)[c]	15.87	2.0060
21	1-Decanol	15.87	2.0059
22	1-Octanol	15.89	2.0059
23	*N*-methylformamide	15.91	2.0059
24	2-Propanol	15.94	2.0059
25	1-Hexanol	15.97	2.0059
26	1-Propanol	16.05	2.0059
27	Ethanol	16.06	2.0059
28	Methanol	16.21	2.0058
29	Formamide	16.33	2.0058
30	1,2-Ethanediol	16.40	2.0058
31	Ethanol–water (1:1)[c]	16.69	2.0057
32	Water	17.16	2.0056
33	10 *M* LiCl aqueous solution	17.52	2.0056

[a]Data taken from Ref. 11.

[b]All data measured at room temperature (23°–24°C). Estimated uncertainties are ±0.02 *G* and ±0.0001 for A_0 and g_0, respectively, relative to the standard dilute aqueous solution of di-*t*-butyl nitroxide for which $A_0 = 17.16$ and $g_0 = 2.0056$.

[c]By volume.

[d]EPA designates a mixture of ethyl ether (diethyl ether), isopentane (2-methylbutane), and alcohol (ethanol).

Reproduced with permission from J. D. Morrisett in "Spin Labeling: Theory and Applications," (L. J. Berliner, ed.), p. 307, Academic Press, New York, 1976.

direction. For example for a rigid oriented nitroxide the observed splitting, T_{obs}, is given by $T_{obs} = [T_{\parallel}^2\cos^2\theta + T_{\perp}^2\sin^2\theta]^{1/2}$ where θ is the angle between the z axis and the magnetic field direction while $T_{\parallel}$ and $T_{\perp}$ are the coupling constants along the *z*- and *x*- or *y*-directions, respectively ($T_{\parallel} = T_{zz}$, $T_{\perp} = T_{xx} = T_{yy}$). Figure 2 depicts the distinct spectral properties obtained for oriented

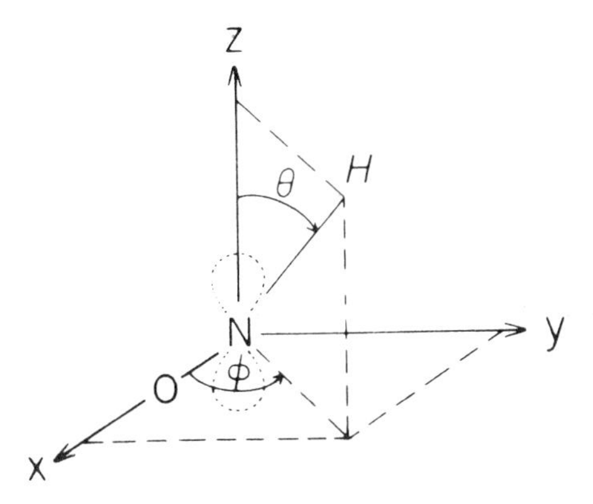

FIGURE 1 The molecular coordinate system and angles which define the external magnetic field direction, H, with respect to the nitroxide principal axis system x, y, and z. With permission from P.C. Jost and O.H. Griffith, *Meth. Enzymol.*, **49G**, 369 (1978).

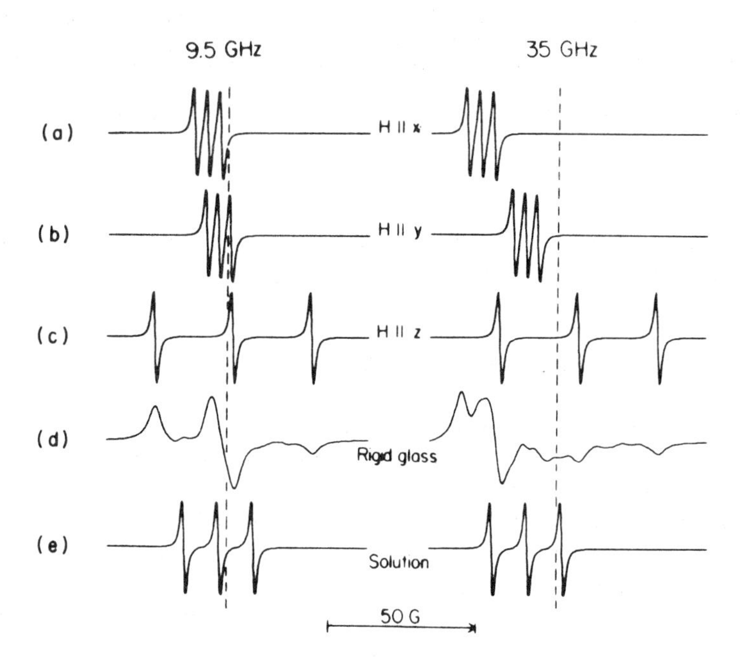

FIGURE 2 Calculated ESR spectra at 9.5 (x-band) and 35 GHz, respectively, for (a–c) the magnetic field along each axis of a rigid oriented nitroxide (d) a random collection of rigid nitroxides (powder spectrum), (e) a random collection undergoing rapid isotropic motion. The principal hyperfine and g-values used were $T_{xx} = 5.9$G, $T_{yy} = 5.4$G and $T_{zz} = 32.9$G; $g_{xx} = 2.0088$, $g_{yy} = 2.0058$, $g_{zz} = 2.0022$. From Ref. 6, p. 460 with permission.

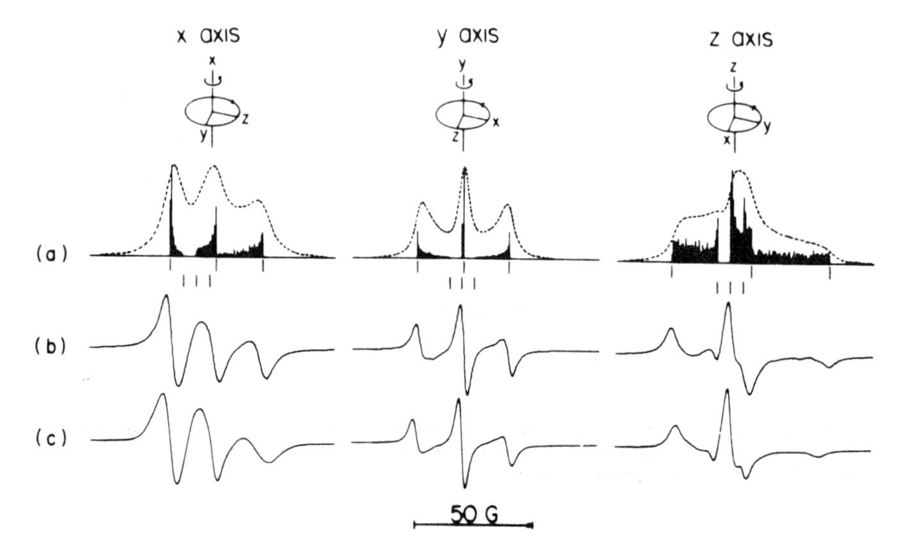

FIGURE 3 Theoretical line shapes and experimental examples of rapid rotation about the three princpal nitroxide axes. (a) Stick spectra and absorption spectra at 9.5 GHz, (b) first derivative spectra of (a), (c) experimental spectra for specific nitroxides: x-axis rotation, 2,2,6,6-tetramethyl-4-piperidinol-1-oxyl dodecanoate trapped in β-cyclodextrin; y-axis rotation, di-*t*-butyl nitroxide trapped in thiourea; z-axis motion, 7-doxyl stearate trapped in γ-cyclodextrin. From Ref. 6, p. 466 with permission.

(rigid) radicals in e.g., a host crystal, while Figure 3 exemplifies what occurs for a nitroxide undergoing anisotropic tumbling.

ORGANIC CHEMISTRY OF SPIN LABELS

Pertinent to the use of nitroxide spin labels in the study of synthetic organic polymers is their versatility as reagents for copolymerization, "labeling" of a preformed polymer or as initiators or terminators of a polymerization reaction. The following brief survey covers the most commonly employed nitroxide precursors followed by a discussion of several new aspects of nitroxide chemistry for future development.

Scheme I depicts the original series of nitroxide precursors as developed by the groups of Rozantsev[3] and Rassat[4] from triacetoneamine XVIII, a compound obtainable from acetone and NH_3.[21] The amino piperidine and hydroxy piperidine nitroxides, XX and XXII, have been used in several spin labeled polymers. During the past ten years a great deal of new nitroxide chemistry has emerged from the laboratory of Keana, who has introduced the

Scheme I

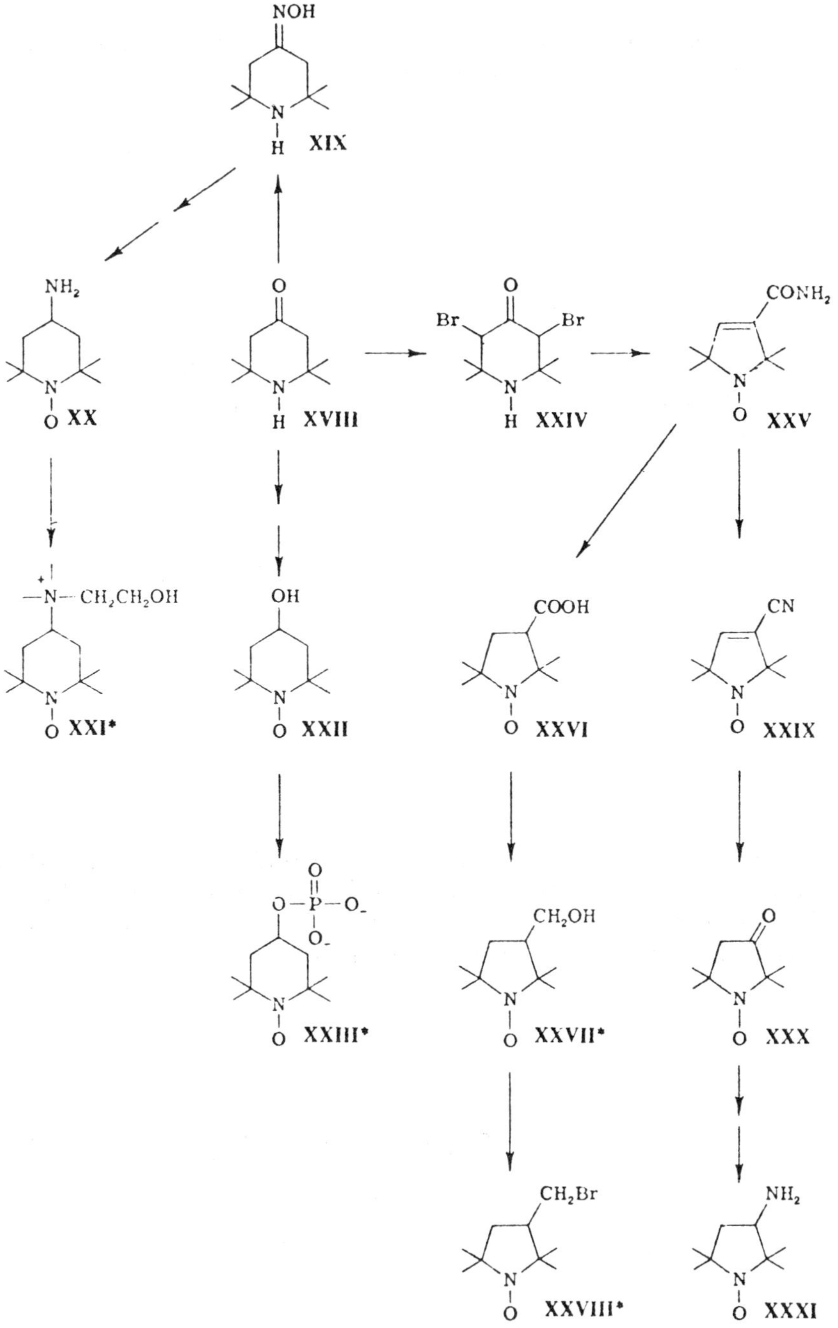

Reaction schemes leading to spin-label precursors.

Reproduced with permission from B. J. Gaffney in "Spin Labeling: Theory and Applications," (L. J. Berliner, ed.), p. 191, Academic Press, New York, 1976.

doxyl, proxyl and azethoxyl nitroxides shown below with the classical piperidine and pyrrolidine nitroxide structures.[19, 20]

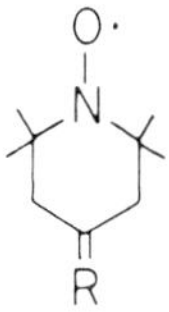

I (A Piperidine Nitroxide)

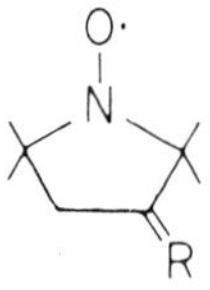

II (A Pyrrolidine Nitroxide)

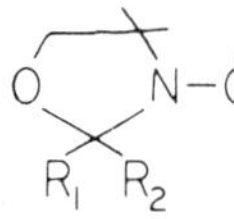

III (A Doxyl Nitroxide)

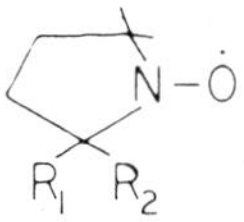

IV (A Proxyl Nitroxide)

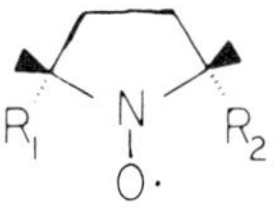

V_c (A cis Azethoxyl Nitroxide)

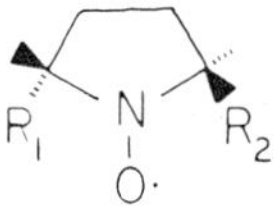

V_t (A trans Azethoxyl Nitroxide)

The doxyl nitroxides are obtained via an oxazolidine intermediate which results from the condensation of 2-amino-2-methyl propanol with a ketone. The oxazolidine is subsequently oxidized to the doxyl nitroxide with m-chloroperbenzoic acid (MCPA) in ether solvent:

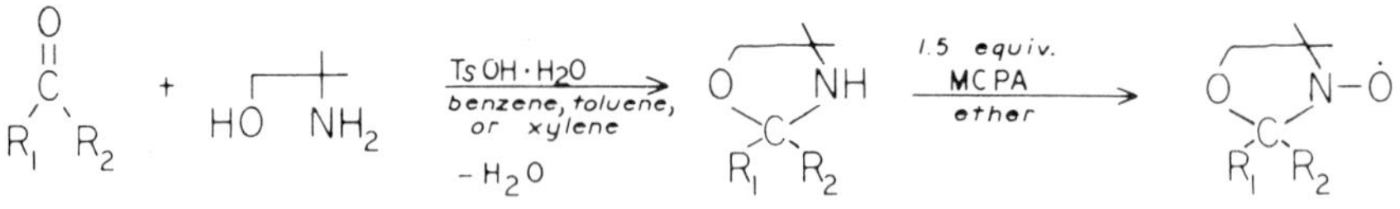

$R_1R_2C{=}O$ + LXXV $\xrightarrow[\text{benzene, toluene, or xylene; } -H_2O]{TsOH \cdot H_2O}$ LXXVI $\xrightarrow[\text{ether}]{1.5\ \text{equiv. MCPA}}$ LXXVII (A Doxyl Nitroxide)

This structure has two outstanding features: 1) the ease of introducing the doxyl group at a ketone group, and 2) its rigid nature as a result of the spiro linkage. A distinct disadvantage is the stereochemistry introduced in the formerly achiral ketone carbon atom. Recently Keana and Lee have introduced a second method of synthesizing doxyl nitroxides from carboxylic acids via the oxazoline, LXXX, which is MCPA oxidized to the oxaziridine, LXXXI, which isomerizes to nitrone LXXXII upon silica gel chromatography, followed by treatment with an organometallic reagent and air oxidation to yield the desired oxazolidine, LXXXIII.

$+ \; HO_2C(CH_2)_4CH_3 \longrightarrow$ LXXX $\xrightarrow{MCPA}$ LXXXI

$\xrightarrow{\text{silica gel}}$ LXXXII (hygroscopic) $\xrightarrow{\text{1. R-metal; 2. air}}$ LXXXIII

$R = CH_3, C_2H_5, n\text{-}C_7H_{15}, CH_2{=}CH\text{-}$

The proxyl nitroxides, IV, based on the pyrrolidine-N-oxyl structures also retain the rigid structure as do doxyl nitroxides, but are both chemically more stable and less polar than the latter type.

These are synthesized from the commercially available nitrone LXXXIV (available from Aldrich Chemical Co., Milwaukee, Wis.) which is converted via a Grignard reaction to the N-hydroxyamine LXXXV, which undergoes *in situ* disproportionation by Cu^{++} catalyzed air oxidation in MeOH to the new nitrone LXXXVII. Reaction with a second organometallic reagent and Cu^{++}-air oxidation yields the desired proxyl compound IV.[22]

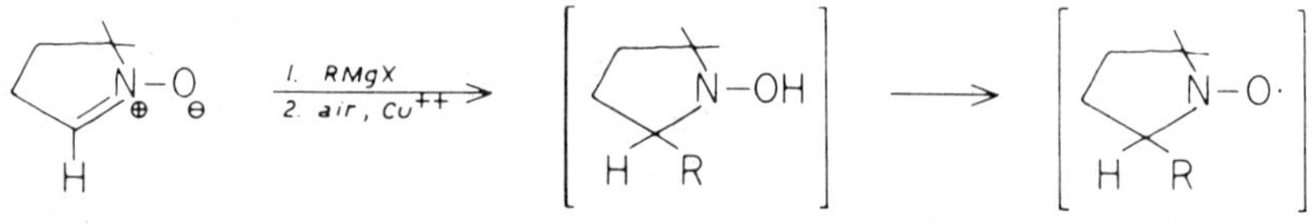

1. RMgX 2. air, Cu^{++}

LXXXIV LXXXV LXXXVI

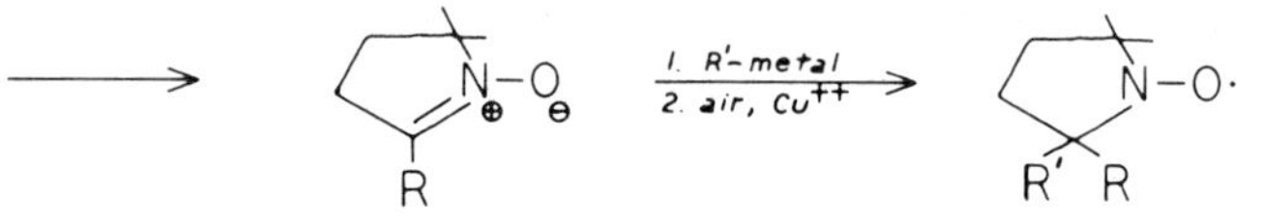

1. R′-metal 2. air, Cu^{++}

LXXXVII IV

Note that either R or R′ may possess remote, protected, functional groups which may be exposed later.

Finally, the azethoxyl nitroxides are the most exciting development by virtue of their minimal steric bulk (minimum perturbation spin labels).[23] Essentially the dialkyl flanked protected nitroxide group is in effect "inserted" into the chain or molecule of interest without the requirement of a complete ring structure as in the earlier labels. Their syntheses parallel those of the proxyl nitroxides with the exception of a different nitrone starting material XCVII. They are also the most chemically stable structures with an additional resistance to reduction over the proxyl nitroxides. The synthetic scheme, which is shown below, affords a mixture of *cis* and *trans* azethoxyl nitroxides, V_c and V_t.

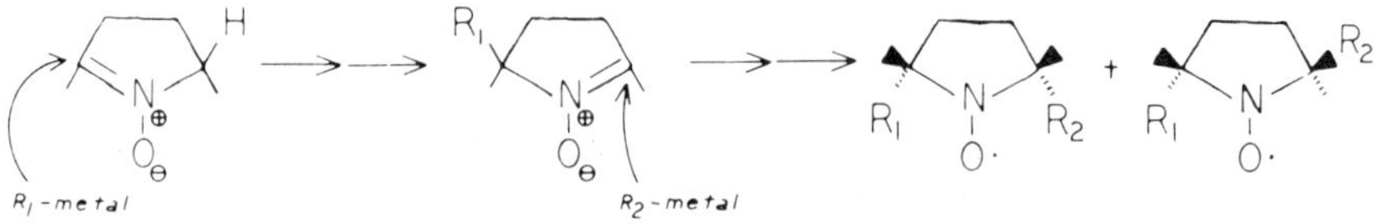

R_1-metal R_2-metal

XCVII XCVIII V_c V_t

Normally the *trans* isomer predominates for steric reasons since R_2 prefers to add *trans* to R_1 on XCVIII. However, by starting with another nitrone, XCIX, the *cis* isomer V_c may be made to predominate.

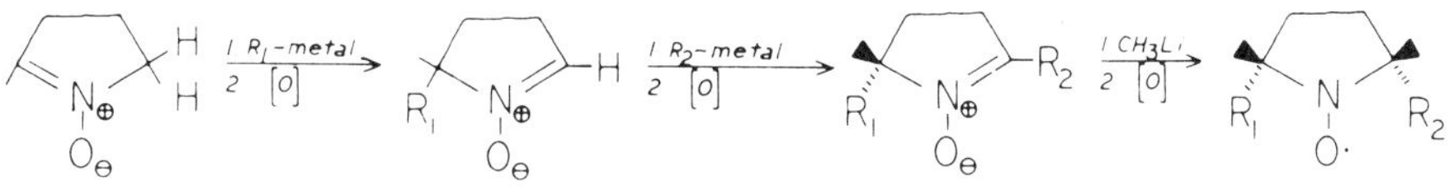

XCIX

cis predominates

Some general guidelines with both proxyl and azethoxyl nitroxide synthesis are:

1) higher yields are obtained by adding the larger alkyl (organometallic) group first,

2) with 2-alkyl nitrones, use Grignard reagents for 2-methyl nitrone, use organolithium reagents if substitution is other than methyl.

Imidazoline nitroxides are formed from a ketone and 2,3-dimethyl-2,3-diaminobutane by a scheme analogous to doxyl synthesis.[24] Over oxidation is not a problem:

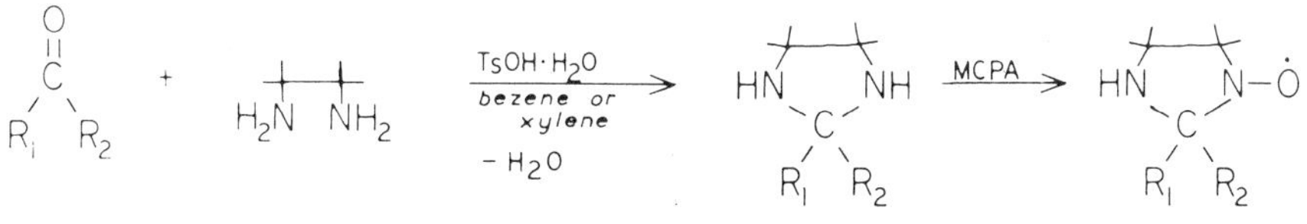

While their ESR spectra become more complicated (5 or 10 lines), various nitronyl nitroxides have been reported where the general functionality begins with an aldehyde as shown below.[25]

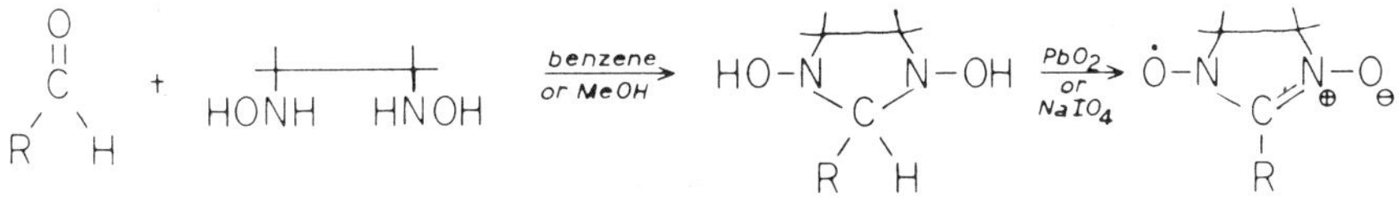

Lastly, with the concern over the ability of nitroxides to inhibit free radical polymerizations, the incorporation of the secondary amine precursor would be the method of choice assuming that it could be selectively oxidized in the final polymer without modifying other groups in the macromolecule.

Another approach to the radical polymerization termination problem might

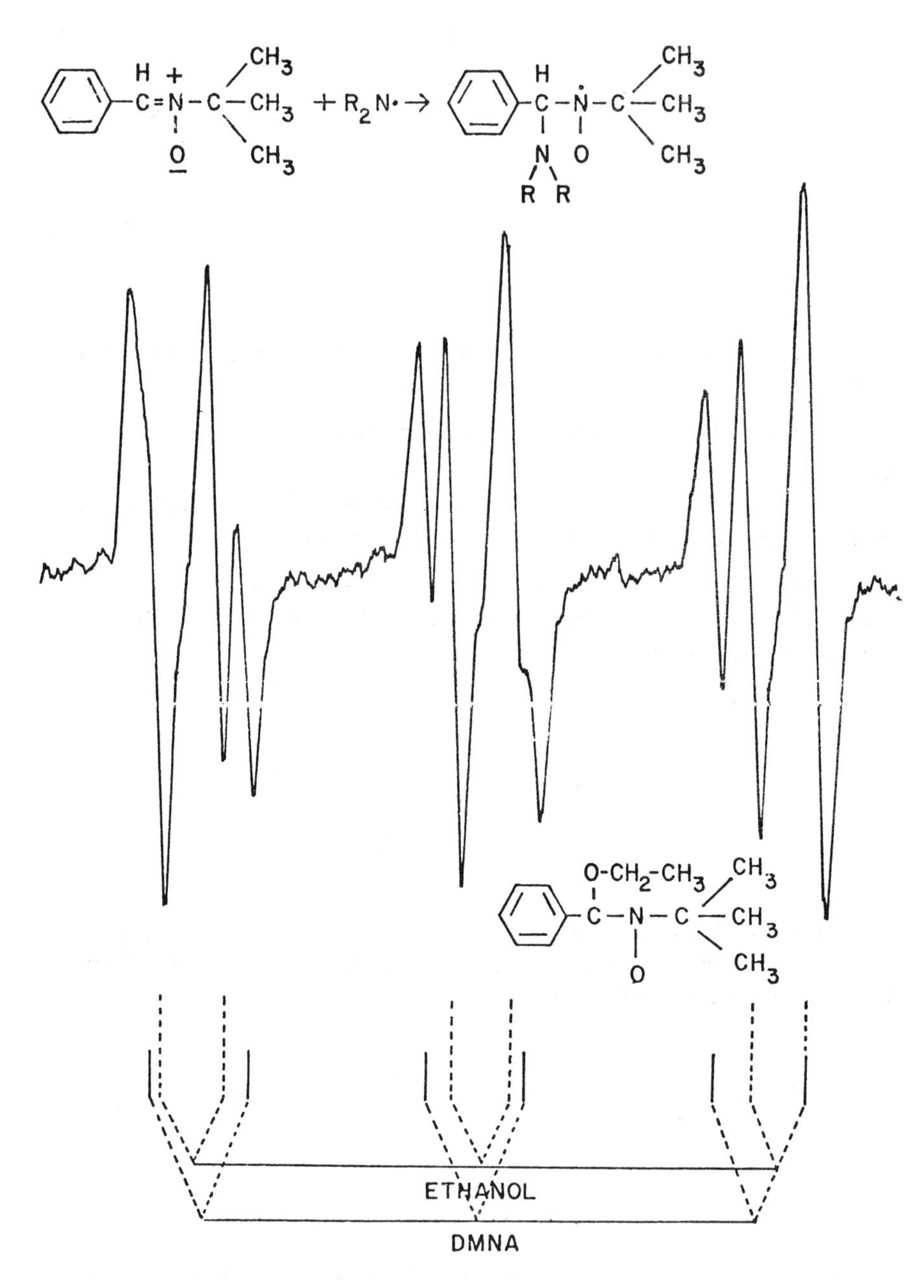

FIGURE 4 ESR spectra of 0.15 M phenyl tertiary butyl nitrone (PBN) adducts formed during the incubation of a 2% dimethylnitrosoamine solution in ethanol for 20 min, 37°C with microsomes and a lipid peroxidation system (0.3 mM NADPH, 1.2×10^{-5} M Fe^{2+}, 2.0×10^{-4} M pyrophosphate). The resultant spectrum reflects the apparent addition of a reactive secondary amine radical. The large a_β^H splitting was 5.75 G. The second splitting $a_\beta^H = 3.35$ G is the ethanol addition product. Adapted from A.N. Sapprin and L.H. Piette, *Arch. Biochem. Biophys.*, **180**, 480 (1977).

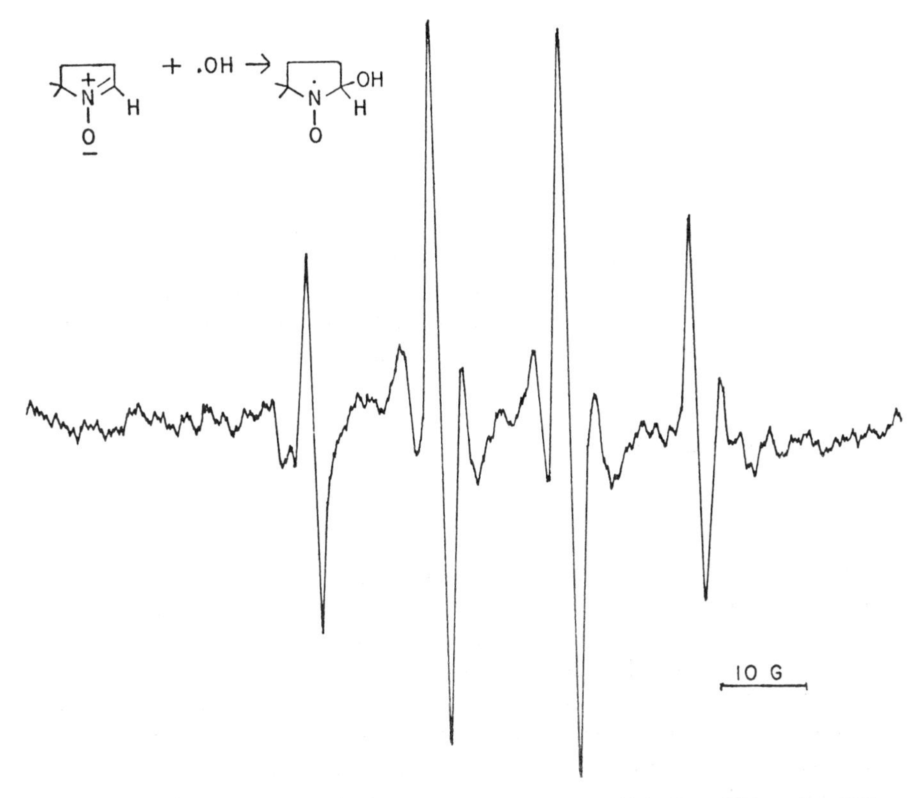

FIGURE 5 ESR spectrum of 5,5′-(dimethyl-1-pyrroline-1-oxide) after addition of an OH· radical across the double bond. Again here, the radical was generated in a microsomal system. Adapted from C.S. Lai and L.H. Piette, *Biochem. Biophys. Res. Commun.*, **78**, 51 (1977).

be through the use and/or incorporation of spin traps. This field has been eloquently developed by Janzen and others.[26] Some examples currently in use as biological spin traps are exemplified in Figures 4 and 5.

SPIN LABELS WHICH ARE NOT NITROXIDES

The polymer matrix has the advantage over a biomolecule in that in some cases, its physicochemical environment is completely aprotic. Thus nominally stable nitroxides and other organic radicals are potential candidates as polymer spin labels.

With a mild diversion from N–O radicals, the sulfur analogue was synthesized many years ago but suffered from the distinct disadvantage that the diamagnetic dimer form was favored.[27]

X = H,H *or* O

This "problem" serves a potential advantage in a hypothetical spin labeled polymer in that the structural constraints associated with various physico-mechanical perturbations of some polymer matrix may be followed by the formation and destruction of the N–S radical signal as the labels come within correct juxtaposition for –S–S– dimer formation.

PROBES OF POLYMER STRUCTURE

Distances

The physical biochemists have been taking greatest advantage of the types of spin-spin interactions discussed earlier in order to map distances in a protein macromolecule. Figure 6 depicts a protein molecule with distinct para-magnetic metal and spin label binding sites, as well as substrate and solvent molecules in chemical exchange with distinct binding sites. Several of these features may also exist in a regular polymer. 1) Interactions between repeating spin labeled sites yields intersite distances and their changes upon structural deformation. For two nitroxides the dstance, r, in Å, is related by

$$r = [(5.56 \times 10^4)/2D]^{1/3}$$

where $2D$ is the dipolar splitting parameter which may be several hundred gauss for, e.g., a 4 to 6 Å separation. 2) Interactions between a spin label and a second metal ion site. The diiminoacetate group, $-N(CH_2COO^-)_2$, which is the chelating component of Chelex 100, a strong polystyrene based deionizing resin, has a high affinity for transition metals such as Mn^{++}, Cu^{++}, Ni^{++}, etc. The phenomenon usually manifests itself as an apparent quenching of the nitroxide spectrum with a $r^{1/6}$ dependence.[12, 13]

Lastly, if the diamagnetic polymer has a characterizable NMR spectrum, distance relationships ($r^{1/6}$) may be obtained from the Solomon-Bloombergen relationship for the case of either a metal or nitroxide spin label.[16]

Accessibility

Another assessment of polymer structure and its physicochemical heter-ogeneity is possible for those macromolecules which are in a dissolved or

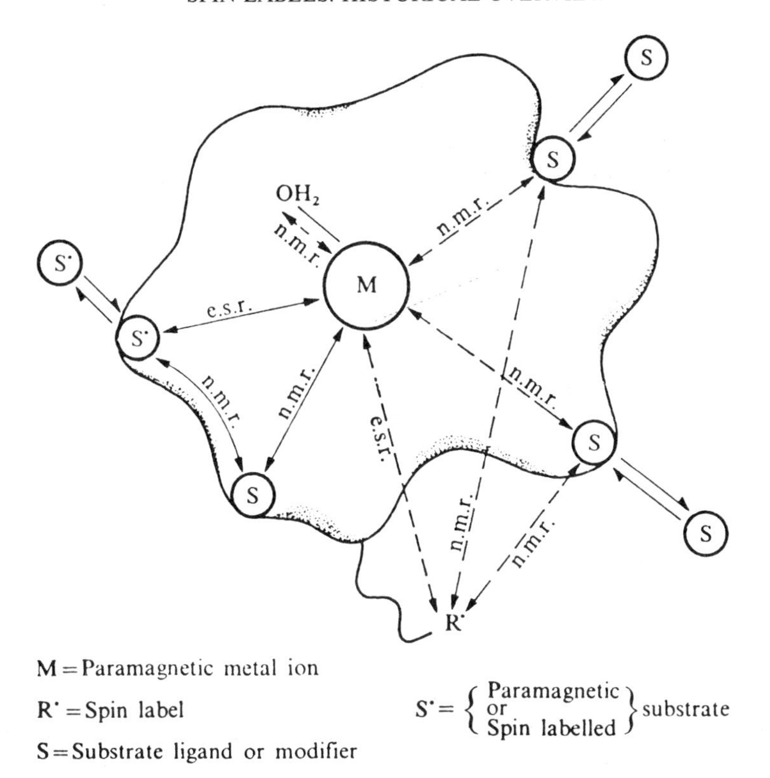

FIGURE 6 Some of the possible interactions involving nitroxides, paramagnetic ions and nuclei that can be studied by magnetic resonance. Note that one can "triangulate" on a site by utilizing two convergent interactions. With permission from R.A. Dwek, "NMR in Biochemistry", Oxford Univ. Press, London, 1973, p. 9.

biphasic solid-solution environment. By virtue of the fact that nitroxides may be reversibly reduced and reoxidized by a variety of soluble reagents, the rate behavior (e.g., multiphasic nature, variation with solvent and solute polarity and charge type) of these redox phenomena may be qualitatively correlated with those regions of the macromolecular structure which are accessible to solute and solvent molecules.

$$>N\dot{-}O \rightleftharpoons >N-O-H \rightleftharpoons >N-H$$

(diamagnetic) (diamagnetic)

An analogous method in biochemistry is hydrogen-deuterium exchange studies on enzymes, polypeptides and polynucleotides as an assessment of structural features such as helix content, buried (interior) residues, etc. While certain pitfalls exist with all techniques, perhaps the great advantage with synthetic organic polymers is, on the one hand, a repeating structural unit, while the structural heterogeneity on the macroscopic scale (end effects, etc.) exemplify one of the problems distinct to a nonglobular macromolecule.

Some oxidizing and reducing agents for nitroxides

Reducing agents[a]	Reoxidizing agents
Ascorbate	H_2O_2
Phenylhydrazine	O_2
Hydroxylamine	MCPA
Thiols	PbO_2
Organolithium	Ag_2O
Na_2S	Cu^{++}
K_2FeCN_6	
$LiAlH_4$	

[a]Doxyl nitroxides tend to irreversibly hydrolyze from the hydroxylamine intermediate.

PROGNOSIS

The future of spin label and ESR applications to physicochemical studies of synthetic organic polymer structure offer great promise. Much of the material in this chapter comes from a more biochemical perspective. While polypeptides (proteins) and polynucleotides (DNA, RNA) are biological polymers, the approaches utilized in physical biochemistry are easily extrapolated to organic polymers, in fact it is this author's assessment that the questions to be answered are more easily addressed in the more structurally homogeneous organic polymer than in an irregular, globular protein or a ribosomal RNA.

ACKNOWLEDGMENTS

LJB is an established investigator of the American Heart Association. His research has been benefitted by the generous assistance of the National Science Foundation, National Institutes of Health, Research Corporation, the American Heart Association and the Ohio State University.

REFERENCES

1. S. Ohnishi and H.M. McConnell, *J. Amer. Chem. Soc.*, **87**, 2293 (1965).
2. A.K. Hoffman and A.T. Henderson, *J. Amer. Chem. Soc.*, **83**, 4671–4672 (1961).

3. E.G. Rozantsev, "Free Nitroxyl Radicals" (translated by B.J. Hazzard), Plenum Press, New York, (1970).
4. Of the many publications of A. Rassat, a principal reference is *Bull. Soc. Chim. Fr.*, **1965**, 3273 (1965).
5. T.J. Stone, T. Buckman, P.L. Nordio, and H.M. McConnell, *Proc. Natl. Acad. Sci. U.S.*, **54**, 1010–1017 (1965).
6. L.J. Berliner (ed.), "Spin Labeling: Theory and Applications", Academic Press, New York, 1976.
7. L.J. Berliner (ed.), "Spin Labeling II", Academic Press, New York, 1979.
8. See chapters by P.L. Nordio (p. 5) and J.H. Freed (p. 53) in Ref. 6.
9. J.S. Hyde, *Meth. Enzymol.*, **49G**, 480 (1978).
10. J.S. Hyde and L.R. Dalton, Chapter 1 of Ref. 7.
11. O.H. Griffith, P.J. Dehlinger, and S.P. Van, *J. Memb. Biol.*, **15**, 159 (1974).
12. J.S. Leigh, Jr., *J. Chem. Phys.*, **52**, 2608 (1970).
13. J.S. Hyde, H.M. Swartz, and W.E. Antholine, Chapter 2 of Ref. 7.
14. Z. Ciecierska-Tworek, S.P. Van, and O.H. Griffith, *J. Mol. Struct.*, **16**, 139 (1973).
15. G.R. Luckhurst, in Ref. 6, p. 133.
16. T.R. Krugh, in Ref. 6, p. 339.
17. J.A. Green, II, L.A. Singer, and J.H. Parks, *J. Chem. Phys.*, **58**, 2690 (1973).
18. J.A. Green, II, and L.A. Singer, *J. Amer. Chem. Soc.*, **96**, 2730 (1974).
19. J.F.W. Keana, *Chem. Rev.*, **78**, 37 (1978).
20. J.F.W. Keana, Chapter 3 of Ref. 7.
21. G. Sosnovsky and Konieczny, *Z. Naturforsch.*, **32b**, 338 (1977).
22. J.F.W. Keana, T.D. Lee, and E.M. Bernard, *J. Amer. Chem. Soc.*, **98**, 3052 (1976).
23. T.D. Lee, B. Birrell, and J.F.W. Keana, *J. Amer. Chem. Soc.*, **100**, 1618 (1978).
24. J.F.W. Keana, R.S. Norton, M. Morello, D. Van Engen, and J. Clardy, *J. Amer. Chem. Soc.*, **100**, 934 (1978).
25. E.F. Ullman, J.H. Osieck, D.G.B. Boobock, and R. Dary, *J. Amer. Chem. Soc.*, **94**, 7049 (1972).
26. E.G. Janzen and B.J. Blackburn, *Accts. Chem. Res.*, **4**, 31 (1971).
27. J.E. Bennett, H. Sieper, and P. Tans, *Tetrahed.*, **23**, 1697 (1967).

DISCUSSION

J. S. Hyde (Milwaukee County Medical Complex, Milwaukee, Wisconsin): What is the "take home" lesson on the differences between doxyl and proxyl labels?

L. J. Berliner: The difference is in the preferred hyperfine axis orientation along the main chain into which the label is incorporated.

J. S. Hyde: What utility does the nitronyl-nitroxide label have?

L. J. Berliner: Obviously much less due to their more complicated ESR spectra (5 or 10 lines). I would guess that only their analytical utility is of interest. Their precursor forms, I believe, can also act as spin traps.

J. S. Hyde: The use of spin-labels to measure polarity is hampered by difficulty in separating motional effects from changes in hyperfine coupling. In principle, measurements at two microwave frequencies can serve to effect this separation.

A. M. Bobst (University of Cincinnati, Cincinnati, Ohio): Why should the azethoxyl radicals be less likely reduced?

L. J. Berliner: Presumably the vicinal alkyl substitution protects the N–O group better than the geminal alkyl substitution in the doxyl or proxyl labels.

B. Rånby (Royal Institute of Technology, Stockholm, Sweden): The spin labels described were synthesized for use in aqueous biological systems. Some of the labels, however, did not appear to be very soluble in water. Could you give the solubility properties in aqueous solution of the labels described?

L. J. Berliner: In fact this is a constant problem in that these are only sparingly soluble in water (unless charged or containing hydroxyl or other protic groups). Usually, however, the concentrations of a normal spin label experiment (10^{-5} to 10^{-4} M) are well within the solubility limits of the label.

J. S. Hyde: What is your understanding of the dynamics of inversion of the angle that the N–O bond makes with respect to the C–N–C plane?

A. T. Bullock (University of Aberdeen, Old Aberdeen, Scotland): The temperature coefficient, da^n/dT, for nitroxides is typically $\sim$1–5 mG K^{-1}. These values are positive for planar nitroxides (5-membered rings) and negative for non-planar nitroxides (6-membered rings).

COMPARISON OF SPIN LABEL AND SPIN PROBE TECHNIQUES

Investigation of Structure and Molecular Dynamics of Polymer Solutions by Spin Probe and Spin Label Techniques

A. L. Buchachenko, A. M. Wasserman, T. A. Aleksandrova, and A. L. Kovarskii

The Institute of Chemical Physics of the Academy of Sciences of the USSR, Moscow, 117334, USSR

Three types of interaction are known to contribute predominantly to the EPR linewidth of spin labels and spin probes: (1) electron-nucleus hyperfine coupling, modulated by molecular rotation and providing information on the frequency, anisotropy, and amplitude of rotation; (2) intermolecular exchange coupling of electrons, modulated by collisions (or, rather, encounters) of radicals and yielding information on the dynamics of translational motion; and (3) intermolecular dipole interaction of electrons which is realized in systems with low molecular mobility and yields information on local concentrations and the static distribution of paramagnetic particles.

In earlier publications dealing with the physical structure and molecular dynamics of isolated macromolecules, polymer solutions, and solid polymers, we scrutinized all three types of interaction. The problem which causes maximum difficulty is that of distinguishing between exchange and dipole contributions to EPR linewidths, since both are linear functions of the concentration of the paramagnetic particles.

The above problem was solved experimentally.[1] We measured the broadening of the EPR lines of the nitroxyl radical TEMPO, for different concentrations of the radical, and the diffusion coefficient of the same radical

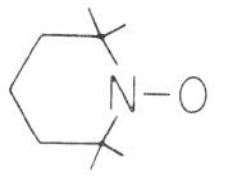

N—O

TEMPO

in liquids (heptane, decalin) and in polymers (polydimethyl siloxane PDMS, natural rubber NR, butadiene-styrene copolymer, polypropylene PP, polyethylene PE, polyformaldehyde PF) at various temperatures. The results are shown in Fig. 1. In systems with high molecular mobility (viz., $D \gtrsim 5 \times 10^{-6}$ cm^2/s) the reduced broadening $\delta H/C$ is a linear function of the diffusion coefficient, coinciding with the theoretical dependence obtained for exchange broadening,

$$\delta H_{ex}/C = 2(3^{-1/2})\gamma^{-1}k_{ex} \tag{1}$$

where

$$k_{ex} = pk_0 = 16\ \pi prD \tag{2}$$

k_{ex} is the exchange rate constant, k_0 is the rate constant of diffusion encounters of radicals, p is the relaxation probability at the radical encounter, and r is the effective radius of spin exchange of radicals. The theoretical straight line in Fig. 1 is plotted for the value of the parameter $pr \approx 1$Å.

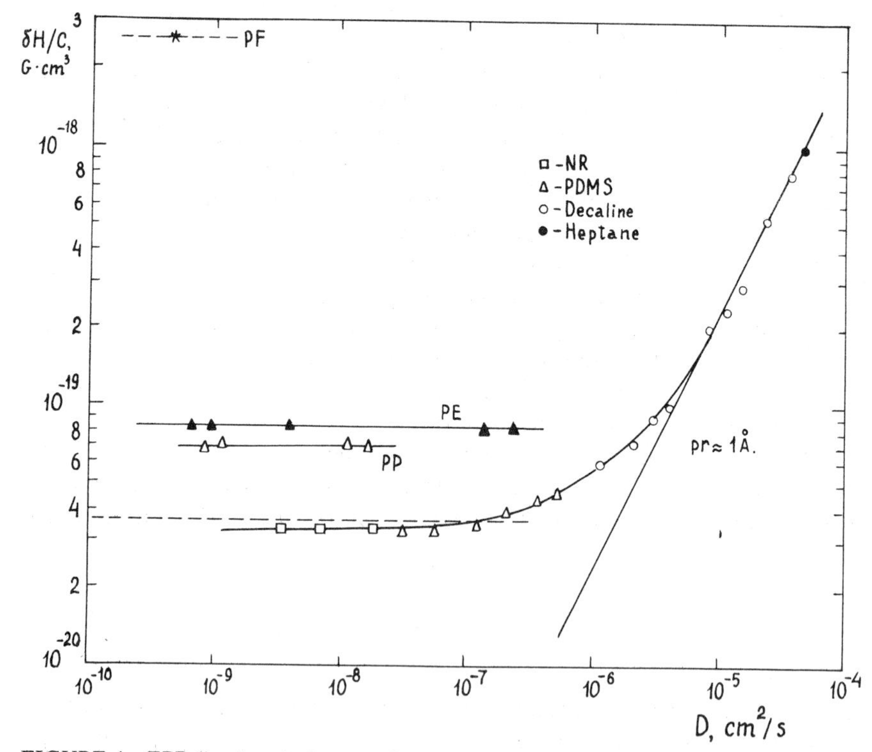

FIGURE 1 EPR line broadening as a function of the translational diffusion coefficient for TEMPO radicals in liquids and polymers. The dashed curve traces the theoretical value of $\delta H/C$ for dipole interaction in a rigid lattice, for statistically distributed radicals.

The exchange rate constants derived from the measured line broadening coincide with the exchange rate constants calculated by means of the expression[2]

$$\Delta W = (\Delta W_0^2 - 8\nu_{ex}^2)^{1/2} \tag{3}$$

from the measured line shifts where ΔW_0 denotes the spacing between hyperfine structure components in a dilute solution, ΔW is the corresponding spacing in concentrated solutions where exchange interaction takes place, and $\nu_{ex} = k_{ex}C$.

In systems with low molecular mobility ($D \lesssim 1 \times 10^{-7}$ cm^2/s) the reduced broadening $\delta H/C$ is independent of the diffusion coefficient and is determined completely by dipole interactions. This is the situation realized in polymers; in the case of amorphous polymers with uniform statistical distribution of radicals, the value of $\delta H/C$ is equal to the theoretically calculated value

$$\delta H/C = (3.6 \pm 0.5) \times 10^{-20} \text{ gauss/cm}^3 \tag{4}$$

In crystalline polymers (polyethylene, polypropylene, and polyformaldehyde), $\delta H/C$ is considerably higher; the reason is that radicals are only localized in the amorphous regions of the polymer, so that their local concentrations, C_{loc}, exceed the averaged concentration, $< C >$, corresponding to the statistically homogeneous distribution of the radicals over the whole volume.

In the case of dipole broadening, we always have $\delta H/C =$ constant, so that

$$\frac{\delta H_{eq}}{\delta H_{noneq}} = \frac{< C >}{C_{loc}} = \frac{V_{loc}}{< V >} = \kappa \tag{5}$$

The ratio of the local concentration to the mean concentration, and the fraction of sample volume, $\kappa = V_{loc}/< V >$ into which the radical cannot penetrate, can be found from the ratio of dipole broadenings of the two systems (those with homogeneous and inhomogeneous distributions of radicals). Table 1 lists the measured values of $\delta H/C$, ratios $C_{loc}/< C >$, and those of κ for a number of crystalline polymers. The values of κ are close to those of the degree of crystallinity of the polymer, which indicates that radicals penetrate only into the amorphous phase of the polymer.

We see, therefore, that the dipole contribution to the EPR linewidth is predominant in the case of low molecular mobility, which enables one to determine the local concentrations of paramagnetic particles, and with this, the statistical parameters of their distribution. The exchange contribution is predominant in the case of high molecular mobility, thus making it possible to obtain exchange rate constants and the coefficients of translational diffusion.

TABLE 1
The values of $\delta H/C$, $C_{loc}/<C>$, and the fraction κ of the polymer inpenetrable for probe radicals

Polymer	% crystallinity, according to IR spectrometry data	$\delta H/C \times 10^{20}$ gauss/cm^3	$C_{loc}/<C>$	κ
Polyethylene	70	8.5 ± 0.4	2.3	57
Polypropylene	47	7.0 ± 0.4	1.9	47
Polyformaldehyde	single crystal	260 ± 40	70	98.6

Let us consider now some applications of the above theory to the study of the physical structure and dynamics of isolated spin labeled macromolecules, polymer solutions, and solid polymers.

1. ISOLATED MACROMOLECULES

Local densities of spin labels and monomer units in a polymer coil were measured[3] in spin labeled poly(vinyl pyridine)

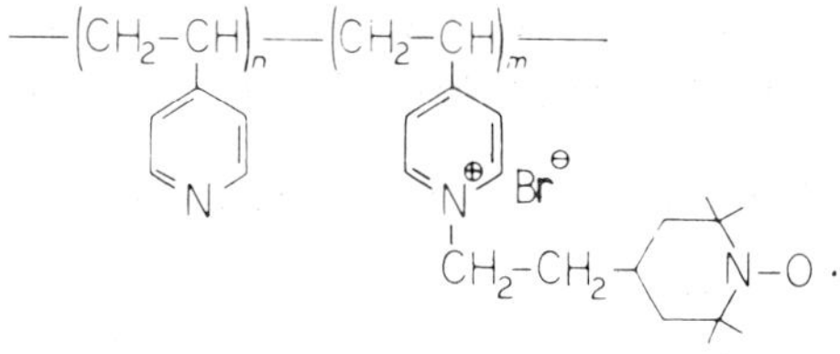

$$—(CH_2–CH)_n—(CH_2–CH)_m—$$

N

$N^{\oplus}$ $Br^{\ominus}$

$CH_2–CH_2$ N–O·

where the molar fraction of spin labels, $\beta = m/(m+n)$, was 0.01, 0.025, 0.05, 0.10, and 0.20. The molecular masses, M_m, of the fractions investigated were 1.5×10^4, 5×10^4, and 25×10^4.

Line broadening and its temperature dependence were studied in polymer solutions in ethanol, at the concentration of 1 wt.%. Diminishing polymer concentration by a factor of 2 to 3 did not change linewidth; this signifies that linewidth is determined, for these concentrations, only by the dipole and exchange interaction of the labels inside a macromolecule. In other words, at these concentrations we are dealing with isolated non-interacting coils of macromolecules.

As follows from the temperature dependence of line broadening (defined as the difference between linewidths in polymers with high and low contents of spin labels), broadening in the range from 0 to 20°C is determined by the intramolecular dipole contribution (Fig. 2), so that we can readily calculate by

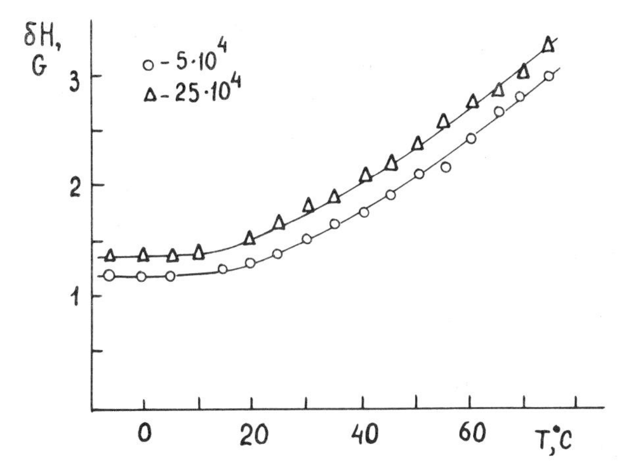

FIGURE 2 Concentration broadening, δH, as a function of temperature for 1% poly(vinyl pyridine) (PVP) ($\beta = 0.20$) in ethanol.

means of eq. (4) the local concentrations of spin labels in polymer coils. The local concentration of labels in a polymer coil corresponds to that of a small volume in the neighborhood of a chosen label; the radius of this local spherical region is approximately 30 Å, and corresponds to the distance at which the dipole contribution to the linewidth of the central label, induced by the label on the periphery of the sphere, is comparable to the linewidth itself, i.e., about 1 gauss. We must emphasize that an elementary volume in which the dipole interaction is significant, is much smaller than the volume of the polymer coil (for instance, for a polymer of $M_m = 25 \times 10^4$, the rms radius of inertia of a macromolecule, given by the Flory-Fox equation of the characteristic viscosity, is found to be ~220 Å).

Local label concentrations, C_{loc}, are related to the local concentration, i.e., density of monomer units, ρ_{loc}, by the formula

$$C_{\text{loc}} = \beta \cdot \rho_{\text{loc}} \tag{6}$$

(it is assumed that labels are distributed uniformly over the length of a macromolecule).

The mean density of monomer units in a coil, $< \rho >$, is calculated from the relationships[4]

$$< \rho > = N/V \cong 0.42\ [\eta]\ M\Phi' \tag{7}$$

where N is the number of monomer units in a macromolecule, $[\eta]$ is the characteristic viscosity, and $\Phi' = 3.1 \times 10^{24}$.

The dipole contribution to the linewidth, and local label concentration, increase as the number of labels per macromolecule increases (Fig. 3). This is an additional confirmation of the conclusion that the linewidth for low polymer concentrations is determined by the intramolecular interaction of spin labels.

The local density of monomer units in a coil, ρ_{loc}, can be found from the known values of C_{loc} by means of eq. (6), and then compared to the mean values, $< \rho >$; these data are listed in Table 2 from which the following conclusions can be drawn.

First, the local concentration of monomer units in a polymer coil depends only slightly on the molecular mass of the polymer; this conclusion is in agreement with the predictions of the statistical theory of macromolecules employing the Gaussian chain model.[5, 6]

Second, the local densities of large macromolecules are greater than the mean values by a factor of 4 to 6.

Third, the local label concentration and local monomer unit density are almost independent of temperature (in the range from 0 to 20°C), i.e., the

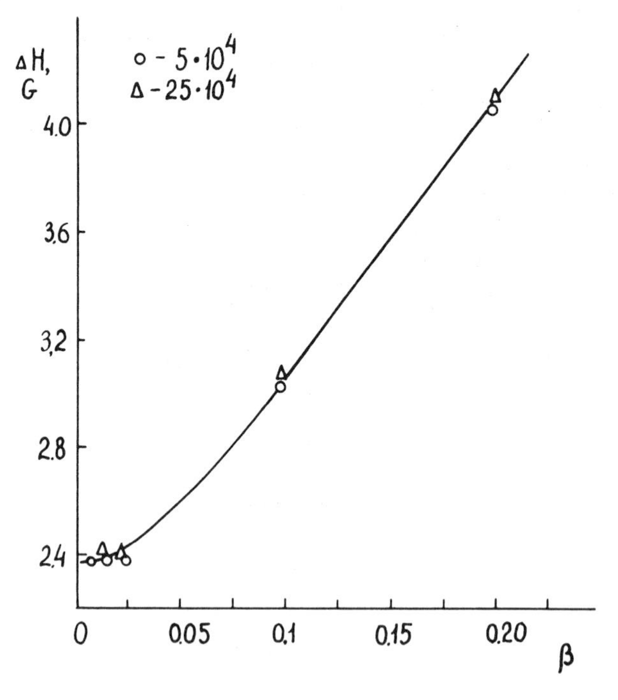

FIGURE 3 Lindwidth of EPR spectra, ΔH, of dilute PVP–ethanol solutions as a function of spin label content in a chain, β, at 20°C.

TABLE 2
Monomer unit density in a polymer coil at 20 °C.

$M_m \times 10^4$	β	$<\rho>$, mol/l	ρ_{loc}, mol/l
25	0.20	—	0.30
25	0.10	—	0.20
25	0.01	0.05	—
5	0.20	—	0.28
5	0.10	—	0.20
5	0.05	0.20	—

distribution of monomer units within a coil, and the coil size depend only slightly on temperature.

Fourth, the local density of monomer units is a function of the number of labels on the polymer chain; as the number of labels is increased by a factor of 2, the local density increases by a factor of 1.5. This effect seems to be due to some modification of the polymer chain as a result of the introduction of spin labels. Unexpectedly, this introduction of large-volume groups into the chain increases the density of monomer units; it is not impossible that electrostatic interactions partially compress the coil thus compensating for its expansion due to the addition of bulky spin labels.

The main contribution to line broadening at high temperatures is produced by the exchange intramolecular interaction. By using eqs. (1) and (2), we can calculate the encounter rate constants and the coefficients of local translational diffusion of spin labels in a coil, D_{loc}; it is necessary to take into account that for the local concentrations included in eqs. (1) and (2), we must use values found earlier from dipole broadening.

Fig. 4 shows D_{loc} of spin labels and probes in the same solution, plotted as a function of temperature. The coefficient of local translational diffusion in a coil are identical for polymers with different molecular mass, and are much smaller than the probe diffusion coefficients. The activation energy of label diffusion is much higher than that of probe diffusion; at the same time, the label diffusion pre-exponential factor is also much higher than that for probes. This partial compensation of the high activation energy by the high pre-exponential factor signifies that intramolecular collisions of spin labels result from a complex cooperative process consisting of elementary stages of displacement of polymer chain segments.

The rotational mobility of labels was investigated in weakly labeled polymers ($\beta = 0.01$) in which dipole and exchange contributions to line broadening are negligible. Fig. 5 illustrates the correlation time of rotational diffusion of spin labels and probes (under identical conditions) as a function of temperature. The correlation time (and activation energy as well) of labels is independent of the polymer's molecular mass, and exceeds that of probe

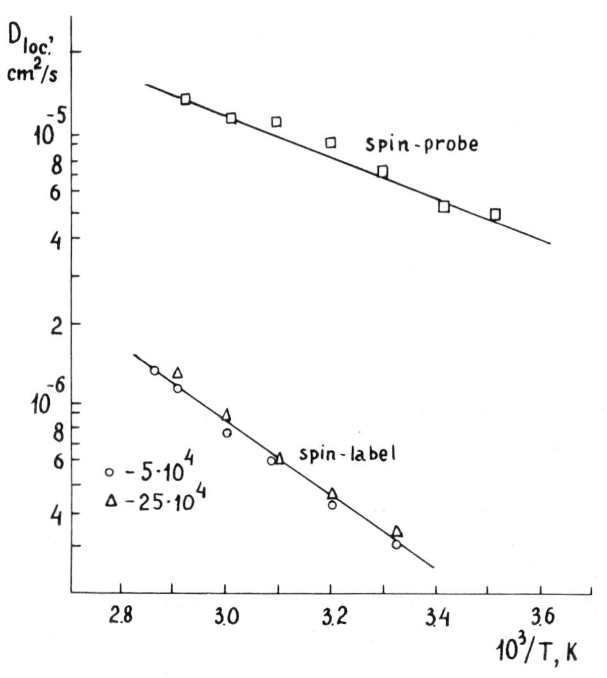

FIGURE 4 Temperature dependence of D_{loc} for labeled PVP, $\beta = 0.20$, and probes in ethanol, polymer concentration 1%.

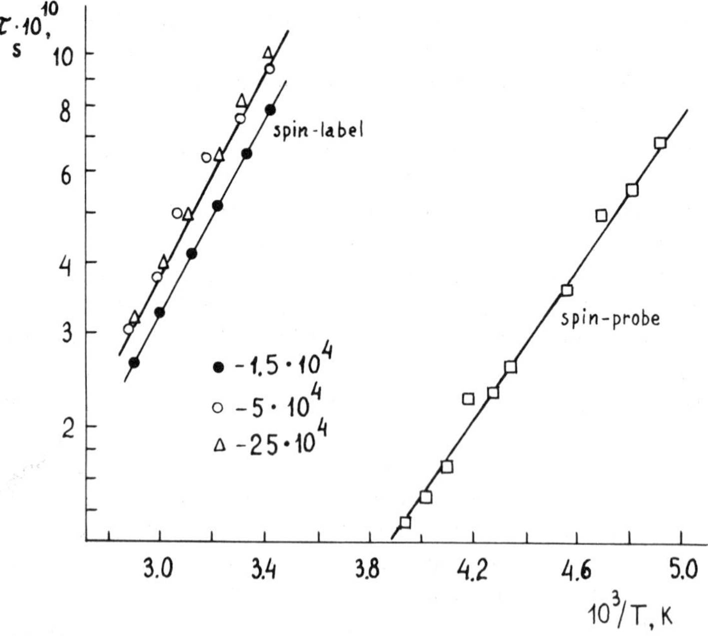

FIGURE 5 Correlation time of rotation as a function of temperature for labels (PVP, $\beta = 0.01$) and probes in dilute polymer solutions in ethanol.

rotation by almost a factor of 20, i.e., by a factor approximately equal to the ratio of the coefficients of local translational diffusion.

In summary, the separation of dipole and exchange contributions to the EPR linewidth of spin labeled macromolecules in dilute solution yields information on the local density of monomer units and the local intramolecular rotational and translational mobility in a polymer coil.

2. INTERACTION OF MACROMOLECULES IN SOLUTION

Our next problem is to describe the behavior of macromolecules due to intermolecular interaction, i.e., the interaction between polymer coils. We have to find the changes in the local density of the intrinsic monomer units of a macromolecule (local density of the host molecule monomer units), in the local density of foreign macromolecules (guest molecules), and the changes produced in the molecular dynamics, both translational and rotational, of the interacting polymer coils.

To achieve this goal, we investigated the dependence of linewidth of spin labeled poly(vinyl pyridine) ($M_m = 25 \times 10^4$, $\beta = 0.2$) on polymer concentration.[7] In order to determine the local density of monomer units in the host molecule, we used dilute solutions of spin labeled polymer (0.5%) in ethanol, with an admixture of non-labeled polymer. It has been demonstrated earlier that the linewidth is determined only by the intramolecular interaction of spin labels; interactions between spin labels incorporated into different macromolecules are ruled out. Line broadening was measured as the difference between linewidths in solutions of heavily labeled ($\beta = 0.2$) and weakly labeled ($\beta = 0.01$) polymer, for identical concentrations of the polymer in question.

In order to separate the dipole and exchange contributions into broadening, we again investigated the temperature dependence of broadening; examples of such dependence for solutions with polymer concentrations of 5.5 and 41% are shown in Fig. 6. Broadening is independent of temperature at all temperatures and is determined by the dipole contribution, so that local concentrations of spin labels in a coil and local densities of monomer units can be calculated by means of eqs. (4) and (6); the results are given in Table 3.

As in isolated macromolecules, the local density of monomer units in host molecules is greater than the mean density of the same units in a coil by a factor of 5 to 6. When the polymer concentration in the solution increases, the local density of host-molecule monomer units rises slightly since the host-molecule coil contracts in the presence of guest molecules. This contraction, however, is negligible; an increase in guest-molecule concentration by two orders of magnitude results in a contraction of the host molecule by not more than 30%.

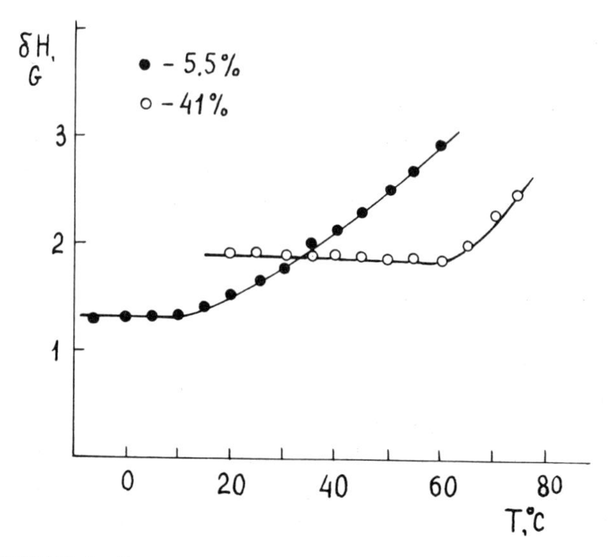

FIGURE 6 Temperature dependence of δH in PVP–ethanol solutions.

These data agree with the results of theoretical computer simulation experiments of polymer solutions; the calculations demonstrate that the local density of host monomer units in a θ-solvent is independent of polymer concentration, and increases slightly in good solvents.[5, 6] The data are also in agreement with the results obtained by means of small-angle X-ray scattering and inelastic neutron scattering,[8, 9] which demonstrate that the molecule's conformation both in a block polymer and in a solution is that of a Gaussian coil with dimensions nearly equal to those of a coil in a θ-solvent. Our latest results show that not only "external" dimensions of a macromolecule but also

TABLE 3
Local density of host-molecule monomer units, ρ_1, in solutions of poly(4-vinyl pyridine) ($M_m = 25 \times 10^4$) in ethanol.

Polymer concentration, wt.%	ρ_1, mol/l
0.5	0.30
1.0	0.30
5.5	0.32
11.3	0.35
25.6	0.35
35.2	0.35
41.0	0.43
54.0	0.43
65.0	0.43

the local density of its monomer units, i.e., the coil microstructure, remain essentially unaltered in high concentration solutions.

In order to determine the changes in the local density of the host-molecule monomer units in a polymer coil, we investigated the linewidths of high concentration solutions of spin labeled polymer ($\beta = 0.20$). Fig. 7 shows the linewidth as a function of polymer concentration: coils in the solution remain isolated at low concentrations (3 to 4%), and linewidth is determined by the intramolecular interaction of spin labels; at higher concentrations the coils begin to overlap, so that the dipole and exchange interaction between spin labels of neighboring molecules produce a substantial contribution to the linewidth, which increases with concentration.

In order to find the molecular contribution to broadening, we measured the difference between linewidth in high concentration solutions of spin labeled polymers with $\beta = 0.2$ and in polymer solutions of the same concentration but with the $\beta = 0.2$ polymer content of only 0.5 wt. % (those where line broadening is due exclusively to intramolecular interaction of spin labels). The molecular broadening measured in this manner contains components from both dipole and exchange origins. To separate these components, we investigated the temperature dependence of intermolecular broadening; one example of this is illustrated in Fig. 8. At low temperatures broadening is

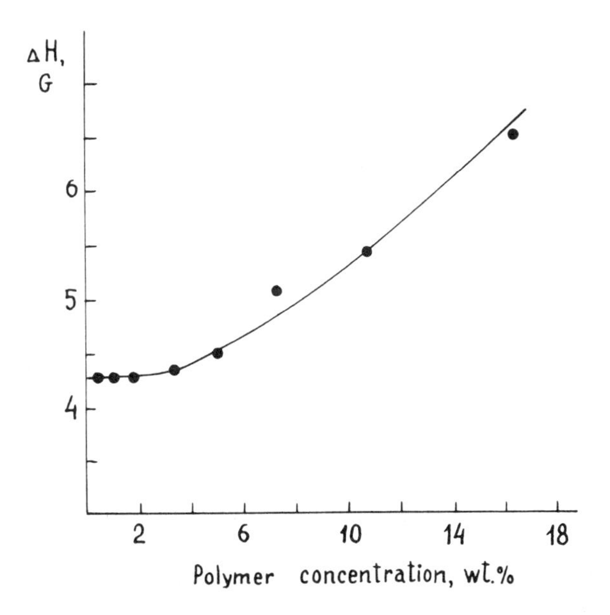

FIGURE 7 Linewidth of EPR spectra, ΔH, as a function of PVP ($\beta = 0.20$) concentration in ethanol at 20°C.

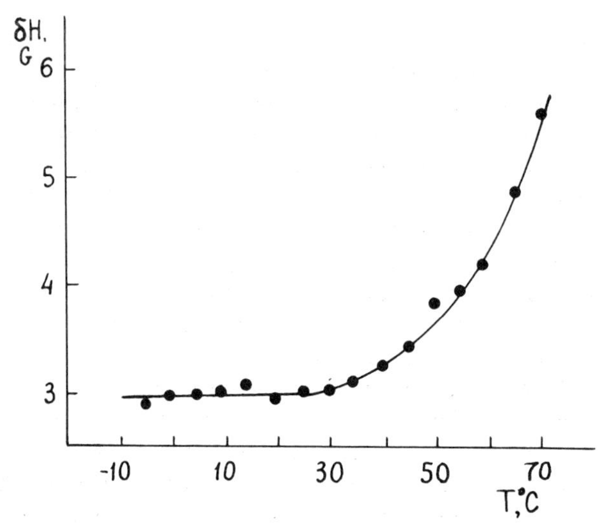

FIGURE 8 Temperature dependence of intermolecular broadening in 10% PVP–ethanol solution.

independent of temperature and is determined by the dipole contribution which yields, via eqs. (4) and (6), the local concentration of spin labels and the local density of monomer units in guest molecules. Fig. 9 demonstrates the behavior of the local density in host molecules (curve 1 and the data of Table 3), that in guest molecules (curve 2), and the total local density of monomer units in the coil (curve 3). When the polymer concentration rises, the local density of guest-molecule monomer units increases sharply; this means that coils of different macromolecules interpenetrate one another and thus overlap. This leaves the local concentrations of host-molecule monomer units, and consequently the host molecule dimensions, practically unchanged. Hence, the monomer units of guest molecules in a host-molecule coil can be regarded as solvent molecules. In other words, an increase in polymer concentration in a host-molecule coil produces a gradual modification of the solvent; ethanol molecules are substituted by the monomer units of guest molecules.

This conclusion is independently supported by measurements of A_{zz} derived from EPR spectra of glassy solutions of nitroxyl radicals (Fig. 10); $A_{zz} = a + 2b$, where a and b denote the isotropic and anisotropic hyperfine coupling constants, respectively. This quantity is sensitive to the polarity of the solvent; normally it is high in aqueous and alcohol solutions and is much smaller in hydrocarbon solvents.[10] The sharp decrease of A_{zz} for spin labels at increasing polymer concentrations (Fig. 10) indicates a change in the solvent in the host-molecule coil; ethanol is being replaced by the monomer units of

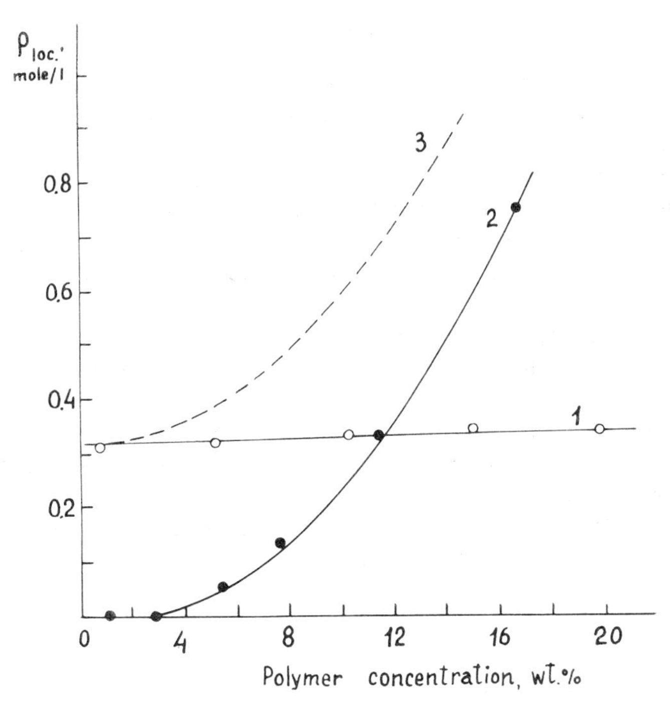

FIGURE 9 Local density of (1) the host-molecule, (2) the guest-molecule, and (3) the total local density of monomer units as a function of PVP concentration in solution.

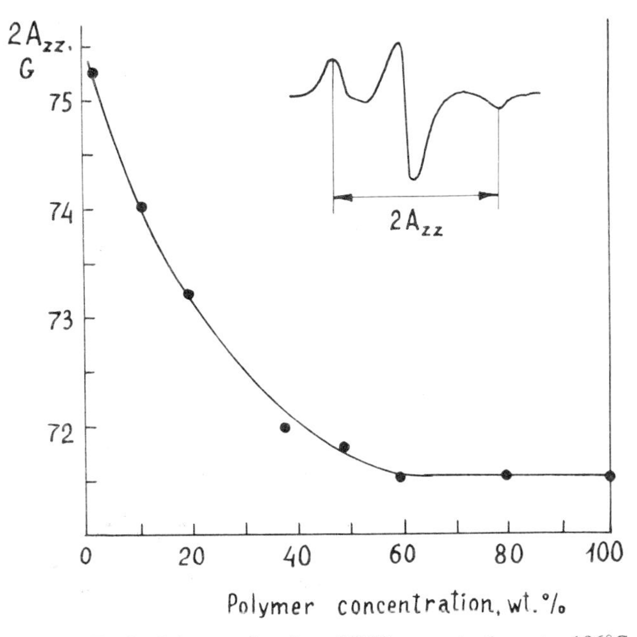

FIGURE 10 $2A_{zz}$ as a function of PVP concentration, at −196°C.

guest molecules. The value of A_{zz} undergoes a change in the same range of polymer concentrations as that where guest molecules penetrate the host-molecule coil.

The changes introduced into the molecular dynamics of the host molecules by guest molecules were investigated in solutions of weakly labeled polymer ($\beta = 0.05$) with added non-labeled polymer, and in solutions of heavily labeled polymer ($\beta = 0.20$), again with non-labeled polymer added. In the first case, the correlation time of rotational motion of spin labels in host molecules was measured, and in the second case, the coefficients of local translational diffusion within the host-molecule coil were found from the exchange contribution to line broadening.

The correlation time for rotational motion is independent of polymer concentration at low concentrations of polymer (Fig. 11), but then increases as polymer concentration rises.

Table 4 lists the activation energies and pre-exponential factors for label rotation, for a number of polymer concentrations in solution.

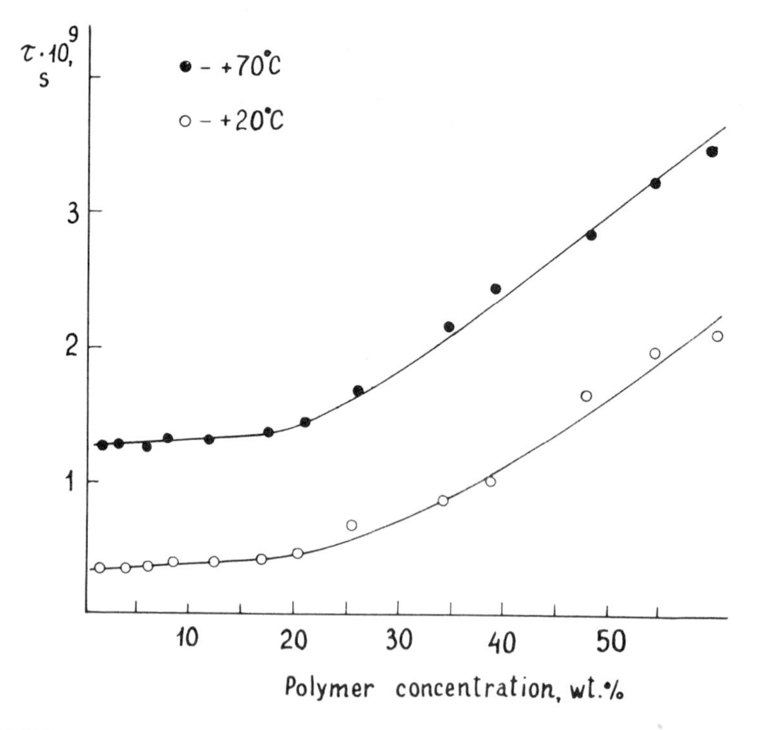

FIGURE 11 Concentration dependence of τ_c of spin labels in PVP solutions ($M_m = 25 \times 10^4$) in ethanol.

TABLE 4
Activation energy, E, and pre-exponential factors, τ_0, for spin label rotation, as functions of concentration of poly(4-vinyl pyridine) in solution.

Polymer concentration in solution	Temperature range, °C	$E \pm 0.5$ kcal/mol	τ_0, s
TEMPO in ethanol	−20 to −70	3.5	1.3×10^{-13}
1.0	−10 to +70	4.7	3.0×10^{-13}
5.5	−10 to +70	4.3	6.0×10^{-13}
25.5	+10 to +85	4.7	5.6×10^{-13}

The rotational motion of the labels is observed to slow down in just that range of concentrations in which the local density of guest-molecule monomer units becomes comparable with the local monomer unit density of host molecules. This means that the concentration dependence of intramolecular mobility, recorded by means of spin labeling techniques and characterizing small-scale motions, is determined by the local density of monomer units of the guest molecules penetrating into the coil of the host macromolecule.

Fig. 12 plots the coefficient of local translational diffusion of spin labels in host-molecule coils as a function of polymer concentration. Similar to rotational diffusion, translational diffusion remains almost constant at low polymer concentrations (up to 15%), but then sharply decreases. The reason for this sharp drop in local translational mobility of spin labels in host-

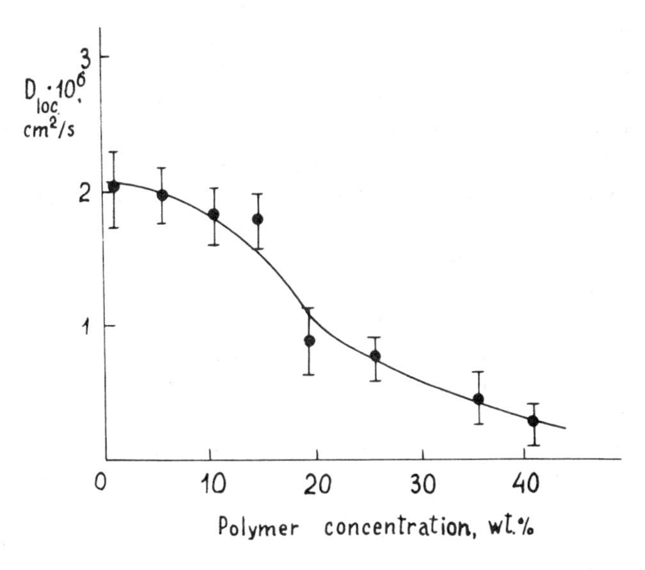

FIGURE 12 D_{loc} as a function of PVP concentration at 70°C.

molecule coils is again the increase in the density of guest-molecule monomer units, accompanying the overlapping of macromolecular coils.

The coefficients of local translational diffusion, characterizing the exchange interaction between spin labels of the host and guest molecules were found from intermolecular broadening measured in high concentration solutions of heavily labeled polymer ($\beta = 0.20$) at high temperature (see Fig. 8). At 70°C and a polymer concentration of 10%, this diffusion coefficient is equal to 1.8×10^{-6} cm^2/s and nearly coincides with the local intramolecular diffusion coefficient (1.2×10^{-6} cm^2/s). Therefore, the local translational diffusion in a compound coil proceeds identically for host-molecule and guest-molecule monomer units, and is determined by the sum of the local densities of the segments.

3. STRUCTURE OF HIGH CONCENTRATION POLYMER SOLUTIONS

When high concentration polymer solutions are investigated, it is of principal importance whether the solution is homogeneous or whether heterophase fluctuations, which are the precursors of phase separation, are formed in the polymer system. To solve this problem, we have studied the behavior of high concentration polymer solutions by means of the spin probe technique.[11–13]

Fig. 13 plots the rotational frequency $\nu = \tau_c^{-1}$ (τ_c being the correlation time of rotational motion) of the spin probe TEMPO as a function of temperature for butyl rubber–toluene systems of varying composition.

In solutions with a polymer concentration of 40, 50, or 55%, the Arrhenius plots of frequency are not linear; as the temperature is decreased, the frequency first decreases exponentially (following Arrhenius behavior), then increases sharply, and then once again drops off exponentially. This anomaly of rotational mobility is due to the phase separation of a homogeneous solution into two phases, one polymer-enriched, and the other solvent-rich. The spin probe is squeezed out into the solvent-rich phase, where the probe rotational mobility is greatly enhanced.

Solutions undergo phase separation (Fig. 13) at 225, 227, and 235°C (these are the positions of the maxima on the curves describing the temperature dependence of the probe rotational frequency). The process of phase separation proceeds, however, in a wide temperature range; this is caused by the gradual variation of the phase composition, in accordance with the phase diagram of the system, when the temperature is lowered.

The fact that phase separation proceeds over a wide temperature range is due to thermodynamic reasons and is not related to kinetic constraints. Indeed, $\ln \nu$ vs. $1/T$ curves are well reproduced whether scanning up or down in temperature. The absence of hysteresis indicates that in the process of phase

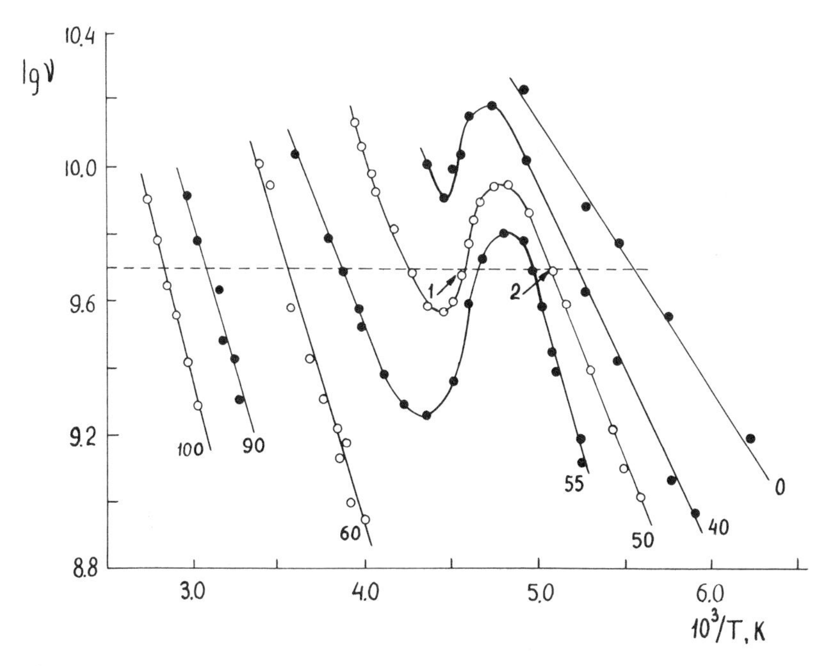

FIGURE 13 Log ν as a function of $1/T$ for probe radical TEMPO in solutions of butyl rubber in toluene. Numbers marking the curves correspond to polymer concentration in the solutions.

separation, the polymer–solvent system is in equilibrium at all temperatures.

The investigation of the exchange broadening of EPR lines proved independently that a new solvent phase is formed and that the spin probe is incorporated into this phase. We demonstrated earlier[14] that in systems with a uniform distribution of radicals there is a one-to-one correspondence of radical rotational frequency to concentration broadening. In other words, concentration broadening must be identical in solutions with uniformly distributed radicals and in those with equal frequencies of probe rotation. Any deviations from this rule point to a nonuniform distribution of radicals; in these cases one can determine the local concentration of spin probes, i.e., a measure of the nonuniformity of the distribution.

We recorded the concentration broadening of TEMPO radicals in solutions of butyl rubber in toluene, under the condition corresponding to a constant rotational frequency $\nu = 5 \times 10^9$ s^{-1} (around the points of intersection of the $\log \nu$ vs. $1/T$ curves in Fig. 13 with the dashed line $\log \nu = 9.69$). The results of these measurements are plotted in Fig. 14. The concentration broadening is found to be the same in all homogeneous systems (including butyl rubber and

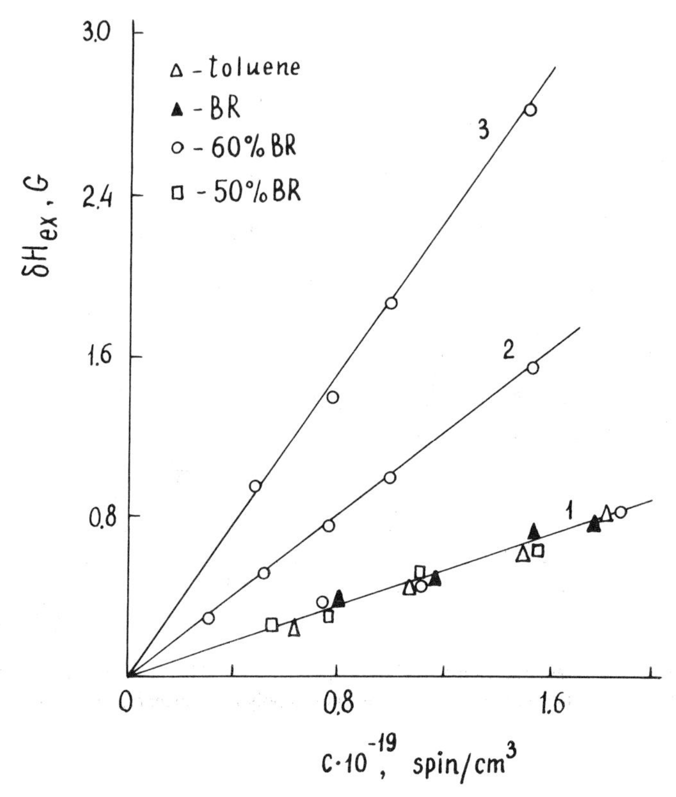

FIGURE 14 Exchange broadening, δH_{ex}, as a function of TEMPO radical concentration for $\nu = 5 \times 10^9\,s^{-1}$. (1) toluene, BR solutions in toluene, and BR; (2) 50% BR–toluene solution, point 1 in Fig. 13; (3) 50% BR–toluene solution, point 2 in Fig. 13.

pure solvent) (curve 1); this rule does not hold for heterogeneous systems only at points 1 and 2. Local concentrations at points 1 and 2 exceed mean values (i.e., the concentration in homogeneous systems) by factors of 2.5 and 5.5, respectively. Therefore, spin probe techniques make it possible to detect phase separations in polymer solutions not only qualitatively but quantitatively as well, by estimating the degree of inhomogeneity.

By recording the phase separation temperatures of polymer solutions by means of the spin probe technique, we can construct phase diagrams for phase separation; examples of such diagrams are given in Fig. 15.[12] An important advantage of the spin probe method exists in one's ability to detect the first stages of formation of a new phase long before phase separation sets in; we have discussed one such example (see Fig. 14). The spin probe method also enables one to detect the new phase formation in very high concentration and

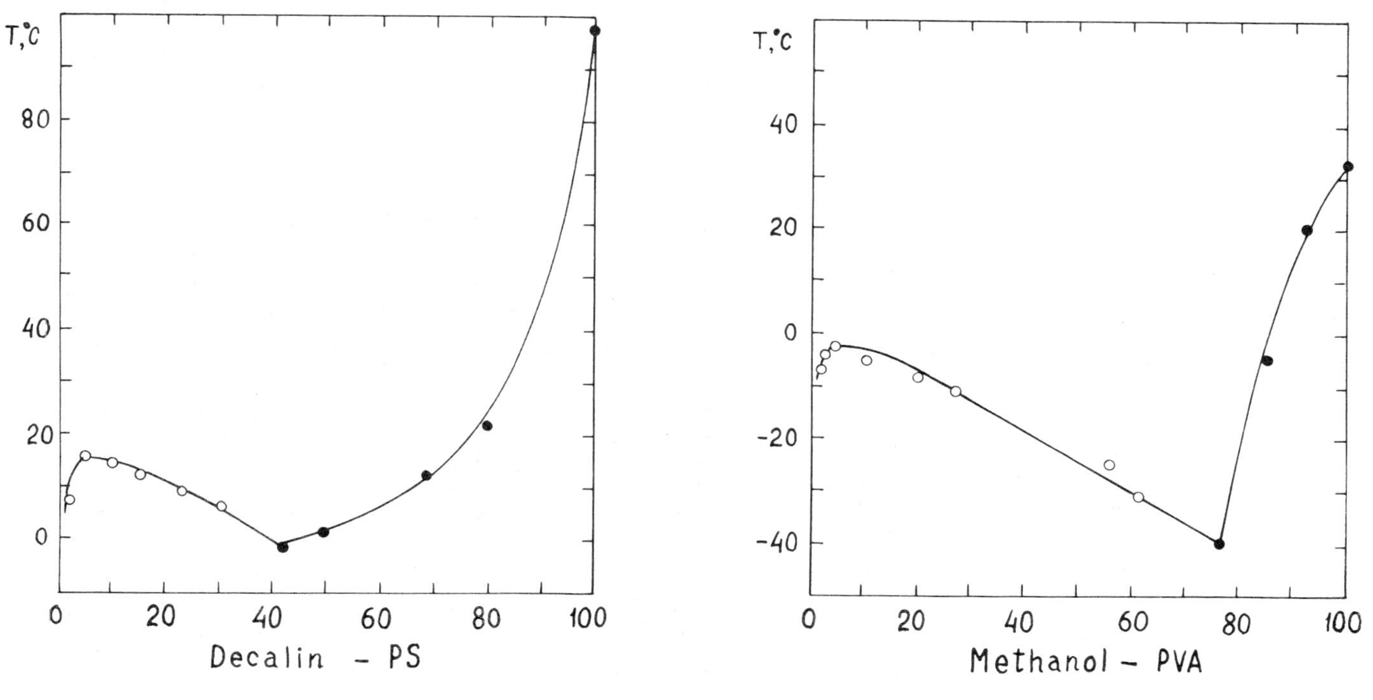

FIGURE 15 Phase diagrams of polystyrene–decalin and poly(vinyl acetate)–methanol systems. ○, phase separation temperatures in solution; ●, glass transition temperatures.

high viscosity polymer solutions, in which kinetic constraints cause extremely slow phase separation not detectable by standard techniques (e.g., light scattering). In principal, this method enables one to follow the kinetics of new phase formation.

A frequently encountered sign of micro-inhomogeneity in polymer systems (solutions, mixtures, etc.) is a superposition of EPR spectra, resulting from spin probes located in various sites of an inhomogeneous polymer system possessing non-identical molecular mobility. This approach has been widely discussed in other publications.[11–13]

Spin labeled polymers can also be employed to study the phase separation of polymer solutions; we have studied the spin labeled poly(vinyl acetate)–methanol system.[13]

A phase diagram of this system was compiled for a wide range of concentrations; it was demonstrated that spin probe and spin label techniques may be essentially complementary when applied to investigations of phase equilibria in polymer solutions.

CONCLUSIONS

The results reported in this paper demonstrate that an analysis of EPR spectra of spin labeled polymers makes it possible to separate the dipole and exchange contributions to intramolecular concentration broadening, and to determine the local density of monomer units and the coefficients of local translational diffusion.

The local density of monomer units characterizes the microstructure of the polymer coils; the dynamics of macromolecules in a solution are characterized by the rotational and translational mobility of the monomer units.

The local density of monomer units in isolated macromolecules (poly(4-vinyl pyridine) in a good solvent, viz., ethanol) is 0.2–0.3 mol/l, substantially exceeding the mean density in the whole volume of a polymer coil. In contrast to the mean density, the local monomer unit density is only weakly dependent on the molecular mass of the polymer. For high molecular mass poly(4-vinyl pyridine) the rotational and translational mobility of spin labels is practically independent of the molecular mass of the polymer, and is smaller by a factor of 10 to 20 than the spin probe mobility in the same solvent.

In changing from a dilute solution to a block polymer, the local density of host-molecule monomer units rises by not more than 20–30%. This means that only a slight "contraction" of polymer coils takes place, while the number and the relative arrangement of host monomer units in a macromolecule in the neighborhood of any monomer unit varies only slightly both in dilute and in high concentration solutions. These results are in agreement with the data obtained by small angle X-ray scattering and neutron scattering studies,[8,9] and demonstrate that not only the "external" dimensions of macromolecules (rms distance between polymer chain terminals and the radius of inertia) but also the "microstructure" of a polymer coil are practically the same in high concentration and dilute solutions.

A comparison of the local density of monomer units and their molecular mobility in high concentration solutions shows that the dependence of the rotational and translational mobility of monomer units on concentration is determined primarily by the increase in the local density of the units belonging to neighboring macromolecules (guest macromolecules) and not by the conformational changes of the host macromolecules.

The application of spin probe and spin label techniques enables one to study the structural micro-inhomogeneity of polymer solutions, to analyze phase separations in them, and to obtain phase diagrams. The degree of inhomogeneity in a system can thereby be not only qualitatively detected but also quantitatively evaluated. An important advantage of these methods lies in the facility to detect the formation of a new phase in very high concentration and in high viscosity systems, i.e., under conditions where standard methods normally fail.

REFERENCES

1. A.M. Wasserman, A.L. Kovarskii, L.L. Yasina, and A.L. Buchachenko, *Teor. i eksper. Khimiya*, **13**, 30 (1977).
2. A. Carrington and A.D. McLachlan, "Introduction to Magnetic Resonance with Applications to Chemistry and Chemical Physics", Harper & Row Publishers, New York, 1967.
3. A.M. Wasserman, T.A. Aleksandrova, and Yu.E. Kirsh, *Vysokomol. Soyed.*, (1979) (in press).
4. V.N. Tsvetkov, V.E. Eskin, and S.Ya. Frenkel, "Struktura macromolekul v rastvorah", Nauka, 1964, (in Russian).
5. T.M. Birshtein, A.M. Skvortsov, and A.A. Sariban, *Vysokomolek. Soyed.*, **19A**, 63 (1977).
6. Yu.Ya. Gotlib and A.M. Skvortsov, *Vysokomol. Soyed.*, **18A**, 1971 (1976).
7. A.M. Wasserman, T.A. Aleksandrova, and Yu.E. Kirsh, *Vysokomol. Soyed.*, (1979) (in press).
8. "Structural Studies of Macromolecules by Spectroscopic Methods", K.J. Ivin, ed., J. Wiley and Sons, 1976.
9. H. Hayashi, F. Hamada, and A. Nakajima, *Macromol. Chem.*, **178**, 827 (1977).
10. A.L. Buchachenko and A.M. Wasserman, "Stable Radicals", Khimiya Publishers, Moscow, 1973 (in Russian).
11. A.M. Wasserman, G.V. Korolev, A.L. Kovarskii, A.I. Malakhov, and A.M. Fokin, *Izv. AN SSSR, Ser. Khim.*, 322 (1973).
12. T.A. Aleksandrova, A.M. Wasserman, A.L. Kovarskii, and A.A. Tager, *Vysokomolek. Soyed.*, **18B**, 326 (1976).
13. T.A. Aleksandrova, A.M. Wasserman, and A.A. Tager, *Vysokomolek. Soyed.*, **19A**, 137 (1977).
14. A.L. Buchachenko, A.L. Kovarskii, and A.M. Wasserman in "Advances in Polymer Science", (Z.A. Rogovin, ed.), p. 33, Halsted Press Wiley, New York, 1974.

Spin-label and Spin-probe Studies of Polyethylene

G. GORDON CAMERON

Department of Chemistry, University of Aberdeen,
Meston Walk, Old Aberdeen AB9 2UE, Scotland

1. INTRODUCTION

Both spin-probe and spin-label techniques are currently in vogue as methods of studying polymer transitions and dynamics. In spin-labelling the paramagnetic molecule is covalently bound to the polymer chain. The synthesis of labelled polymers, however, is not always a straightforward operation, especially if it is desired to attach the label at specific points on the polymer chain. Polymer samples containing spin probes, on the other hand, can usually be prepared by straightforward physical mixing because in the probe experiment the paramagnetic species is present as a "free" or "guest" molecule in the polymer matrix at concentrations of 10 to 100 ppm. The spin-probe method therefore appears to be the more attractive. Nevertheless, in favourable circumstances spin-label studies are capable of yielding more information. Polyethylene provides a particularly suitable subject for contrasting the advantages and disadvantages of the two methods and for showing how they may be complementary.

In common with the vast majority of such studies, the paramagnetic species employed were nitroxides, chosen in this case to ensure that both label and probe were structurally identical. For reasons discussed elsewhere in this volume, nitroxides are eminently suited for this type of work.[7]

2. SYNTHESIS OF SPIN-LABELLED POLYETHYLENE AND THE SPIN-PROBE

Pure polyethylene is a rather difficult polymer to label because it contains no functional groups and is chemically fairly inert. Recourse was therefore made to copolymers containing 0.5 to 2 per cent of carbonyl groups of the type

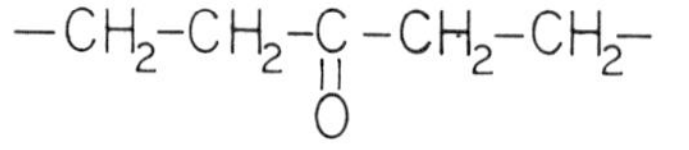

$-CH_2-CH_2-C(=O)-CH_2-CH_2-$ copolymer of ethylene and carbon monoxide (PE/CO)

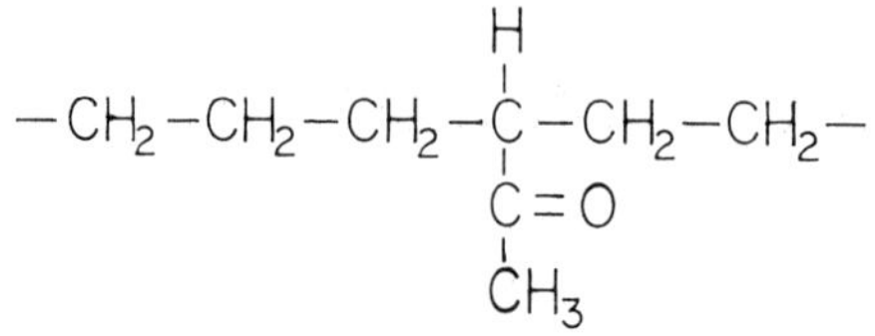

$-CH_2-CH_2-CH_2-CH(C(=O)CH_3)-CH_2-CH_2-$ copolymer of ethylene and methyl vinyl ketone (PE/MVK)

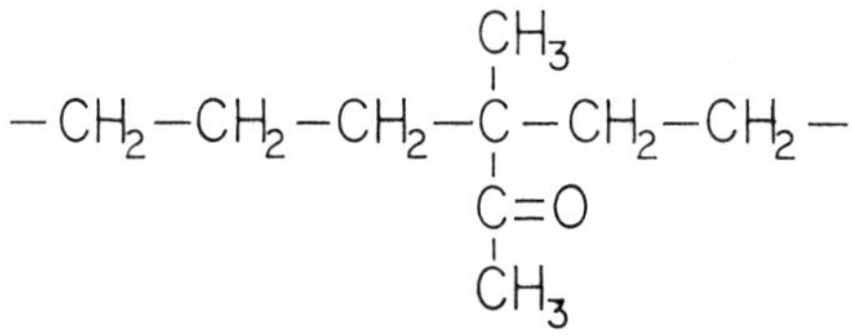

$-CH_2-CH_2-CH_2-C(CH_3)(C(=O)CH_3)-CH_2-CH_2-$ copolymer of ethylene and methyl isopropenyl ketone (PE/MIPK)

These polymers were all prepared by radical polymerisation and may be regarded as low-density polyethylene.[1] The carbonyl groups provided the reactive functions for labelling via the Keana synthesis[2] with 2-methyl-2-amino-propan-1-ol.

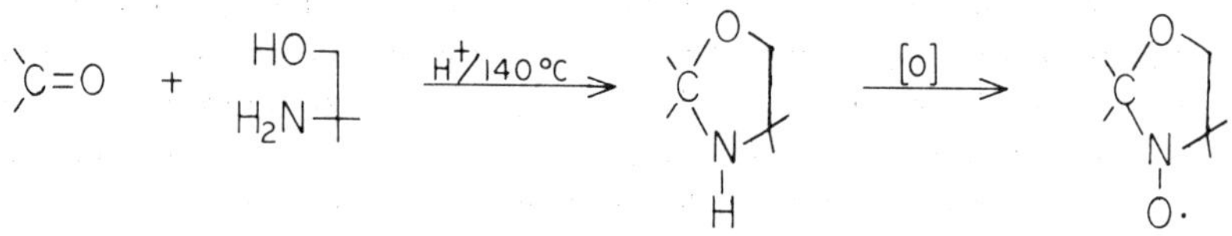

$$>C=O \;+\; HO-CH_2-C(CH_3)_2-NH_2 \xrightarrow{H^+/140\,°C} \text{oxazolidine (N–H)} \xrightarrow{[O]} \text{oxazolidine (N–O·)}$$

The three polymers therefore carried labels of the type

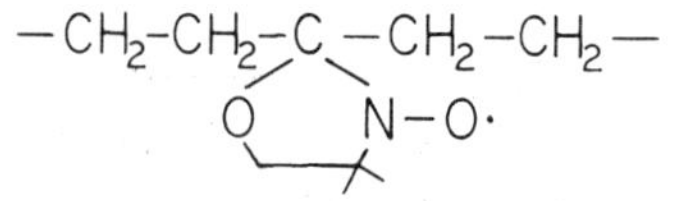

$-CH_2-CH_2-C(-CH_2-CH_2-)$ with oxazolidine ring (O, N–O·) for PE/CO[3] **I**

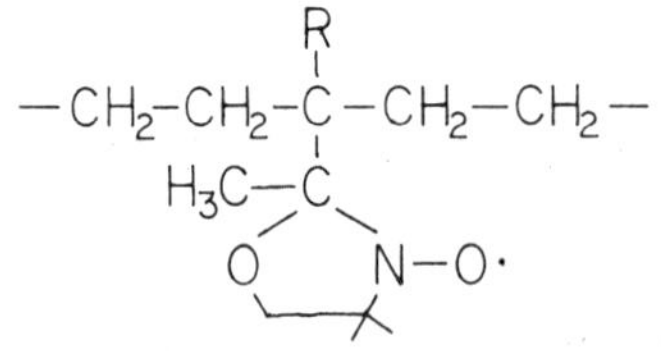

$-CH_2-CH_2-C(R)(-CH_2-CH_2-)-C(H_3C)$ with oxazolidine ring (O, N–O·) for PE/MVK (R=H) and for PE/MIPK ($R=CH_3$)[4] **II**

To avoid spin-exchange contributions to the line-width only a fraction of the carbonyl groups were labelled in this manner so that the final spin concentrations lay in the range one radical per 50,000 to 235,000 monomer units.

In the PE/CO label (I) one of the carbon atoms of the oxazolidine ring is part of the backbone. This imposes some restriction on the mobility of this label compared with the labels (II) on the other two polymers where the oxazolidine ring is a pendant group. These differences in mobility are reflected to a degree in the e.s.r. spectra of the polymers.

All three polymers had the same nitrogen isotropic coupling constants and the same isotropic g-values:[3, 4]

$$a_N = 40.0 \pm 0.3 \text{ MHz and } g_{iso} = 2.00570 \pm 0.00007$$

The spin-probe was 2,2,5,5-tetramethyl-3-oxazolidinyloxy (TMOZ)

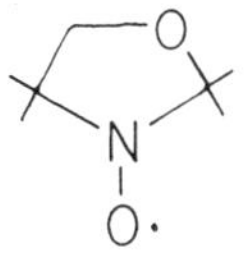

prepared from acetone by the Keana synthesis above. The e.s.r. spectrum in solution had the characteristic three narrow lines of equal intensity with $a_N = 40.8 \pm 0.1$ MHz and $g_{iso} = 2.0056 \pm 0.001$. At 291K the centre line-width was 1.94 ± 0.03 MHz and no proton hyperfine structure was observed.[3] Other groups of workers[5,6] have conducted spin-probe investigations on polyethylene using the probe 2,2,6,6-tetramethyl-4-hydroxypiperidin-1-oxyl benzoate (BzONO).

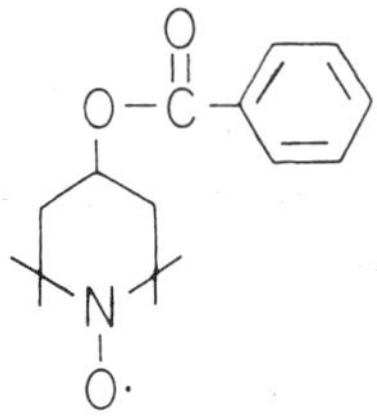

The BzONO molecule is somewhat larger than the TMOZ molecule and this difference results in different values of transition temperatures obtained with the two probes as will be discussed later.

3. CALCULATION OF CORRELATION TIMES FOR NITROXIDE RADICAL TUMBLING

For correlation times in the range $10^{-9} < \tau_c < 10^{-6}$ s the e.s.r. spectrum of the nitroxide label or probe is characteristic of a solid or powder. A marked change occurs, however, when

$$\tau_c^{-1} \sim 4/3\ \pi\ [A_{zz}^{-1/2}\ (A_{xx} + A_{yy})]$$

where the A's are the principal values of the hyperfine coupling tensor. For dialkyl and saturated cyclic nitroxides the value of τ_c in this critical region is about 3×10^{-9} s and the spectrum for correlation times shorter than this is of the typical motionally-narrowed form. In many spin-label and probe investigations on polymers the observed correlation times straddle this critical region and two distinct ranges can be identified—"slow" tumbling $10^{-9} < \tau_c < 10^{-6}$ s and "fast" tumbling $5 \times 10^{-11} < \tau_c < 10^{-9}$ s. The method of calculating correlation times depends upon the range into which τ_c falls, that is, whether the spectrum is of the solid or motionally-narrowed type.

In the fast tumbling region τ_c is generally calculated from equations which relate τ_c to the linewidth parameter T_2. These equations are discussed in full elsewhere in this volume and further elaboration here is unnecessary.[7] Values of τ_c for labelled polyethylenes in the molten state and in solution, and for spin-probed polyethylenes at temperatures above T_{50G}, were calculated by this means. Unresolved couplings to protons in the TMOZ bring about inhomogeneous broadening of the spectral lines and for accurate calculations appropriate corrections must be made. In this work a simulation technique, as used in an earlier investigation on polystyrene,[7] was employed to produce a plot of "true" versus "observed" linewidths. The values of τ_c calculated from the true linewidths were 5–10 percent lower than those calculated from the uncorrected data. Other workers[8, 9] have used the method of "additional broadening" to allow for the effects of unresolved proton coupling[10] in calculations of correlation times. The magnitudes of the corrections in this work are comparatively small but in other systems considerable errors may be introduced by calculating correlation times from uncorrected linewidths. Errors in the values of τ_c generate further errors in calculations of energies of activation.[11]

In the region of "slow" tumbling ($\tau_c >$ ca.10^{-9} s) the e.s.r. spectrum is no longer given by a simple superposition of Lorentzian lines. Correlation times in this range are observed for most spin-labelled polymers in the bulk or solid state and for spin-probed polymers at temperatures below T_{50G}. For many polymer studies calculations of correlation times in this range are therefore of considerable importance. Such calculations, however, are not always straightforward because the spectra are affected in subtle and complex ways by the motions of the radicals and the magnetic spin interactions. The most rigorous approach to this problem is via spectral simulations, involving well-established theoretical principles,[12] which require accurate values of the magnetic tensors A and g. For most applications in polymer chemistry, however, simplified methods of estimating τ_c are adequate. Some of these simplified methods are briefly reviewed below. All are based on a relatively straightforward analysis of the outer extrema. Usually the inward shift of these extrema, with the onset of molecular motion, from the "rigid limit" positions is exploited. The central region of a nitroxide e.s.r. spectrum is, of course, also very sensitive to dynamic effects, but it is also very sensitive to deviations of the magnetic parameters from axial symmetry and hence general methods based on this region have not been developed.

Correlation times in the range $10^{-9} < \tau_c < 10^{-7}$ s may be derived from the theoretical dependence of τ_c on the parameter κ.[13]

$$\kappa = \frac{H(\tau) - H(\tau \to 0)}{H(\tau \to \infty) - H(\tau \to 0)} \times 100 \tag{1}$$

In equation 1, $H(\tau)$, $H(\tau \to 0)$ and $H(\tau \to \infty)$ are the magnetic field values corresponding to the maxima of the first derivative absorption line ($m_N = +1$) at the correlation time τ_c, at the "free" ($\tau_c < 10^{-11}$ s) and the rigid limit ($\tau_c > 10^{-7}$ s), respectively. The parameter κ, which is shown as a function of τ_c in Figure 1, is weakly dependent on the g and A values, and on the intrinsic linewidth (δ) over the range 1 to 4G. Although based on an arbitrary jump

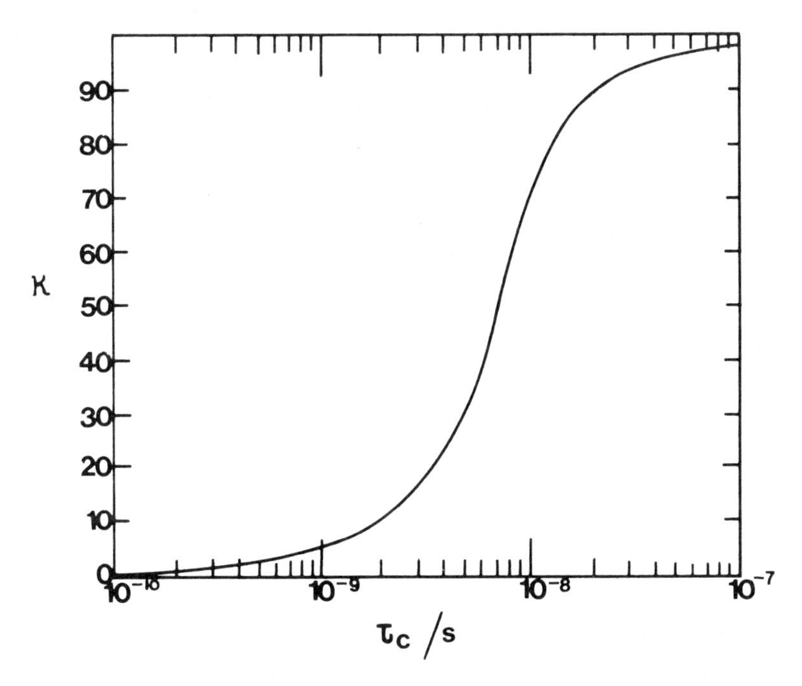

FIGURE 1 The parameter κ as a function of correlation time. (From ref. 13). (Reproduced by permission of the North-Holland Publishing Company).

model for radical reorientation, it has been claimed that the method is also applicable to a Brownian diffusion model.[8] This method may be useful for estimating correlation times which are just outside the rapid tumbling region but which are too short to be estimated by the other methods discussed here.

Goldman, Bruno and Freed[14] have shown that for correlation times in the range $10^{-8} < \tau_c < 10^{-6}$ s the correlation time is related to the change in extrema separation by the expression

$$\tau_c = a(1 - S)^{b} \tag{2}$$

In equation 2, S is the ratio A'_{zz}/A_{zz} where $2A'_{zz}$ is the extrema separation at the correlation time τ_c and $2A_{zz}$ is the rigid limit value. The parameters a and b depend on the diffusion model and on the intrinsic line width δ. Simulation studies show that for a given value of A_{zz} the value of S is insensitive to changes in the other A-tensor components or in the g-tensor components. The magnitude of A_{zz} does affect the value of S but this dependence can be allowed for by a simple scaling procedure without appreciable loss of accuracy. The method becomes very insensitive when $\tau_c > 10^{-7}$ s where $(1 - S) < 3\%$, and in the region of $\tau_c > 10^{-8}$ s the results are very sensitive to the choice of δ. The reader is referred to the original literature[12, 14] for a more detailed account of this method. When using this and other methods for determining τ_c the choice of an appropriate diffusion model can be facilitated by measuring the relative shifts ΔH_+ and ΔH_- of the low- and high-field extrema with respect to the rigid limit positions.[15] The value of ΔH_- is much more sensitive to the character of the motion than is ΔH_+. This can be seen in Figure 2 which shows the parameter $R(=\Delta H_-/\Delta H_+)$ as a function of the low-field shift ΔH_+ for Brownian rotational diffusion and large jump diffusion.

By numerical solution of diffusion-coupled Bloch equations McCalley, Shimshick and McConnell[16] calculated nitroxide e.s.r. spectra as a function of τ_c. From these simulations the dependences of τ_c on the inward shifts ΔH_+ and ΔH_- of the low- and high-field lines respectively were derived as shown in Figure 3. From these reference curves correlation times in the range $10^{-8} < \tau_c < 10^{-6}$ s may be estimated. This method does not appear to be oversensitive to variations in g- or A-tensors or to the intrinsic linewidth. Thus, for correlation times in the region 10 to 20 ns a variation in Δg by 0.0015 shifts ΔH_+ by ~0.3G, a 4G variation in A_{zz} produces ~0.1G variation and a 30% change in δ produces ~0.1G variation.[17] In the present work the Brownian diffusion model was found to be appropriate and the method of McConnell *et al.* was employed to calculate correlation times in the slow tumbling region.

In all three of these methods it is necessary to establish the position of one or both extrema at the rigid limit. For labelled biopolymers this is often achieved by extrapolation of solution spectra to infinite viscosity[16] but such a method is inappropriate for solid polymers and usually an extrapolation based on

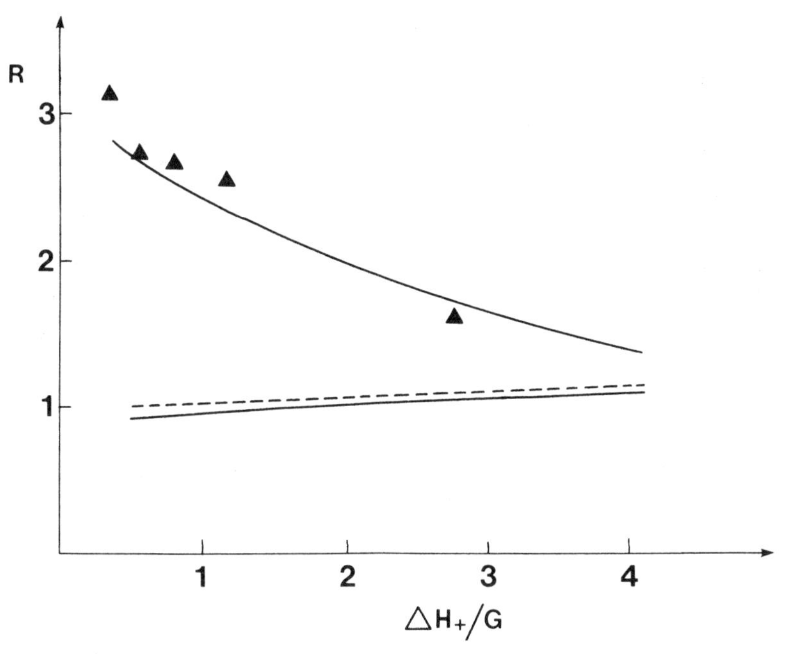

FIGURE 2 The parameter R as a function of the shift ΔH_+. The upper full line and the triangles correspond to Brownian rotational diffusion and the lower full and dotted lines to arbitrary jump tumbling (——) $A_{\parallel} = 34.3$ G, $A_{\perp} = 6.5$ G, $\Delta g = 0.0051$; (- - -) $A_{\parallel} = 32.0$ G, $A_{\perp} = 6.0$ G, $\Delta g = 0.0053$; (▲) $A_{\parallel} = 30.8$ G, $A_{\perp} = 5.8$ G, $\Delta g = 0.0053$. (From ref. 15). (Reproduced by permission of the North-Holland Publishing Company).

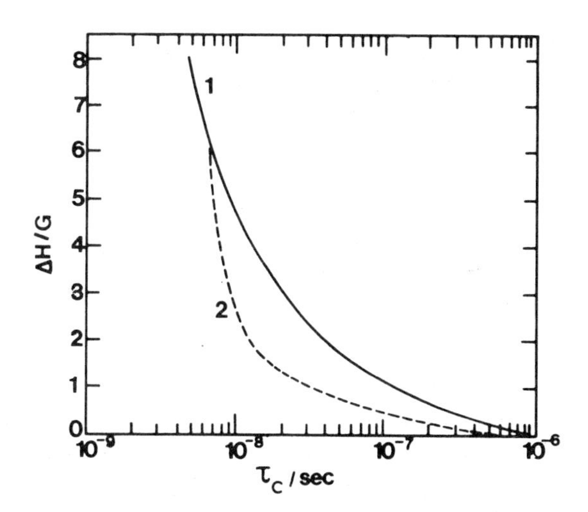

FIGURE 3 Inward shifts of (1) high-field and (2) low-field extrema of derivative spectra as a function of τ_c calculated for $g_z = 2.0024$, $g_x = 2.0090$, $g_y = 2.0060$, and $A_{zz} = 34.3$ G, $A_{xx} = 6.8$ G, $A_{yy} = 6.29$ G, and intrinsic linewidth = 2.73 G. (From ref. 16). (Reproduced by permission of the North-Holland Publishing Company).

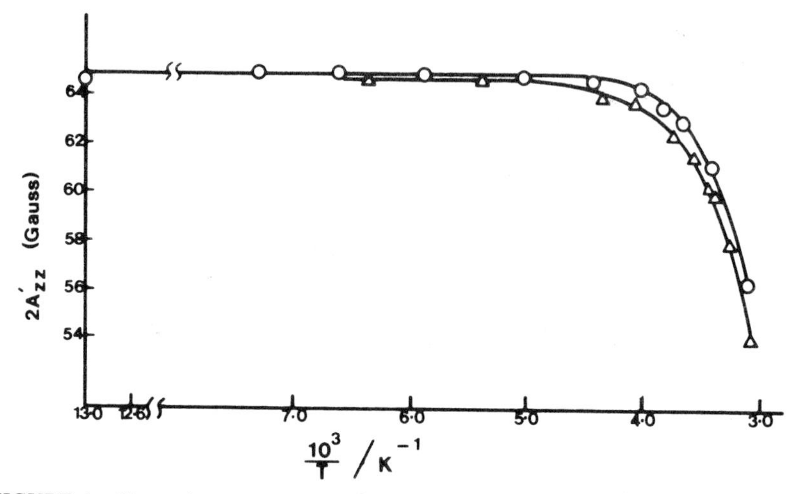

FIGURE 4 Change in extrema separation ($2A'_{zz}$) with temperature for spin-labelled polyethylene PE/CO. ○, amorphous; Δ, annealed. (From ref. 3). (Reproduced by permission of the Pergamon Press).

temperature is required. In the case of spin-labelled polyethylene the rigid limit value $2A_{zz}$ was established by plotting A'_{zz} versus reciprocal temperature[3, 4] (Figure 4): there is an asymptotic approach to the rigid limit value as $1/T$ increases. A similar procedure was found to be adequate in earlier experiments with labelled polystyrene[18] but in more recent work with labelled polyacrylates it was found that the rigid limit had not been reached at 77K and it was necessary to extrapolate to zero K by a least-squares curve-fitting procedure to a polynomial in T.[19]

Finally, a method which may be capable of extending the range of measurable correlation times to $\tau_c < 5 \times 10^{-6}$ s has been devised by Mason and Freed.[20] The technique is based on the lifetime broadening of nitroxide spectra near the rigid limit and requires the determination of the dimensionless parameters W_i which are given by

$$W_i = \Delta_i / \Delta_i^r \tag{3}$$

where $i = l$ or h, corresponding to the low- or high-field lines respectively; Δ_i values are the half-widths at half-height and Δ_i^r are the corresponding values at the rigid limit. Plots of $(W_i - 1)$ as a function of τ_c are available[20] and these plots can be represented in the form

$$\tau_c = a(W_i - 1)^{-b} \tag{4}$$

The coefficients a and b are dependent on the diffusion model and on the intrinsic linewidth δ. W_i, like S, is insensitive to deviations from axial A and g tensors and is virtually independent of A_{zz} over the range 27 to 40G. Unfortunately, there are various contributions to Δ_i^r which are rapidly averaged with the onset of molecular motion resulting in a decrease in Δ_i. When this is the case it is necessary to estimate a value of δ and hence effective inhomogeneous widths $\Delta_i^{er} (< \Delta_i^r)$ from the relationships

$$2\Delta_{\mathrm{l}}^{er} = 1.59\delta$$
$$\text{and } 2\Delta_{\mathrm{h}}^{er} = 1.81\delta \tag{5}$$

Equation 3 then becomes

$$W_i = \Delta_i/\Delta_i^{er} \tag{3'}$$

The correct value of δ gives the same value of τ_c from both extrema. As a further check another value of τ_c may be obtained (when feasible) from S.

4. RELAXATIONS IN BULK SPIN-LABELLED POLYETHYLENES

Of the three spin-labelled polyethylenes the PE/CO polymer proved to be the most useful because the label remained reasonably stable well above the melting point of the polymer and correlation times could be measured over a wide temperature range as shown in the map in Figure 5.[3] The labels on PE/MVK, and even more so on PE/MIPK polymers, decomposed rapidly at temperatures above *ca.* 295K in the bulk and hence the correlation map for these polymers (Figure 9) covers a much more restricted temperature range.[4]

Relaxations in labelled PE/CO

The correlation map for the bulk polymer in Figure 5 shows two well defined transition temperatures, the upper of which corresponds to the melting temperature T_m. In the temperature region below 350K, the lower transition temperature, it seems likely that the β-relaxation, usually attributed to cooperative segmental motion in the amorphous phase, is being observed. The energy of activation for motion in this region lies in the range 29 kJ mol^{-1} for "amorphous" polymer to 24 kJ mol^{-1} for the annealed sample (see Table 1). These figures are in reasonable agreement with comparable data (24 to 33 kJ mol^{-1}) from proton NMR relaxation studies covering a similar temperature range. It follows that the discontinuity in the correlation map at 350K corresponds to T_α above which the α-relaxation also occurs. This relaxation is

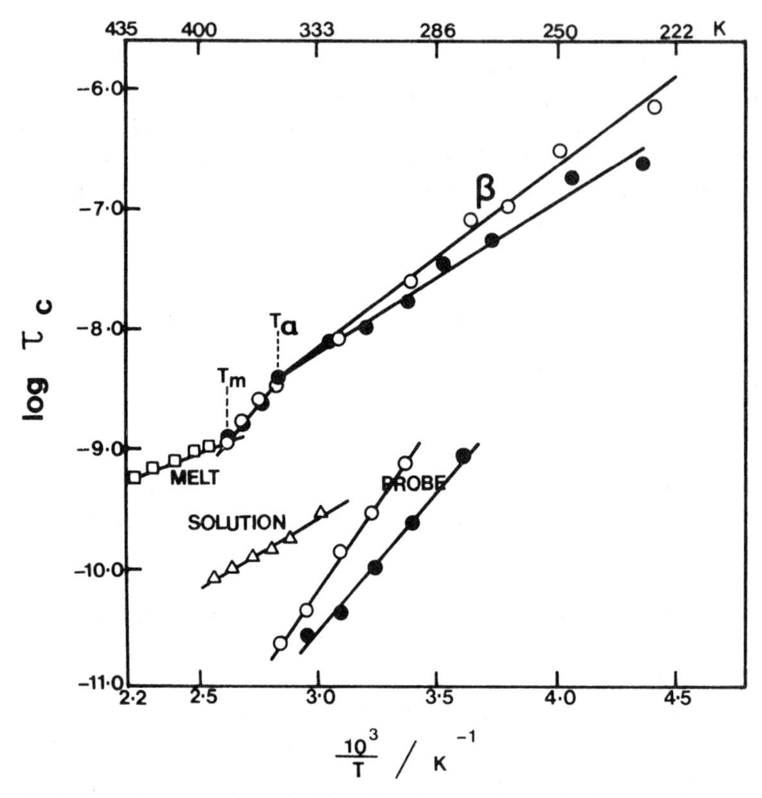

FIGURE 5 Correlation map for spin-labelled and spin-probed polyethylene PE/CO. Spin-probe experiments in bulk polymer only. ○, amorphous; ●, annealed. τ_c in sec. (From ref. 3). (Reproduced by permission of the Pergamon Press).

attributed to large-scale reorientations of polymer chains within crystalline regions.

In the β-relaxation region annealing the sample, by holding it at its melting point for 10 minutes and cooling slowly, brought about shorter correlation times and lower energies of activation. This can be explained on the basis that annealing allows the reformation of crystalline regions from which irregularities such as labelled segments are excluded. Therefore, after annealing, the label is in a more amorphous environment than before where motion is generally less restricted. Annealing had no effect on relaxations in the α-region or the melt.

Data for the molten polymer are discussed and compared later with data for the polymer solutions.

The glass temperature T_g of semicrystalline polymers has been the subject of

TABLE 1
Activation energies for spin-label and spin-probe relaxations in polyethylenes

Region	Temp. range, K	Polymer		Activation energies, kJ mol^{-1}			Ref.
				Amorphous	Melt or Solution	Annealed	
β-transition	220–350	PE/CO	label	28.6±2.0		23.9±1.5	3
	275–350		probe	54.2±1.9		44.6±1.8	3
β-transition	220–330	PE/MVK PE/MIPK	label	25.1±1.2		18.3±1.8	4
	283–383		probe	44.1±2.0		46.1±2.0	25
α-transition	350–380	PE/CO	label	48.0±1.9		47.3±5.0	3
Melt	380–450	PE/CO	label		15.2±1.4		3
—	123–203	PE/MVK	probe	6.90±0.5		6.86±0.5	25
—	113–193	PE/MIPK	probe	5.02±0.5		5.19±0.5	25
Solution (in xylene)	330–390	PE/CO	label		22.9±1.9		3
Solution (in xylene)	293–367	PE/MVK PE/MIPK	label		27.6±1.9		4

some controversy over the years. The particular case of polyethylene has been reviewed by Boyer[21] who has suggested that the spin-label and probe methods may provide some clarification in such controversial cases. Unfortunately our labelled polyethylenes provide little assistance here. To define the amorphous T_g it is necessary to identify in the correlation map a discontinuity T_β similar to T_α and T_m. All that can be said is that T_β lies below 222K. The method of McConnell *et al.* for calculating τ_c from the extrema separation is insensitive when τ_c is greater than *ca.* 10^{-6} s. Therefore, to calculate values of τ_c for PE/CO below 220K some other technique must be considered. It may be possible to extend the calculations of correlation times into the region where the extrema have reached the rigid limit positions by adopting the method of Mason and Freed.[20] Alternatively, more sophisticated instrumental techniques such as saturation studies[22] may be employed to gain access to this region of very slow tumbling. We hope to conduct such investigations in the future. The temptation to refer to Figure 4 and associate T_β with the region of the curve at which $2A'_{zz}$ begins to decrease significantly should be avoided.

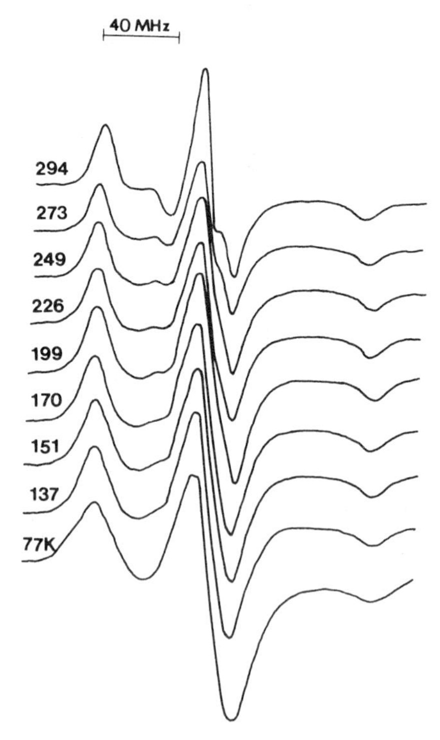

FIGURE 6 Solid state spectra of labelled polyethylene PE/CO in the temperature range 77 to 294 K. (From ref. 3). (Reproduced by permission of the Pergamon Press).

This region does not represent a discontinuity or the sudden onset of molecular motion. Even at temperatures on the horizontal part of this curve, where the extrema separation equals the rigid limit value, the subtle changes in the spectral line shapes with temperature indicate the existence of some motion (see Figure 6). The problem of defining T_g is referred to again when the spin-probe experiments are considered.

The spectra of the labelled PE/CO polymer in the temperature range 320–350K shown in Figure 7 merit further comment. The extrema, particularly the low-field peaks, show pronounced subsplitting at these temperatures. This could arise if these spectra comprise a motionally-narrowed type superimposed upon a solid-state spectrum. Such composite spectra have been reported in spin-probe studies by other authors.[5,23] If this interpretation is

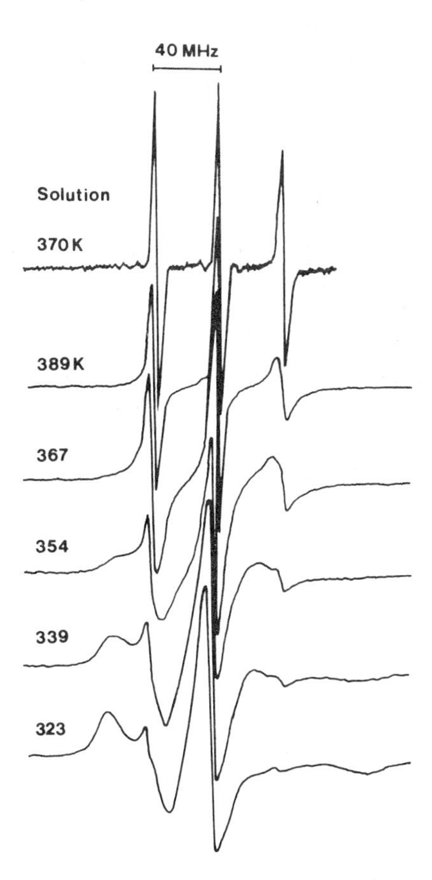

FIGURE 7 Solid state spectra of labelled polyethylene PE/CO in the temperature range 323 to 389 K and solution spectrum (3% in xylene) at 370 K. (From ref. 3). (Reproduced by permission of the Pergamon Press).

correct then it means that certain of the labelled sites enjoy a much higher degree of motional freedom than the others; the more mobile labels would have values of $\tau_c < 10^{-9}$ s, which is at least an order of magnitude smaller than τ_c for the remaining labels. It should be recalled, however, that the axis system of the oxazolidine nitroxide label I is fixed with respect to the polymer with the $2p\pi$ orbital on the nitrogen atom (the z-axis) parallel to the direction of the extended backbone (see Figure 8). Therefore, crankshaft-type rotations, which are expected to be the main form of macromolecular motion at these temperatures and which occur predominantly in the x–y plane, can average out the anisotropy between the x- and y-tensor elements only. The z-axes of the labels, however, remain effectively fixed until large-scale reorientations can occur. This markedly anisotropic type of motion could give rise to or at least contribute to the subsplitting in the spectra of the labelled PE/CO polymer. Although spectral simulation studies are necessary to furnish a more conclusive interpretation of these spectra it may be significant that this feature was absent from the spectra of the polymers carrying the motionally freer labels of type II.

Relaxations in labelled PE/MVK and PE/MIPK

The correlation map for these polymers is shown in Figure 9. Because of the poor stabilities of the labels in these polymers only the region of the β-relaxation was observed.[4] There is no significant difference in behaviour between the labelled PE/MVK and PE/MIPK polymers over the temperature range 220–330K, but the correlation times and activation energies are lower for these polymers than for PE/CO over the same temperature range (Figure 9 and Table 1). The lower motional freedom of the label in PE/CO is qualitatively obvious from inspection of the e.s.r. spectra of this polymer and of

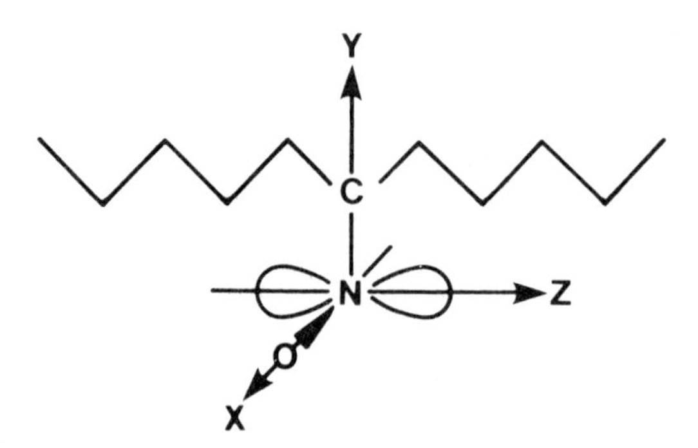

FIGURE 8 Axis system for oxazolidine nitroxide spin label on PE/CO.

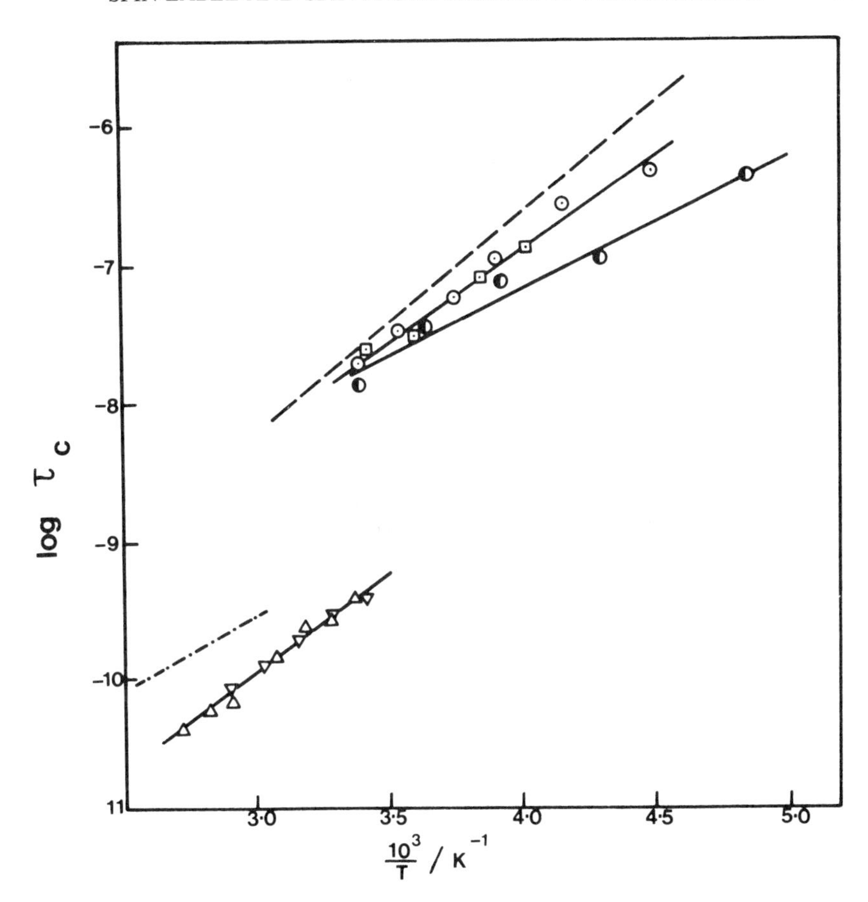

FIGURE 9 Correlation map for spin-labelled polyethylenes PE/MVK and PE/MIPK. ⊙, amorphous PE/MVK; ⊡, amorphous PE/MIPK; ◐, annealed PE/MVK; Δ, PE/MVK in xylene solution (5% wt/vol) ∇, PE/MIPK in xylene solution. – – – amorphous PE/CO and –·– PE/CO in xylene solution from Figure 5. τ_c in sec. (From ref. 4). (Reproduced by permission of the American Chemical Society).

PE/MVK in Figure 10. The temperature at which the latter assumes the motionally-narrowed form is *ca.* 320K (Figure 10c) but at this temperature the spectrum of the labelled PE/CO polymer is still of the solid-state type and comparable narrow lines do not develop until *ca.* 365K (Figure 7). It is also noteworthy that the low-field subsplitting discussed above was not in evidence in the spectra of labelled PE/MVK or PE/MIPK at any temperature in the range studied. This observation gives some support to the hypothesis that anisotropic motion of the label contributes to the subsplitting in the spectra of labelled PE/CO because in the other two polymers the labels should tumble

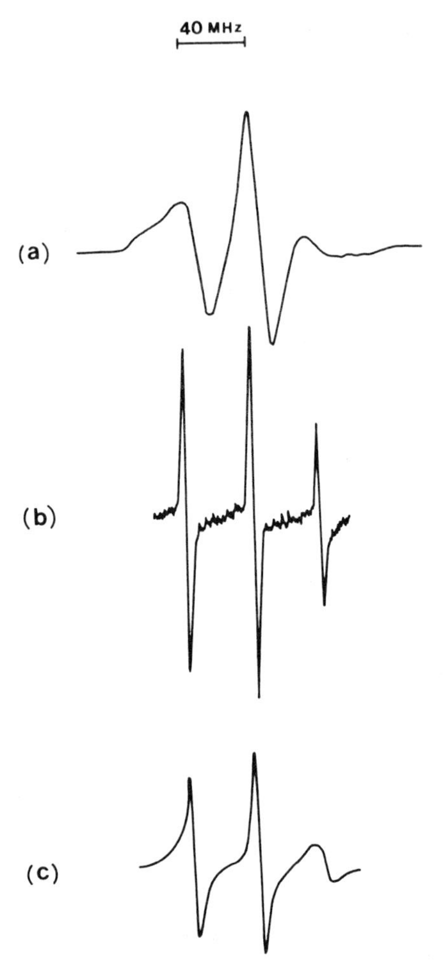

FIGURE 10 E.s.r. spectra of spin-labelled polyethylene PE/MVK. (a) solid polymer at 299 K, (b) xylene solution at 296 K, (c) solid polymer at 321 K. (From ref. 4). (Reproduced by permission of the American Chemical Society).

more isotropically by rotations about the C_1C_2 axis in co-operation with crankshaft motion.

Relaxations in solution and in molten PE/CO

The correlation map in Figure 5 also includes data for the labelled PE/CO polymer in solution (3 per cent in xylene) which yield an activation energy of 22.9 ± 1.9 kJ mol^{-1}. The molecular weight of this polymer was *ca.* 100,000 and for reasons discussed elsewhere[7] it is assumed that relaxation in solution can be attributed to local segmental modes with a negligible contribution from whole-molecule rotation. Therefore, the activation energy for rotation in the "isolated" polymer molecule can be obtained by subtracting from 22.9 kJ mol^{-1} the activation energy (9.0 ± 0.3 kJ mol^{-1}) for viscous flow of xylene. This yields 13.9 ± 1.2 kJ mol^{-1} as the energy barrier to internal rotations in the polymer, a figure which is remarkably close to the 12 kJ mol^{-1} barrier to internal rotations in ethane molecules in the gas phase.[24] In the molten state, where whole-molecule tumbling is also insignificant, the overall activation energy for relaxation (from Figure 5) is 15.2 ± 1.5 kJ mol^{-1}. But the energy of activation for viscous flow of molten polyethylene[25] is 46–61 kJ mol^{-1} and clearly viscous drag of the type exerted by solvent on polymer segments does not operate in the bulk, molten polymer. In other words, the energy barrier to segmental rotation in the molten polymer is not much greater than the internal barriers to rotation in an isolated polymer chain. This is in general accord with Bueche's hypothesis[26] that in the melt chain segments are surrounded by equivalent segments and no hydrodynamic shielding is operative. Thus, the reason why relaxations in the melt are slower than in solution is not because the energy barriers are greater in the former, but because the pre-exponential factor for rotations in the melt (1.5×10^{10} s^{-1}) is lower than in the isolated molecule (9.5×10^{10} s^{-1}).

Correlation times for the labelled PE/MVK and PE/MIPK in xylene solution were the same within experimental error as shown in Figure 9. At a given temperature, however, the correlation times for these polymers are about half of the value for PE/CO. Again it is evident that the nitroxide label has greater motional freedom when pendant to (as in II) rather than integrated (as in I) with the polymer backbone. The greater motional freedom of the labels II over I is due, however, to a higher pre-exponential factor for rotation of the former because the energy barrier to rotation of the labels in solution (in contrast to the situation in the bulk polymers) is higher in PE/MVK and PE/MIPK than in PE/CO (Table 1). The higher pre-exponential factor for labels II is a reflection of their greater conformational freedom.

If the temperature of the xylene solution drops below 333K the dissolved polyethylene comes out of solution as a cloudy, gelatinous precipitate. When this occurred the lines in the e.s.r. spectrum of labelled PE/CO broadened abruptly signifying a marked increase in τ_c. The e.s.r. spectra of the other two labelled polymers in contrast showed no obvious change on precipitation and the values of τ_c lay on the same Arrhenius plot as those from solution spectra. This indicates that the labels II have as much mobility in the precipitated swollen state as in solution while the movements of label I are greatly hindered on precipitation. This somewhat surprising effect must also have its origin in the greater motional freedom of label II compared with I.

5. SPIN-PROBE EXPERIMENTS WITH POLYETHYLENE

The most striking difference between the spin-probe and spin-label results with the three polyethylene samples is the very high mobility of the nitroxide probe compared with the labels. It is clear that in the bulk polymer the frequency of probe rotation is much higher than that of the α- or β-process in the polymer. The greater mobility of the probe is immediately apparent from the correlation map for PE/CO in Figure 5: at a given temperature the τ_c value for the probe in the *solid* polymer is lower, over most of the temperature range covered, than τ_c of the labelled polymer *in solution*. Another contrasting feature between the two experiments is highlighted by considering the temperature range 200–230K. In this region there is remarkably little change in the spectra of the labelled polymers; the extrema separation changes very little and alterations in the line shapes are comparatively minor. For the TMOZ probe, on the other hand, this temperature range represents the critical region alluded to earlier; below 200K the spectra are typical of the slow-motion powder variety with extrema separation of 60–65G, but above 230K the spectra assume the motionally-narrowed form with extrema separation of 30–35G (see Figure 11).[27] Evidently in this temperature range there is a transition in the polymer to which the probe is highly sensitive but which is much more difficult to detect by the spin-label experiment. At higher temperatures, say beyond T_α, the spin-probe is tumbling very rapidly and the resulting narrow-line spectrum is uninformative. In this region, however, the e.s.r. spectra of the spin-labelled polyethylene PE/CO are still amenable to analysis. Thus, the two different experiments can be complementary.

Plots such as Figure 11, of extrema separation versus temperature from spin-probe investigations have been employed frequently in the recent past as a means of defining transition temperatures, particularly the glass transition, in polymers. In this connection it has now become customary to quote T_{50G},[5,6] the temperature at which the separation between the spectral extrema becomes 50G. An extrema separation of 50G corresponds closely to the mid-point of the transition between the slow-motion and fast-tumbling types of spectra and at this point the correlation time for probe tumbling is *ca.* 3×10^{-9} s. Although T_{50G} is not identical with T_g the two temperatures are related, though not linearly.[6] Generally T_{50G} is greater than T_g but the exact relationship depends, among other factors, on the size of the probe, the smaller probes giving lower values of T_{50G}.[5,28] The value of T_{50G} for PE/MVK from Figure 11 is 220K which is substantially lower than the figure of 277–280K recorded by Kumler and Boyer[6] for the probe BzONO in polyethylene. Part of this discrepancy can be attributed to the much smaller size of the TMOZ probe compared with BzONO, but Kumler and Boyer found that different samples of polyethylene gave different T_{50G} values with BzONO. They suggested that these variations could have arisen from the different preparative histories of the samples.[6]

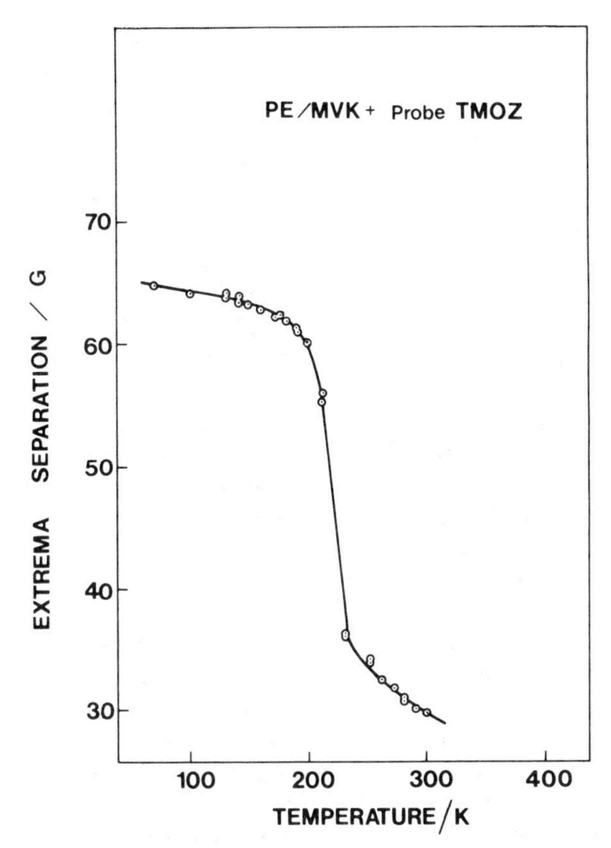

FIGURE 11 Extrema separation as a function of temperature for polyethylene PE/MVK containing the probe TMOZ.

In an earlier spin-probe study of PE/CO it was tentatively suggested that at T_{50G} the probe might be responding to the γ-relaxation,[3] that is to T_γ (usually quoted as *ca.* 150K) rather than to the amorphous phase glass transition T_g. However, T_γ from mechanical relaxation measurements does not appear to be sensitive to variations in sample crystallinity[29] and on hindsight it seems more probable that T_{50G} is associated with the glass transition in polyethylene as in most other polymers. If this is so T_{50G} values for polyethylene are of considerable interest because there is some controversy surrounding the amorphous T_g of polyethylene. Boyer has argued convincingly[6, 21] for a T_g value of 197 ± 10K. His argument was based in part on dilatometrically determined T_g values of ethylene-propylene copolymers, for which there is no ambiguity, extrapolated by the Gordon-Taylor equation to 100 per cent polyethylene. The argument was reinforced by spin-probe determinations of

T_g (using a T_{50G}–T_g calibration plot). The latter points fitted quite well to the Gordon-Taylor extrapolation. Unfortunately, there is as yet no T_{50G}–T_g calibration available for the probe TMOZ and the only contribution Figure 11 makes to this controversy is to indicate that T_g lies below 220K! This agrees, incidently, with the conclusions from the spin-label investigations.

An alternative method of using the spin probe is to measure correlation times over a range of temperatures then to plot these as a correlation map. If the temperature range spans a transition temperature then, in ideal circumstances, this will be revealed as a discontinuity T_d (as for T_m and T_α in Figure 5). Such a correlation map for PE/MVK containing TMOZ is shown in Figure 12. Assuming that the data points lie on two straight lines, these intersect at 268K. Although some uncertainty surrounds the value of T_d, because of lack of points in its close vicinity, it is unlikely to be as low as 220K, the value for T_{50G}. A decision as to which of the two, T_d or T_{50G}, is the more significant in pinpointing T_g for polyethylene and other controversial polymers must await a more detailed investigation. It is quite possible, however, that reliable T_g values from T_d measurements will require the establishment of a T_d–T_g calibration plot as in the case of T_{50G}. Whether or not T_d is sensitive to probe size also remains to be seen.

When comparing transition temperatures from spin-probe or spin-label studies with those obtained by more conventional techniques such as torsion pendulum measurements, dilatometry, etc., the very different time scales of the experiments should not be overlooked. The e.s.r. experiment is concerned with frequencies of the order 10^7 Hz, whilst in many mechanical methods the relevant frequency is of the order 1 Hz. Frequency and temperature in such measurements are often related by an Arrhenius equation

$$f_{max} = B \exp(-\Delta H_a/RT) \tag{6}$$

in which f_{max} is the frequency at which a mechanical or dielectric loss peak in a polymer is a maximum for a temperature T, and ΔH_a is an apparent enthalpy of activation. A modified Arrhenius equation, the Williams-Landel-Ferry equation,[30] is also appropriate to many polymer systems. For polymers such as polyethylene with low values of ΔH_a, the observed value obtained for a transition temperature is therefore very sensitive to the frequency range of the method employed. Thus, at T_d in Figure 12 the value of τ_c is 10^{-8} s and since τ_c is related to f_{max} by $\tau_c^{-1} = 2\pi f_{max}$, the corresponding though fictitious value of f_{max} is 1.59×10^7 Hz. Assuming that the above Arrhenius equation is valid and taking ΔH_a for polyethylene[6] as 18 kcal mol^{-1}, the values of T_d for frequencies of 10^2 and 1 Hz are 197K and 179K respectively. In the absence of calibration data, these values are more relevant in comparisons with transitions observed by mechanical techniques. If, for the reasons given

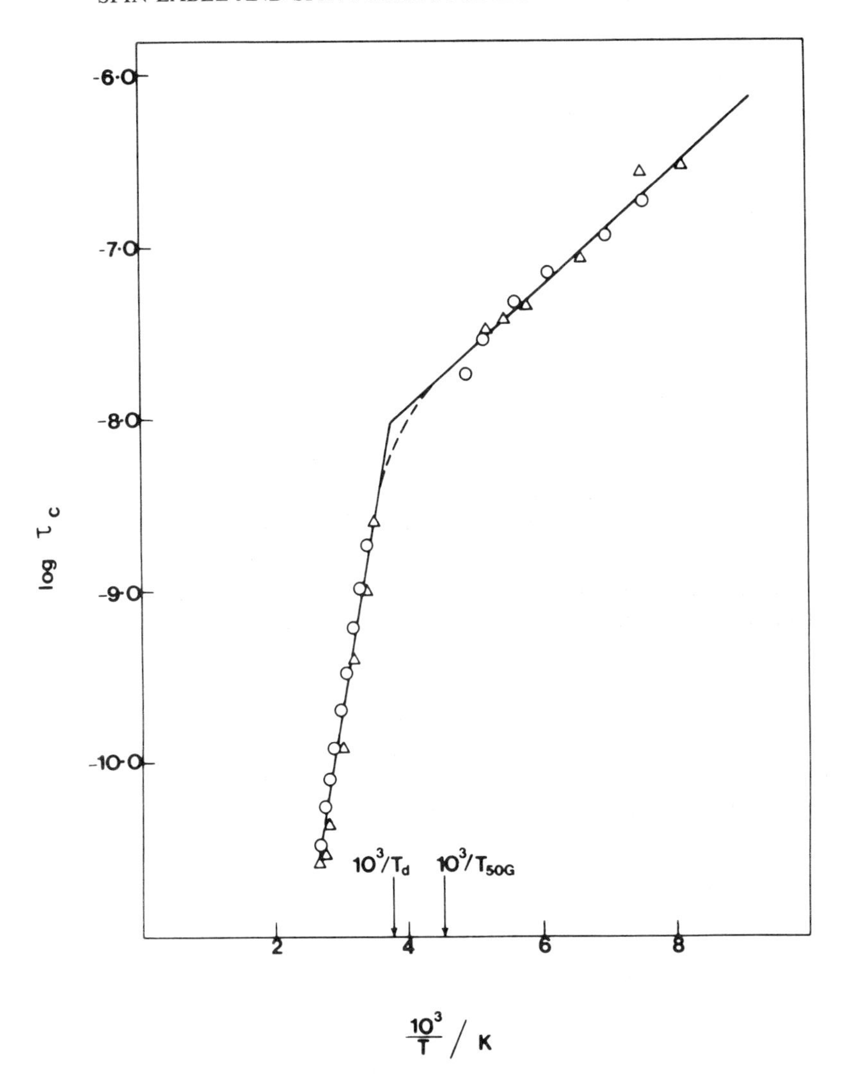

FIGURE 12 Correlation map of PE/MVK containing the spin-probe TMOZ. Δ, annealed; ○, amorphous. τ_c in sec.

earlier, T_d is taken as the amorphous glass temperature at $\sim 10^7$ Hz then a value of T_g in the range 175–200K is indicated. While the approximations and assumptions involved in reaching these figures must be acknowledged it is noteworthy that they are not far removed from Boyer's estimate.

It has already been noted that the activation energies for relaxation of the labelled polyethylene are in accord with other data for the β-process in polyethylene. The values for the activation energy of probe tumbling (Table 1) over the relevant temperature range are about twice what one would expect for the β-relaxation. This would suggest that the motions of the probe and the surrounding macromolecules are not linked in a simple manner. Although the probe enjoys greatest mobility when the polymer chains are undergoing large-scale reorientations, the energies of probe tumbling cannot be equated, as was the case with the label in PE/CO, with those governing polymer chain dynamics. For this reason it is difficult to interpret the very low activation energies for probe dynamics at temperatures below T_{50G}.

6. CONCLUSIONS

From this study of polyethylene we conclude that the spin-probe and spin-label techniques may in favourable circumstances complement one another. At low temperatures, where the motion of the spin-label is very slow and may be difficult to analyse, the spin-probe method may provide information. Conversely, in a higher temperature range, e.g., above the α-transition in polyethylene, the very rapid tumbling of the spin-probe can result in a narrow-line e.s.r. spectrum which is comparatively uninformative, but here the slower motion of the spin-label can yield spectra which may be subjected to a line-width analysis. The main value of the spin probe probably lies in its sensitivity to polymer transitions, notably the glass transition, but in a study of the energetics or absolute rates of macromolecular motions the spin-label experiment is likely to prove the more reliable of the two.

REFERENCES

1. We are indebted to Professor J.E. Guillet for his donation of these polyethylene samples.
2. J.F.W. Keana, S.B. Keana, and D. Beetham, *J. Amer. Chem. Soc.*, **89**, 3055 (1967).
3. A.T. Bullock, G.G. Cameron, and P.M. Smith, *Eur. Polym. J.*, **11**, 617 (1975).
4. A.T. Bullock, G.G. Cameron, and P.M. Smith, *Macromolecules*, **9**, 650 (1976).
5. G.P. Rabold, *J. Polym. Sci., Part A-1*, **7**, 1203 (1969).
6. P.L. Kumler and R.F. Boyer, *Macromolecules*, **9**, 903 (1976).
7. Contribution from A.T. Bullock in this volume.
8. A.M. Wasserman, T.A. Alexandrova, and A.L. Buchachenko, *Eur. Polym. J.*, **12**, 691 (1976).
9. J. Labský, J. Pilăr, and J. Kálal, *Macromolecules*, **10**, 1153 (1977).
10. A.N. Kuznetsov, A.Y. Kolkov, V.A. Livshits, and A.T. Mirsoian, *Chem. Phys. Lett.*, **26**, 369 (1974).
11. G. Poggi and C.P. Johnson, *J. Magn. Res.*, **3**, 436 (1970).
12. J.H. Freed in "Spin Labelling: Theory and Applications" Ed. L.J. Berliner, Academic Press, N.Y., (1976), 53.
13. A.N. Kuznetsov, A.M. Wasserman, A.U. Volkov and N.N. Korst, *Chem. Phys. Lett.*, **12**, 103 (1971).
14. S.A. Goldman, G.V. Bruno, and J.H. Freed, *J. Phys. Chem.*, **76**, 1858 (1972).

15. A.N. Kuznetsov and B. Ebert, *Chem. Phys. Lett.*, **25**, 342 (1974).
16. R.C. McCalley, E.J. Shimshick, and H.M. McConnell, *Chem. Phys. Lett.*, **13**, 115 (1972).
17. E.J. Shimshick, private communication. We are grateful to Dr. Shimshick for providing numerical data from ref. 16.
18. A.T. Bullock, G.G. Cameron, and P.M. Smith, *J. Polym. Sci., Polym. Phys. Ed.*, **11**, 1263 (1973).
19. A.T. Bullock, G.G. Cameron, and V. Krajewski, *J. Phys. Chem.*, **80**, 1792 (1976).
20. R.P. Mason and J.H. Freed, *J. Phys. Chem.*, **78**, 1321 (1974).
21. R.F. Boyer, *Polymer*, **17**, 996 (1976).
22. S.A. Goldman, G.V. Bruno, and J.H. Freed, *J. Chem. Phys.*, **59**, 3071 (1973).
23. N. Kusumoto, M. Yonezawa, and Y. Motozato, *Polymer*, **15**, 793 (1974).
24. E.B. Wilson, *Adv. Chem. Phys.*, **2**, 367 (1959).
25. H. Schott and W.S. Kagham, *J. Appl. Polym. Sci.*, **5**, 175 (1961).
26. F. Bueche, "Physical Properties of Polymers", Interscience, N.Y., 1962.
27. A.T. Bullock, G.G. Cameron, and N.K. Reddy, unpublished results.
28. R.F. Boyer and P.L. Kumler, *Macromolecules*, **10**, 461 (1977).
29. K.-H. Illers, *Kolloid Z.Z. Polym.*, **252**, 1 (1974).
30. M.L. Williams, R.F. Landel, and J.D. Ferry, *J. Amer. Chem. Soc.*, **77**, 3701 (1955).

DISCUSSION

A. M. Bobst (University of Cincinnati, Cincinnati, Ohio): In the ordered state of nucleic acids the spin label displays a smaller activation energy, than it does after the swelling of the nucleic acids. This oddity has so far not been explained. You seem also to observe an oddity when measuring activation energies on polyethylene. What do the activation energies here really reflect?

G. G. Cameron: The oddest feature concerning energetics that we observe is, in my view, the very high activation energy for probe tumbling above T_{50G}. Exactly what the significance of these very high values is I cannot say, but they are certainly much higher than the energy of activation of the β-relaxation in the polymer.

L. J. Berliner (Ohio State University, Columbus, Ohio): Are polarity corrections necessary when using the slow motion approximation calculations? These are necessary in lipid-membrane order parameter calculations.

G. G. Cameron: Not in the case of a non-polar medium such as bulk polyethylene. In general, however, it is important that the rigid limit parameters be determined under the same conditions of polarity as obtained for the measurement of τ_c.

W. G. Miller (University of Minnesota, Minneapolis, Minnesota): I am skeptical of your tentative interpretation of anisotropic motion rather than a superposition of a fast and a slow component in polyethylene at some temperatures. This is discussed in some detail in my contribution to this volume. Do you object to the interpretation that spin labels may be moving differently in the amorphous region than those in the crystalline region, resulting in a superposition spectrum?

G. G. Cameron: Not entirely. See my reply to Sohma below.

W. G. Miller: I feel that one must be very careful in interpreting breaks in Arrhenius plots. Such apparent breaks can come from using an inappropriate method to extract the correlation times from the spectra, as has been shown in the case of membrane studies.

J. Sohma (Hokkaido University, Sapporo, Japan): You observed complex spectra at higher temperatures. These spectra have small peaks corresponding to those from an isotropic sample. You do not interpret these spectra as the superposition of the spectral components from the mobile and the immobile parts of the sample. I have good reason to attribute such spectra to such a superposition. Could you tell us the reasons why you do not attribute these spectra to superposition?

G. G. Cameron: My interpretation of these spectra is not quite as positive as you imply. I accept the point that at higher temperatures the complexity of the spectra from labelled PE/CO may result from a superposition of motionally-narrowed and broad-line spectra. It should not be overlooked, however, that the label on PE/CO cannot rotate independently of the main chain. If crankshaft rotation is the main form of motion, then the z-tensor elements will be less readily averaged than the x- and y-tensor elements. This could contribute to the complexity of the spectra under discussion.

N. Kusumoto (Kumamoto University, Kumamoto, Japan): You showed the temperature variation of extrema separation, and said that the T_{50G} is attributable to the initiation of "crankshaft" motion. How did you identify this temperature, T_{50G}, as resulting from "crankshaft" motion?

G. G. Cameron: I think I said that T_{50G} (for spin probe experiments) is related to the glass transition and hence to the molecular motions, including segmental rotation, which are associated with this transition.

L. J. Berliner: Concerning the sharp line component, could this be due to just a very mobile minor component of the whole PE sample? Since it probably represents only 0.1–0.2% of the total spins, its significance to the overall spectral interpretation may be small. Also, if you increase the microwave power you can saturate out those lines if they really are due to a mobile component.

G. G. Cameron: I agree with all these points.

R. F. Boyer (Midland Macromolecular Institute, Midland, Michigan): According to the conventional view of polyethylene morphology, one would expect to find your nitroxide label in both the loose loops and in the cilia, but probably not in the crystalline regions. The loops, being anchored on both ends to crystallites, would be under greater restraint than the cilia which have one free end. This situation, having two amorphous regions, might explain some aspects of your spectra. One other comment that I'd like to make is that Cowie and McEwen [*Macromolecules*, **10**, 1124 (1977)] present convincing evidence that the T_g for amorphous PE is 197–200 K.

L. J. Berliner: An interesting study would be to follow the effect of increasing spin probe concentration on the temperature transitions observed, that is, could the probe be acting as a classical solute?

G. G. Cameron: This is possible.

N. Kusumoto: Have you ever seen in your spin-label work in polyethylene a small narrowing taking place below T_{50G}? In the slide you showed I did not see this small narrowing in the spin-label case. This is, however, very popular in the spin-probe work.

G. G. Cameron: Not in spin-label experiments but we have seen this effect in spin-probe measurements.

Spin Label and Probe Studies of Relaxations and Phase Transitions in Polymeric Solids and Melts

PERTTI TÖRMÄLÄ

Institute of Materials Science, Tampere University of Technology, P.O.B. 527, SF-33101 Tampere 10, Finland[a] *and Deutsches Kunststoff-Institute, D-6100 Darmstadt, Federal Republik of Germany*[b]

GUNTER WEBER[c]

Deutsches Kunststoff-Institut, D-6100 Darmstadt, Federal Republik of Germany

and

J. JOHAN LINDBERG

Department of Wood and Polymer Chemistry, University of Helsinki, Meritullinkatu 1, Helsinki, Finland

1. INTRODUCTION

The studies of molecular relaxations in polymeric systems have aroused large interest among polymer scientists during the last decades [1]. Especially during the last years the importance of these studies has been emphasized while it has been shown that molecular motions contribute to most of the important properties of the polymers. We could mention e.g., mechanical properties (tensile, flexural, compressive, impact and yield strength, creep, stress relaxation and fracture phenomena), thermal properties (glass transition T_g, melt transition), electrical properties (conductivity), solid state reactivity and diffusivity, melt rheology, crystallization and solution properties [see e.g., refs. 1–6].

Traditionally molecular dynamics of polymers have been studied by means of dielectric and mechanical loss measurements [1]. During the last years a multitude of new relaxation measurement techniques have been developed.

[a] Present address where the correspondence should be sent.

[b] Visiting researcher, January 1977—July 1978.

[c] *Present address: Ingenieurabteilung Angewandte Physik, Bayer AG, D-4150 Krefeld 11, Federal Republik of Germany.*

These are based on such measurement methods as calorimetry, dilatometry, electrical conductivity, electron spin resonance (ESR), luminescence depolarization and quenching, neutron spectroscopy, nuclear magnetic resonance (NMR) and ultrasonic discharge [see e.g., refs. 7–21]. These techniques together allow relaxation measurements of polymers in the frequency range of 10^{-6}–10^{12} Hz.

The spin label and probe methods which are based on ESR measurements have been applied in studies of synthetic polymers during the last ten years. The progress in this field has been largely due to the synthesis of a large amount of new stable nitroxide free radicals [22] and to a detailed analysis of their electronic structures [23–24].

In spin label and probe experiments the free radicals are bonded to the polymer molecules by means of primary bonds and secondary forces, respectively. The difference between the conceptions, probe and label is vague in the case of polymers while, as we will see later, the behavior of probes approaches rapidly with increasing probe molecular weight (M_w) the behavior of labels.

Tables 1–4 summarize structural data of some of the most important nitroxide labels and probes which have been applied in studies of polymeric solids and melts.
Tetramethylpiperidine, tetramethylpyrroline, and dimethyloxazolidine derivatives are the most usual cyclic label and probe radicals.

The application of nitroxide labels and probes is based largely on the analysis of their motional state. This can be characterized quantitatively by means of the line width analysis of ESR spectra which leads to the calculation of the rotational correlation time τ_R. Figure 1 shows as an example ESR spectra of probe XXI in poly(ethylene oxide) (PEOX) at different temperatures [23].

Several theories which were developed by teams of Kivelson [24], Freed [25], McConnell [26], and Kuznetsov [27] allow the calculation of τ_R in the range of 10^{-7} s–5×10^{-11} s.

Accordingly the following convenient equations are usually applied in calculation of τ_R:

$$\frac{W_m}{W_0} = 1 - \frac{\tau_R}{\sqrt{3}\pi\Delta\nu(0)}(c_1 m + c_2 m^2) \tag{1}$$

where $m = \pm 1$, $\Delta\nu(0)$ = the width of the middle line (in MHz) and c_1 and c_2 depend on radical properties [24,25],

$$\tau_R = a(1 - A'_{ZZ}/A_{ZZ})^b \tag{2}$$

TABLE 1
Spin label and probe radicals: piperidine derivatives

X	N:O	M_w	Method	X	N:O	M_w	Method
$>C(H)H$	I	156	P[a]	$>C(H)-CH_2-COOH$	XI	—	L
$>C=O$	II	170	P	$>C(H)NH_2$	XII	171	L, P
$>C(H)OH$	III	172	L[b], P				
$>C(H)-O-C(=O)-(CH_2)_n-H$, n = 2	IV	228	P	$>C(H)-N(H)-C(=O)-(CH_2)_n-H$, n = 1	XIII	213	P
n = 15	V	410	P	n = 2	XIV	227	P
n = 17	VI	438	P	n = 3	XV	241	P
				n = 15	XVI	409	P
$>C(H)-O-C(=O)-C_6H_3(R_1)(R_2)$, $R_1 = R_2 = H$	VII	276	P	$>C(H)-N(H)-$(4,6-dichloro-1,3,5-triazin-2-yl)	XVII	319	L, P
$R_1 = H, R_2 = NO_2$	VIII	321	P				
$R_1 = R_2 = NO_2$	IX	366	P				
$>C(H)-O-(Si(CH_3)_2-O)_2-$(2,2,6,6-tetramethylpiperidine N–O·)	X	474	P				

[a]Spin probe
[b]Spin label

TABLE 2
Spin label and probe radicals: pyrroline derivatives 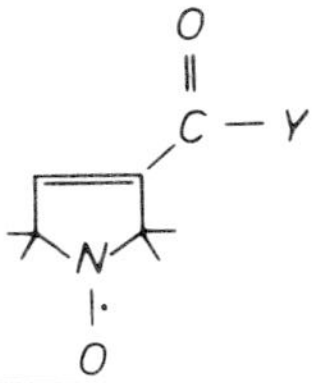

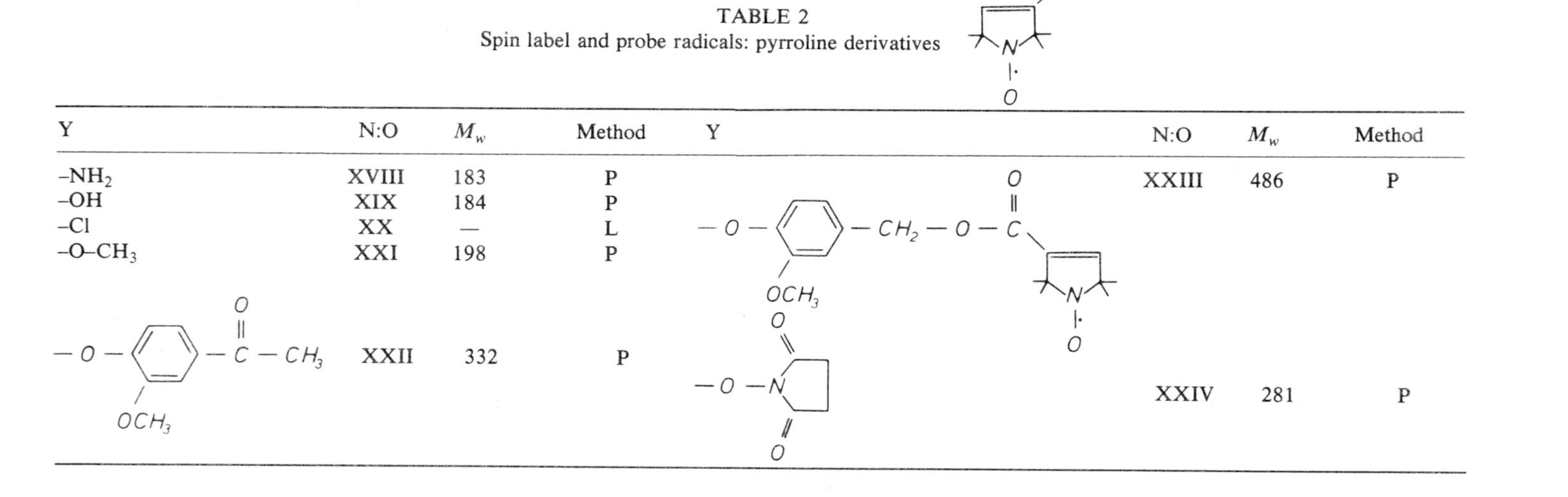

Y	N:O	M_w	Method	Y	N:O	M_w	Method
–NH_2	XVIII	183	P	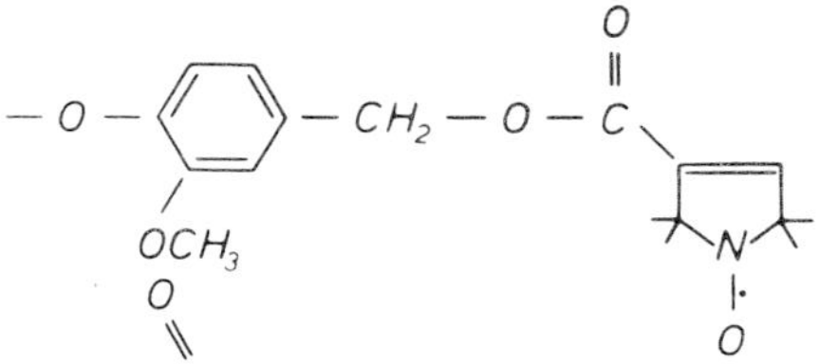	XXIII	486	P
–OH	XIX	184	P				
–Cl	XX	—	L				
–O–CH_3	XXI	198	P				
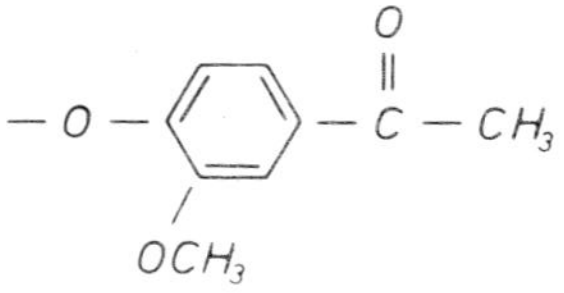	XXII	332	P	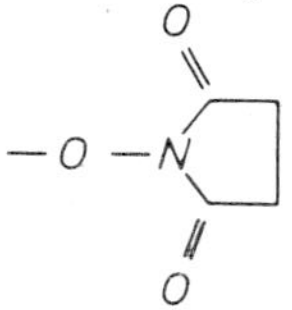	XXIV	281	P

TABLE 3
Spin label and probe radicals: oxazolidine derivatives

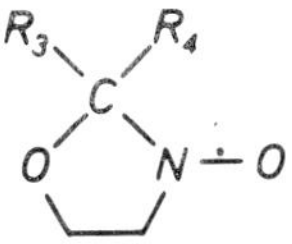

R_3	R_4	N:O	M_w	Method
$-H$	$-H$	XXV	144	P
$-CH_2-CH_2\sim$	$-CH_2-CH_2\sim$	XXVI	—	L
$-H$	$-CH(CH_2-CH_2\sim)(CH_2-CH_2\sim)$	XXVII	—	L
$-CH_3$	$-CH(CH_2-CH_2\sim)(CH_2-CH_2\sim)$	XXVIII	—	L

TABLE 4
Spin label and probe radicals: other radicals

Radical	N:O	M_w	Method
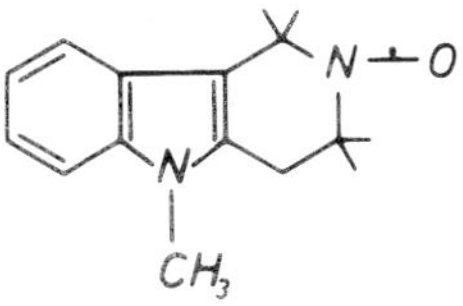 N—O, N, CH_3	XXIX	257	P
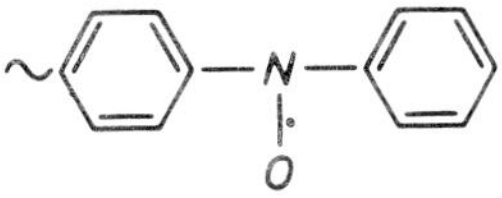 $\sim C_6H_4-N(\dot{O})-C_6H_5$	XXX	—	L
 $t-Bu-N(\dot{O})-t-Bu$	XXXI	144	P

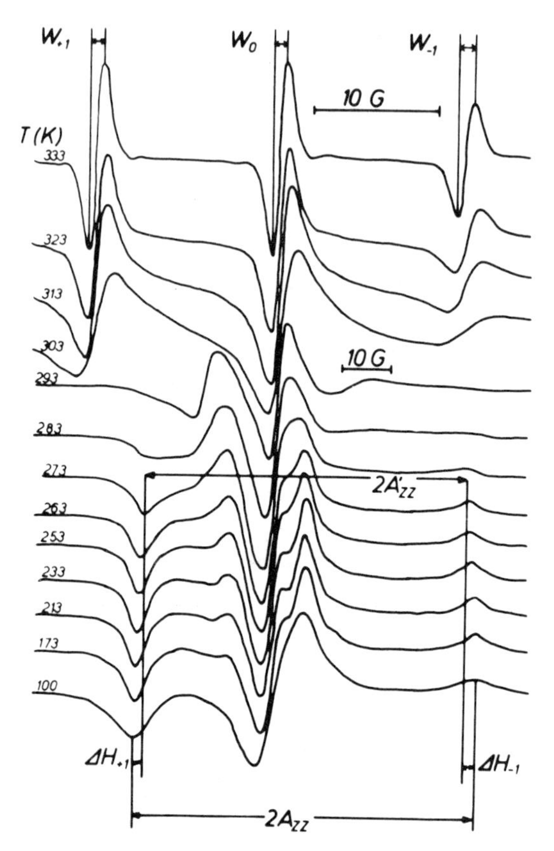

FIGURE 1 ESR spectra of the probe radical XXI in PEOX of $\bar{M}_n = 4000$ at different temperatures [23]; $2A_{zz}$ = the peak-to-peak separation of the rigid limit spectrum, $2A'_{zz}$ = the peak-to-peak separation at higher temperatures, W_i $(i = \pm 1, 0)$ = the line-widths of the hyperfine components.

where a and b are constants which depend on the selected diffusion model [25],

$$\Delta H_{\pm 1} = k_{\pm 1} \times \tau^{-2/3} \tag{3}$$

where $\Delta H_{\pm 1}$ are the shifts of the outer spectral lines and $k_{\pm 1}$ are constants [26],

$$\kappa = \frac{H(\tau_R) - H(\tau_R \to 0)}{H(\tau_R \to \infty) - H(\tau_R \to 0)} \tag{4}$$

where $H(\tau_R)$, $H(\tau_R \to 0)$, and $H(\tau_R \to \infty)$ are the positions of the maximum of the low field derivative line for the correlation time τ_R, for free rotation, and for the rigid state [27].

In polymer solids and melts there exists an intimate contact between polymer and additive (label or probe) molecules. Therefore the motional freedoms of labels and probes are sensitive to the motions of their environment and they become an indirect source of information on the dynamics of the surrounding polymer segments.

The early studies of Rozantsev [22] and Rabold [28] showed that the motions of polymer segments strongly contribute to the dynamic state of nitroxide probes.

In the early seventies Bullock, Cameron, and Smith in Aberdeen [29] and we in Helsinki [30] demonstrated that when τ_R values are converted into the corresponding effective frequency values of $f_{\text{eff}} = (2\pi\ \tau_R)^{-1}$ it is possible to establish correlations between dynamic data of ESR measurements and of other relaxation measurements.

The usefulness of this principle is shown in Figure 2 where dynamic ESR

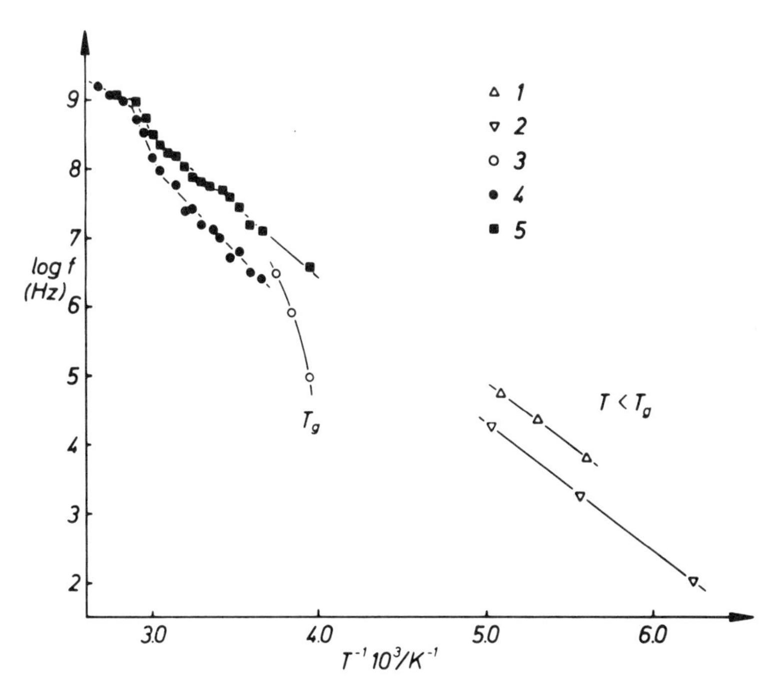

FIGURE 2 Relaxations of PEOX. Dielectric data: (1) and (2) [1], (3) (M_w of PEOX = 3×10^4 g mol^{-1}) [1]; ESR data: (4) label XX ($\bar{M}_n$ of PEOX = 2×10^4 g mol^{-1}) [23], (5) probe XXI (M_n of PEOX = 2×10^4 g mol^{-1}) [23].

data of medium M_w PEOX ($\bar{M}_n$ = 22,000) are compared with dielectric relaxation data of PEOX. Figure 2 shows that labels and probes in PEOX respond to the merged T_g and $T<T_g$ relaxation at high temperatures. The frequency difference between covalently bonded labels and free probes is only about one decade which clearly shows the local nature of the relaxation process.

A noticeable number of reports concerning spin label and probe studies of synthetic polymers have been published during the last years. Because this method is indirect the basic mechanisms which generally determine radical—polymer relationships are still only vaguely known.

2. T_g, T_{50G} CORRELATION

The first attempts to obtain a general interpretation of radical—polymer relationships are based on the so-called T_g, T_{50G} correlation.

When the ESR spectra of label and probe radicals in polymers are measured

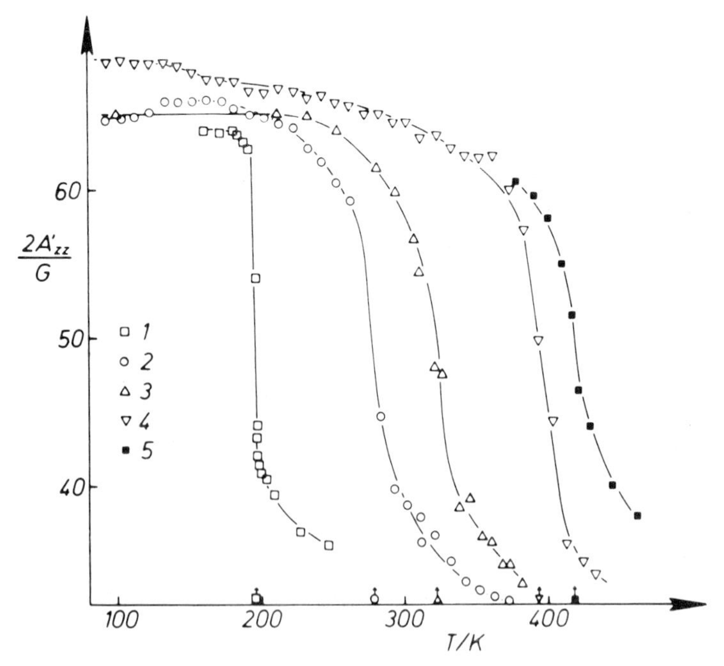

FIGURE 3 Plots of extrema separation ($2A'_{zz}$) *vs.* temperature in the ESR spectra of radical VII in poly(dimethyl siloxane) (PDMS) (1) [31], high density polyethylene (HDPE) (2) [28], polyisobutylene (PIB) (3) [32], poly(vinyl chloride) (PVC) (4) [28], and polycarbonate (PC) (5) [31]. T_{50G} values are given by arrows.

over a wide temperature range and the separations of the outer hyperfine extrema ($2\,A'_{ZZ}$) are given as a function of temperature sigmoidal curves are obtained. T_{50G} is the temperature at which the sigmoidal curves show the sharp break [28]. Figures 3 and 4 show that both the structure of polymer matrix and of radicals contribute to the value of T_{50G}.

The examination of Figure 3 shows qualitatively that T_{50G} increases with increasing T_g as Rabold first reported [28]. Kumler and Boyer gave for the correlation between T_g and T_{50G} of probe VII (BzONO) a semiquantitative, two-parameter expression

$$T_{50G} = T_g \Big/ \left(1 - \frac{0.03\,T_g}{\Delta H_a}\right) \tag{5}$$

where ΔH_a is the apparent activation energy for the T_g relaxation process [31].

We measured T_{50G} values of BzONO radical in some selected amorphous polymers [33]. The results are given in Figure 5 as a function of the dynamic

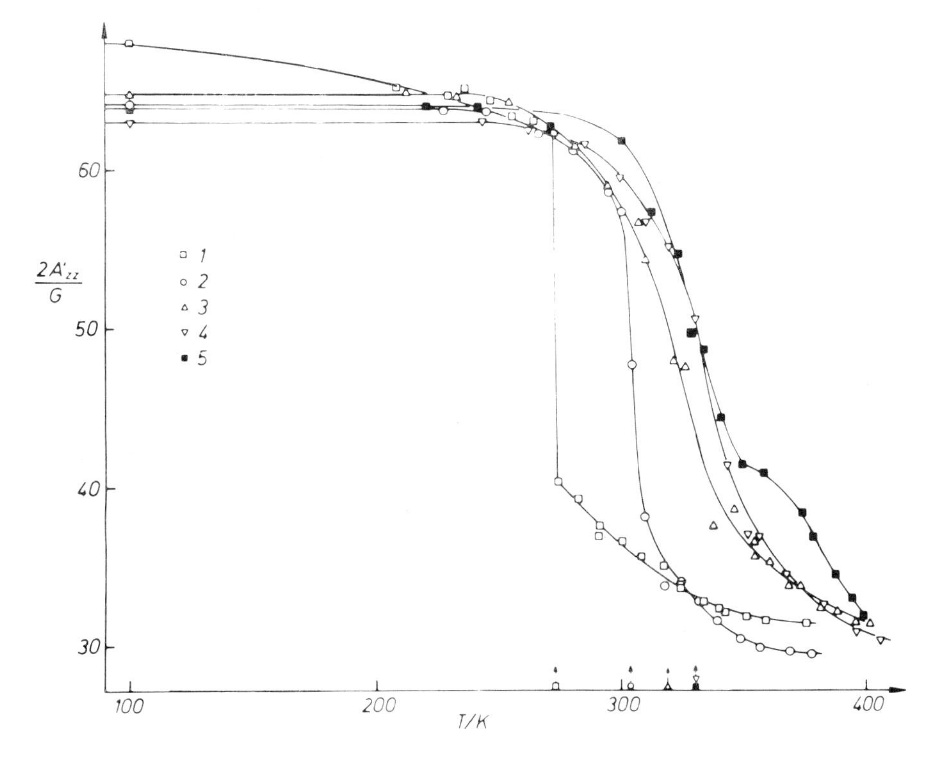

FIGURE 4 Plots of extrema separation ($2A'_{zz}$) *vs.* temperature in the ESR spectra of radicals III (1), XVIII (2), VII (3), XXII (4), and XXIII (5) in PIB. T_{50G} values are given by arrows. Reprinted by permission from Ref. [32].

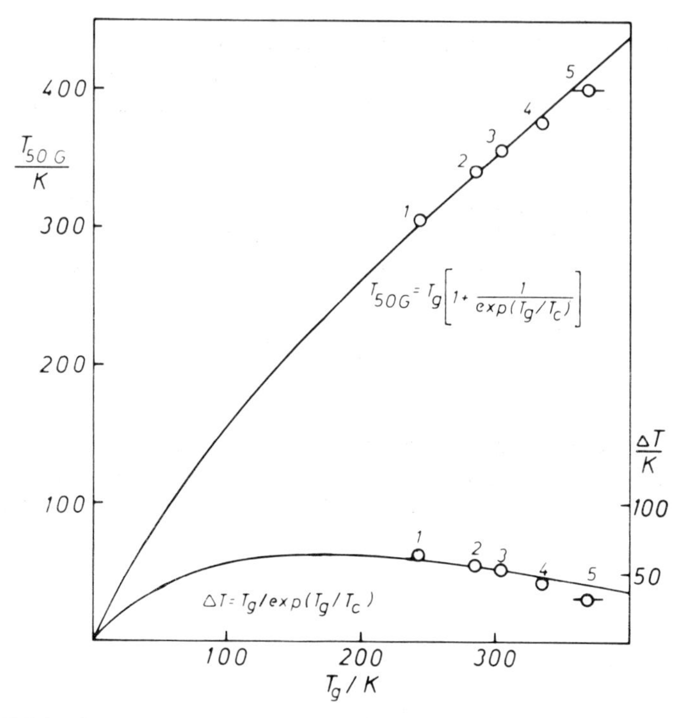

FIGURE 5 Calculated and measured T_{50G} values of probe VII in some amorphous polymers as a function of the dynamic T_g. Experimental values: (1) PIB, (2) poly(ethyl acrylate) (PEA), (3) poly(methyl acrylate) (PMA), (4) poly(vinyl acetate) (PVAC), and (5) poly(vinyl chloride) (PVC) [33]. (———) The drawn curves calculated according to equations (6) and (7).

(10^2 Hz) glass transition temperature (T_g) whose values were obtained from literature [1] or from manufacturer.

The temperature shift $\Delta T = T_{50G} - T_g$, which originates from the frequency difference between low frequency T_g and high frequency T_{50G} values, is also given in Figure 5.

T_g,ΔT data of Figure 5 obey fairly good the one parametric equation:

$$\Delta T = T_g/\exp\,(T_g/T_c) \tag{6}$$

where T_c is a correlation temperature. Accordingly the following expression describes the T_g, T_{50G} relationship:

$$T_{50G} = T_g \left[1 + \frac{1}{\exp\,(T_g/T_c)}\right] \tag{7}$$

A value of $T_c = 173$ K is obtained by best fit method. The corresponding calculated curves are also given in Figure 5. Within experimental error, there is a satisfying agreement between experimental and calculated results. This shows that in the studied T_g range the motions of bulky probe radicals are correlated to motions of polymer segments undergoing T_g relaxation.

When T_{50G} values of Figure 4 are given as a function of probe M_w it is seen that T_{50G} rapidly approaches a limiting value T^l_{50G} with increasing probe M_w. This behavior seems to be general as can be seen from Figure 6 where T_{50G} data of different probes in some polymers are given as a function of probe M_w [32]. The existence of T^l_{50G} values may indicate that the dynamic state of large flexible probes is determined straightforwardly by the dynamic state of the polymer matrix.

Figure 6 shows that in polymers with high T_g (such as PC and PDMPO) T^l_{50G} is attained by smaller M_w values as in polymers with low T_g (such as PIB). A logical conclusion is that in high T_g polymers probes are responding at T_{50G} to local mode motions while in low T_g polymers they respond to co-operative motions. This hypothesis is supported by the following frequency comparisons.

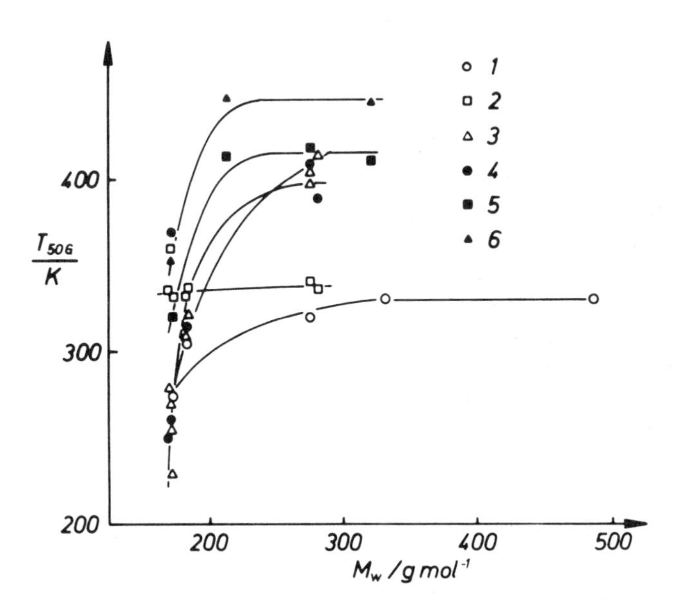

FIGURE 6 T_{50G} *vs.* probe molecular weight. (1) PIB, (2) PVDF, (3) PS, (4) poly-2,6-dimethylphenylene oxide (PDMPO), (5) PC, (6) poly-4,4′-isopropylidene-2,2′,6,6′-tetrachloro-1,1′-diphenyl carbonate ($PCCl_4$). Reprinted by permission from Ref. [32].

It has been proposed that the rotational frequency of nitroxide radicals is at T_{50G} the order of magnitude 10^7–10^8 Hz [31,33]. A more accurate frequency value can be deduced by the examination of the validity limits of the τ_R calculation methods. The rapid rotation theories give relevant τ_R values when ESR spectra are symmetrical and the hyperfine components do not overlap. According to Kuznetsov [27] the upper limit of τ_R is then $\sim 4 \times 10^{-9}$ s. The slow rotation theory of Freed is valid when the outer lines of the unsymmetrical ESR spectra do not converge to the motionally narrowed spectrum ($\tau > 10^{-8}$ s) [25]. When $2A'_{ZZ}$ is 50 G the outer lines clearly converge to the motionally narrowed spectrum although the hyperfine components overlap still considerably as can be seen e.g., from Figure 1. Therefore the value of τ_R corresponding to T_{50G} seems to lie between 4×10^{-9} s–10^{-8} s when $2A'_{zz}$ is the order of magnitude of 65 G. From these τ_R limits can be calculated a practical mean frequency $f = 3 \pm 1 \times 10^7$ Hz corresponding to T_{50G}.

Figure 6 shows that T_{50G} values have attained T^l_{50G} in the case of PIB, PVDF, PDMPO, PC, and $PCCl_4$ (chlorinated polycarbonate). These T^l_{50G} values can be compared with general T_g relaxation data of different polymers by calculating from literature the temperatures at which the frequency of T_g relaxation is 3×10^7 Hz and giving these temperatures as also T^l_{50G} values as a function of low frequency T_g (10^2 Hz). Figure 7 shows such a comparison for 12 different polymers. In addition to T^l_{50G} values of PIB, PVDF, PDMPO and PC, a value of 410 K for polyamide (PA) 610 was extrapolated from literature data [30].

Figure 7 shows that T^l_{50G} values of PIB, PVDF, and PA 610 coalesce with T_g relaxation data. T^l_{50G} data vs. low frequency T_g data gave by best fit method using equation (7) a mean T_c value of 214 K (T^l_{50G} values of PC and PDMPO were omitted from this calculation because they clearly do not represent T_g). Figure 7 shows also the calculated correlation line which was constructed by means of equation (7) using the T_c value of 214 K. The observed good correlation between measured and calculated results indicates that equation (7) describes generally the temperature shift between glass transitions at low and high frequencies. Figure 7 demonstrates also the importance of segmental relaxations in radical–polymer relationships while T^l_{50G} coalesces with temperatures of glass transition only if T_g and local segmental relaxations are coalesced at this temperature [32].

3. EXTENDED RELAXATION STUDIES OF THERMOPLASTICS

The above discussion shows that the examination of T_{50G} values alone do not suffice for the general characterization of radical—polymer interrelationships. The comparison of dynamic data of radicals and polymers over wide

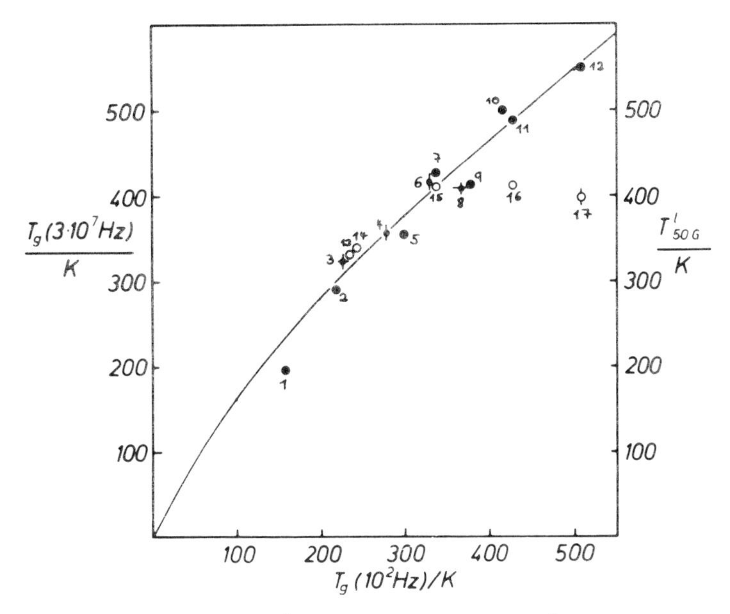

FIGURE 7 High frequency (3×10^7 Hz) T_g *vs.* low frequency (10^2 Hz) T_g. (●): (1) PDMS, (2) Natural Rubber (NR), (3) PIB, (4) polypropylene (PP), (5) PMA, (6) PVAC, (7) PA 610, (8) PVC, (9) Poly(ethylene terephthalate) (PETP), (10) poly(methyl methacrylate) (PMMA), (11) PC, (12) PDMPO; (○) ESR parameter T'_{50G}: (13) PIB, (14) PVDF, (15) PA 610, (16) PC, (17) PDMPO. (———) The drawn curve calculated according to equation (7).

temperature regions seems to be one way to general conclusions. The comparisons of dynamic data can be done conveniently by means of Arrhenius plots as was already demonstrated in Figure 2. We have applied this technique in the following discussion to clarify radical—polymer interrelationships in several thermoplastics.

3.1. Polyisobutylene (PIB)

The relaxation map of PIB is given in Figure 8. It shows two prominent relaxations: T_g and $T \ll T_g$ (δ process = rotations of CH_3 groups) [1,17]. Additionally Stoll and Pechhold have proposed the existence of two $T < T_g$ processes (γ and γ') [34]. From Figure 8 can be seen that T'_{50G} is in good agreement with the loss maximum temperatures of dielectric and mechanical relaxation measurements and with T_1 minima of NMR measurements when the frequency differences are taken into account. Evidently the large, flexible probes respond to the merged $T_g + T<T_g$ relaxation. Small probe I rotates at high

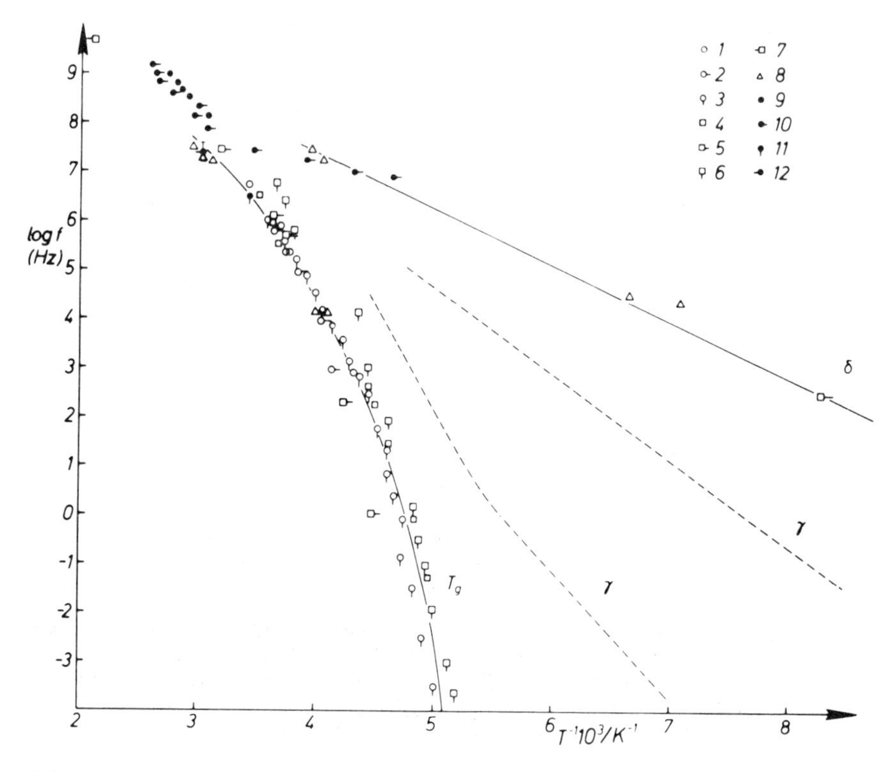

FIGURE 8 Relaxation map of PIB. Dielectric data: (1) [1], (2) [17], (3) [34]; mechanical data: (4) [1], (5) [17], (6) [34], (7) [35]; NMR data: (8) [17]; ESR data: (9) probe I [36], (10) probe I [37], (11) probe VII [38], (12) T^{l}_{50G} [32].

temperatures more rapidly than polymer segments. It shows a discontinuity temperature T_d at ~300 K. When $T<T_d$, E_a of probe I is 10 ± 2 kJ mol^{-1}. This value is in good agreement with E_a of δ relaxation in PIB (9.5 kJ mol^{-1}) [39]. Therefore radical I seems to respond to δ relaxation at low temperatures ($T<290$ K) while the more bulky probe VII responds still at 290 K to the merged $T_g + T<T_g$ relaxation. This comparison supports the proposition of Kumler and Boyer that probes of different size may in certain conditions respond to different relaxations [40].

3.2. Poly(ethylene oxide) (PEOX)

We have made a series of dynamic studies of PEOX by means of spin label XX and probes XVIII and XXI [13,23,30,41]. The measurements were

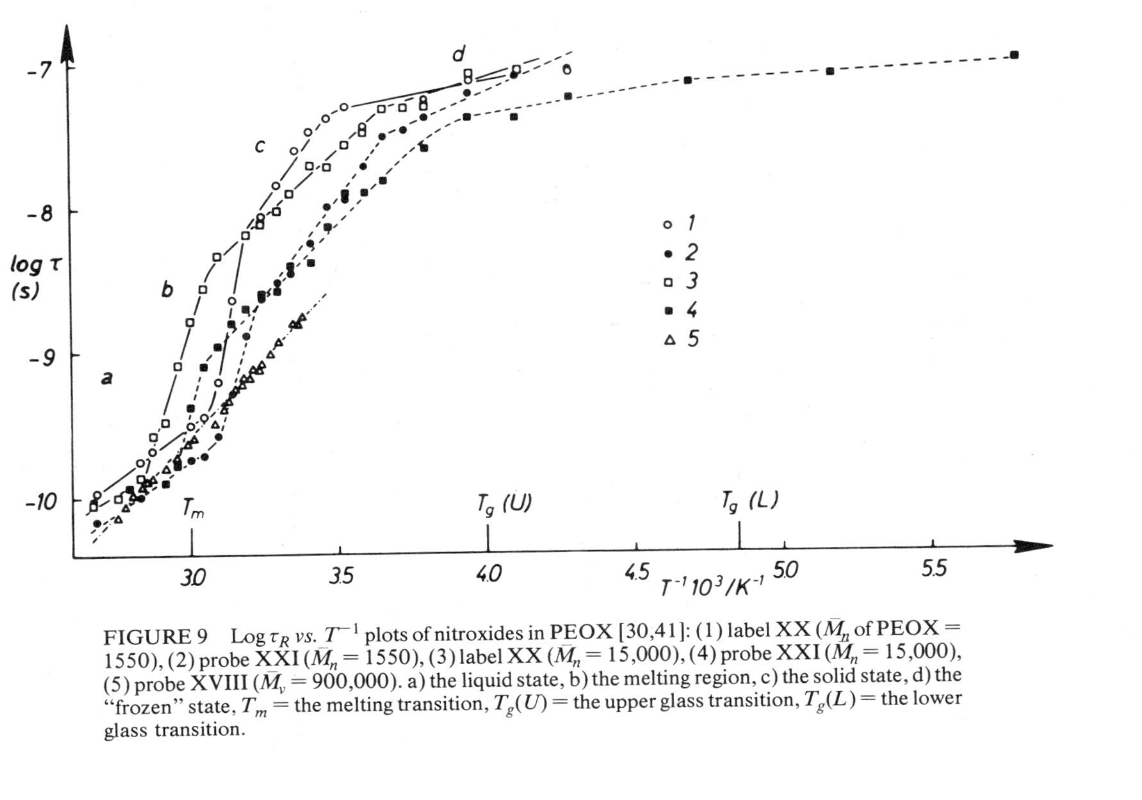

FIGURE 9 Log τ_R *vs.* T^{-1} plots of nitroxides in PEOX [30,41]: (1) label XX ($\bar{M}_n$ of PEOX = 1550), (2) probe XXI ($\bar{M}_n = 1550$), (3) label XX ($\bar{M}_n = 15{,}000$), (4) probe XXI ($\bar{M}_n = 15{,}000$), (5) probe XVIII ($\bar{M}_v = 900{,}000$). a) the liquid state, b) the melting region, c) the solid state, d) the "frozen" state, T_m = the melting transition, $T_g(U)$ = the upper glass transition, $T_g(L)$ = the lower glass transition.

made over a wide temperature region and polymers of different molecular weights were studied (M_w = 200–900,000).

The labels and probes showed in polymers of medium M_w (1000–22,000) four different motional regions (Figure 9): a) the liquid state, b) the melting region, c) the solid state, and d) the "frozen" state. The frozen state was later tentatively divided in two regions of different E_a values [30].

The studies of Lang *et al.* [42] and of Shapiro *et al.* [43] have confirmed the existence of the above proposed motional regions. These results indicate that PEOX displays several relaxation processes which are activated at different temperatures. Figure 9 shows also the low frequency temperatures of the three important transitions T_m (the melting transition), T_g (U) (the upper T_g), and T_g (L) (the lower T_g).

It is seen that the melting region (320–340 K) of medium M_w PEOX is characterized by an abrupt decrease of τ_R. Such a discontinuity is strongly reduced in PEOX whose end groups have been esterified [44] and it is practically absent in the high M_w sample [41]. Therefore we concluded that the transition during melting is caused by the dissipation of strain forces generated by –OH end groups in the amorphous phase.

The discontinuity temperatures (T_{d_1} between regions *c* and *d* and T_{d_2} inside the region *d*) are correlated to T_g (U) and T_g (L) transitions. While label radicals are bonded to polymer chains they evidently follow intimately the motions of polymer chains. T_d values of label radicals are about 20 degrees higher than the corresponding low frequency T_g (U) and T_g (L) values. This is not surprising when the high frequency of ESR measurements is taken into account. On the other hand, probe XXI shows T_d values which are about 20 degrees lower than those of labels. This manifests the effect of size difference between small and large radicals as we already saw in the T^{l}_{50G} discussion.

Figure 9 shows that τ_R values of labels and probes decrease with increasing polymer M_w. This behavior is analoguous to the change of frequencies of T_g relaxation as a function of polymer M_w as measured by means of dielectric relaxation [1].

On the other hand, the E_a values of labels and probes in solid PEOX (region *c*) (and in the high M_w sample also in the polymer melt) attain at $\bar{M}_n > 10{,}000$ a constant value ~40 kJ mol^{-1}. This value is in good agreement with E_a of $T < T_g$ relaxation in PEOX (35 kJ mol^{-1}) [11]. Therefore we conclude that labels and probes respond in the solid and melted PEOX to the merged $T_g + T < T_g$ relaxation when $T > T_{d_1}$.

3.3. Natural Rubber (NR)

Kovarskii *et al.* studied the high temperature rotations of several probes (I, III, IV, IX, XIV, XVI and XXIX) in natural rubber [45]. We have used these comprehensive τ_R data to analyze radical–polymer relationships in natural rubber. Figure 10 shows τ_R *vs.* T^{-1} data of Kovarskii *et al.*

A tendency towards a limiting τ_R value with increasing M_w of probes can be concluded from this figure. This behavior is analogous to the T_{50G}–M_w correlation. We converted τ_R values of Figure 10 to the corresponding effective frequencies (f_{eff}) and found that f_{eff} values approach with increasing probe M_w the correlation frequencies of the merged $T_g + T < T_g$ relaxation of NR [46].

The following analytical expression was developed to reduce the f_{eff} values of different probes to a single master curve (the data of the stiff radical XXIX were excluded from this examination) [46]:

$$f_{red} = \frac{f_{eff}}{\exp\left[\dfrac{E_0 e^{-M_w/M_s}}{RT}\right]} = f_0 \exp\left(-\frac{E_0}{RT}\right) \tag{8}$$

where E_0 is the activation energy of the merged T_g and $T < T_g$ relaxation (32.6

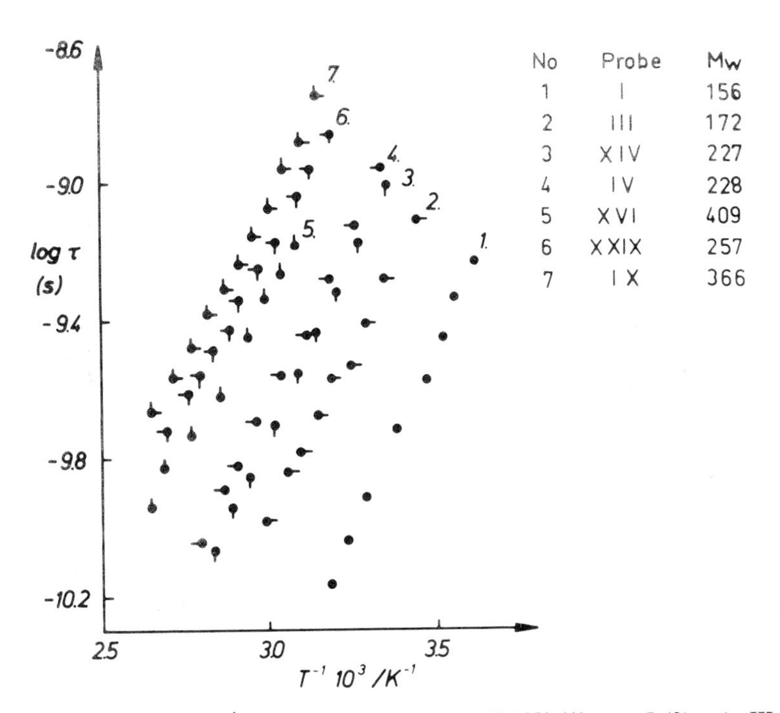

FIGURE 10 Log τ_R *vs.* T^{-1} plots of different probes in NR [45]. (1) probe I, (2) probe III, (3) probe XIV, (4) probe IV, (5) probe XVI, (6) probe XXIX, (7) probe IX.

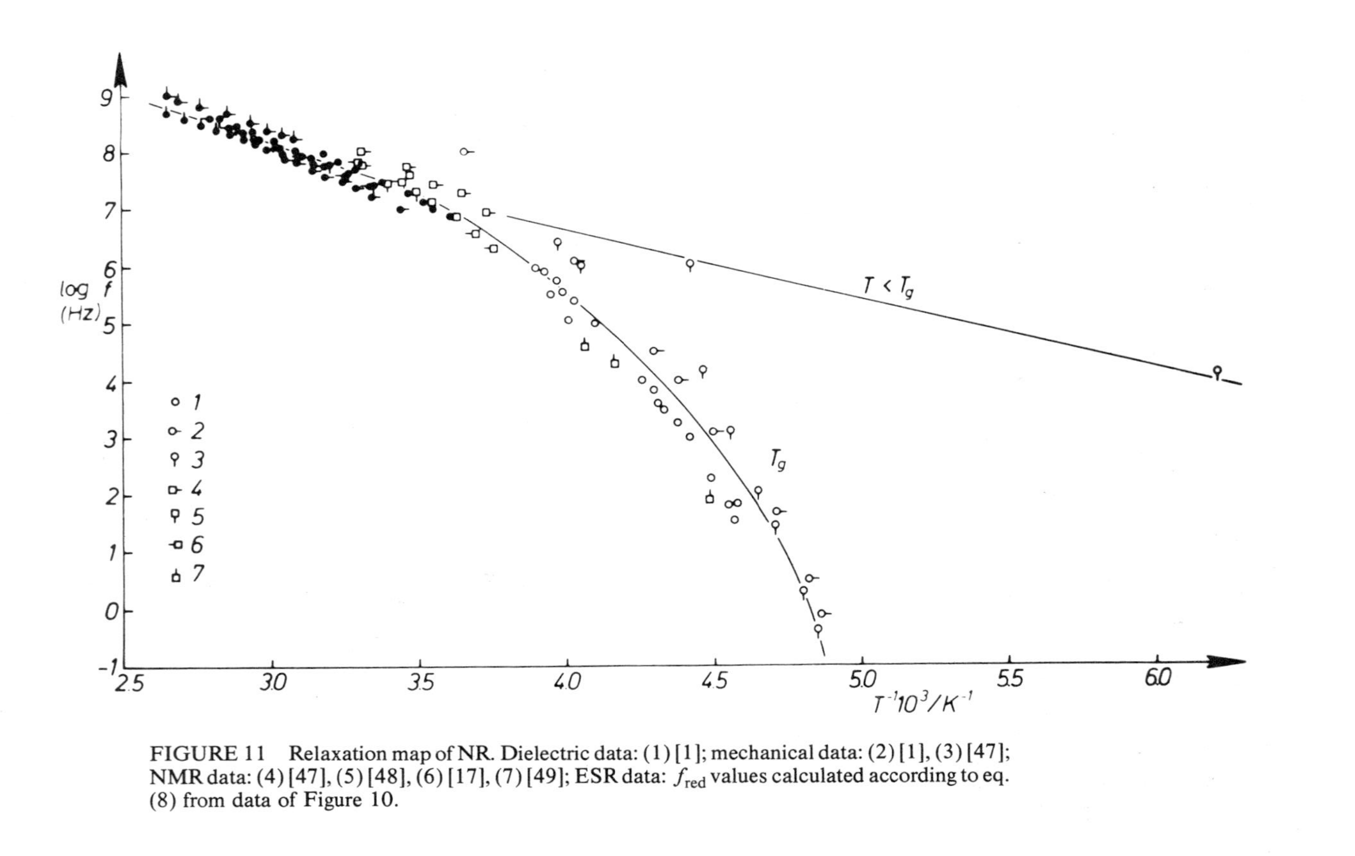

FIGURE 11 Relaxation map of NR. Dielectric data: (1) [1]; mechanical data: (2) [1], (3) [47]; NMR data: (4) [47], (5) [48], (6) [17], (7) [49]; ESR data: f_{red} values calculated according to eq. (8) from data of Figure 10.

kJ mol^{-1}) [47], f_0 is the corresponding pre-exponential factor (2×10^{13} Hz) and M_s is a correlation parameter (114 g mol^{-1}).

In Figure 11 the f_{red} values of radicals I, III, IV, IX, XIV and XVI are compared with dielectric [1], mechanical [1,47], and NMR [17,47,48] relaxation data.

A good correlation between f_{red} values and other relaxation data can be seen. The parameter M_s is interesting as it may represent the size of the smallest rotating polymer segment which interacts with probes. Accordingly the coupling of radical rotations with rotations of polymer chains may take place via multiplets of M_s. As a consequence the dynamic state of large, flexible radicals whose motions are coupled with motions of several M_s units, approach with increasing probe M_w the dynamic state of polymer chains. Since the obtained M_s value corresponds to a segment of ~1.7 monomer units in length and on the other hand T_g and $T < T_g$ relaxations are merged in the high frequency region, the T_g relaxation can be interpreted as a coordinated motion of a group of short segments.

3.4. Polyethylene (PE)

Polyethylene shows multiple relaxations related to motions both in the crystalline and amorphous phases. These relaxations are often sensitive to morphology and thermal history. Therefore there exists still widespread controversy about the molecular interpretation of relaxations in PE.

The relaxations of PE can be divided briefly into 3 groups: α, β, and γ processes. Each group can be divided into several single relaxations which are related to various structures of segments in the ordered and disordered regions:

α (α_1, α_2, α'); Motions of structural defects in the crystalline phase [11,47].

$\beta[T_g(U), T_g(L)]$; The double glass transition: $T_g(L) = 195 \pm 10$ K, $T_g(U) = 240 \pm 20$ K [50].

γ; Multiple local mode motions related to the various structures of segments in the disordered regions of crystalline phase (low temperature branches of γ group) and of amorphous phase (high temperature branches) [11].

In Figure 12 are compiled results of dielectric [47,51–53], mechanical [47], and NMR [54–56] relaxation measurements. Additionally are given f_{eff} values which were calculated from dynamic ESR data of several label and probe radicals. The examination of Figure 12 leads to the following observations concerning radical–polymer relationships in PE:

a) The relaxation data of all radicals (labels and probes) are located in the regions of γ and $T_g(L)$ relaxations.

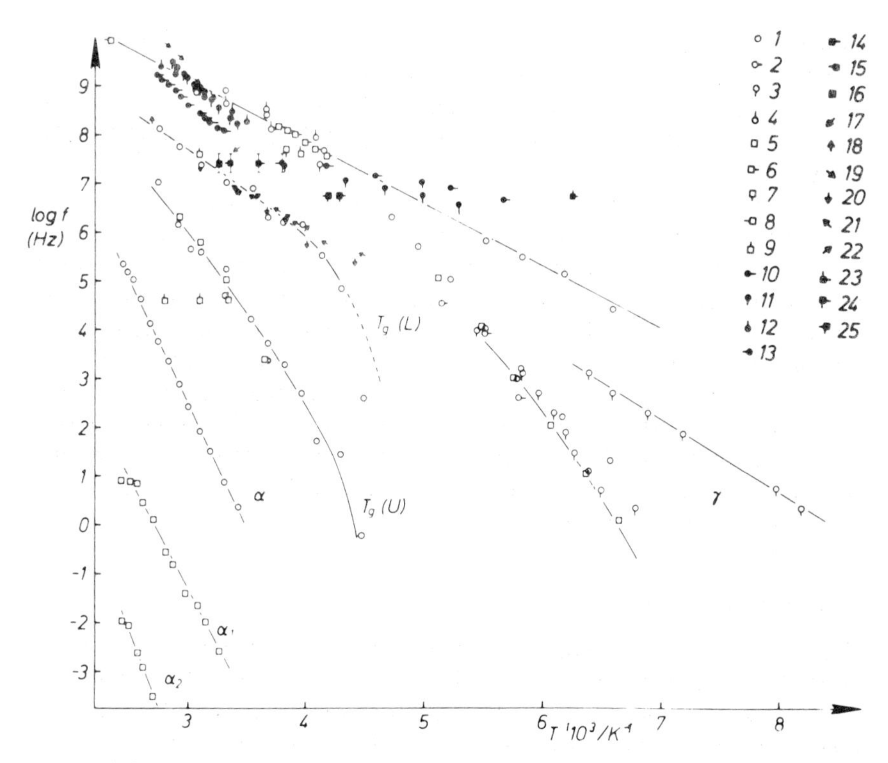

FIGURE 12 Relaxation map of PE. Dielectric data: (1) [47], (2) [51], (3) [52], (4) [53]; mechanical data: (5) [47]; NMR data: (6) [54], (7) [55], (8) [56], (9) [49]; ESR data: (10) probe I [37,57], (11) probe I [37,57], (12) probe I [58], (13) T_{50G} of probe III [28], (14–16) T_{50G} values of probe VII [28,31], (17) probe VII [38], (18) probe X [59], (19) probe XXV [60], (20) label XXVI [60], (21) label XXVII [61], (22) label XXVIII [61], (23–25) T_u (the "unfreezing" temperature) values of probes I, III, and VII, resp. [12].

b) The motional state of probes approaches with increasing probe M_w the dynamic state of polymer segments showing $T_g(L)$ relaxation. Accordingly the rotational frequencies of large probe X ($M_w = 474$) and of labels XXVII and XXVIII (which can be regarded as "macroprobes") practically coalesce with the frequencies of $T_g(L)$ relaxation when $T > 240$.

c) The effective activation energy E_a is inversely proportional to M_w of radicals. This behavior seems to be characteristic for nitroxides tumbling in different polymers. We have proposed that this tendency originates from the differences of temperature dependency of E_a of different radicals [46]. With increasing M_w of probes E_a seems also to approach the activation energy of $T_g(L)$ relaxation (30–35 kJ mol^{-1} in the high frequency region [47]) while small probe XXV has an E_a value of 54 kJ mol^{-1} [60], larger probe VII, 46 kJ mol^{-1} [28], and labels XXVI–XXVIII show E_a values of 25–29 kJ mol^{-1} [60–61].

3.5. Poly(vinylidene fluoride) (PVDF)

Figure 13 shows the T^l_{50G} value of PVDF from Figure 6 in comparison to the schematic relaxation map which is constructed according to literature data [10,62,63].

This relaxation map follows the general character for a semicrystalline polymer showing $T\alpha_c$, $T_g(U)$, $T_g(L)$ and $T < T_g(L)$ processes. T^l_{50G} is in a good agreement with the merged $T_g(L) + T < T_g(L)$ relaxation which evidently shows the coupling of radical motions with segmental relaxations of PVDF chains at high temperatures.

3.6. Polypropylene (PP)

Figure 14 summarizes dielectric [17,64], mechanical [17,35,65,66], and NMR [17,66–69] relaxation data of isotactic PP in addition to f_{eff} values which are calculated from literature ESR data [12,28,31,45,70].

Figure 14 indicates that the dynamic state of probe radicals approaches the state of polymer chains with increasing probe M_w. This is clearly seen also from Figure 15 where the temperature T_s at which f_{eff} of different probes

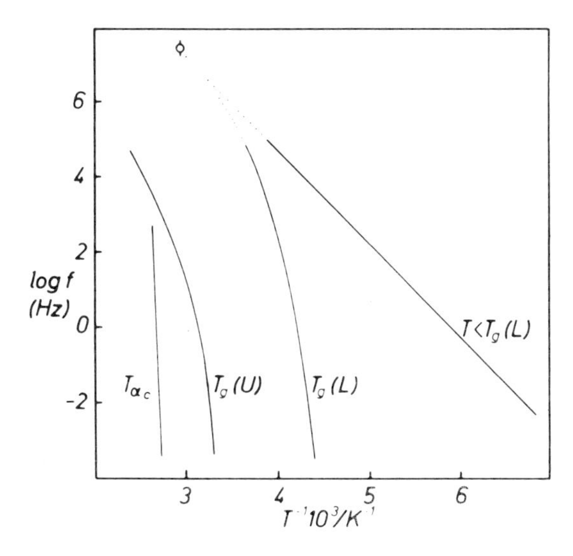

FIGURE 13 Schematic relaxation map of PVDF [10,62,63]. (○) T^l_{50G} value of Figure 6.

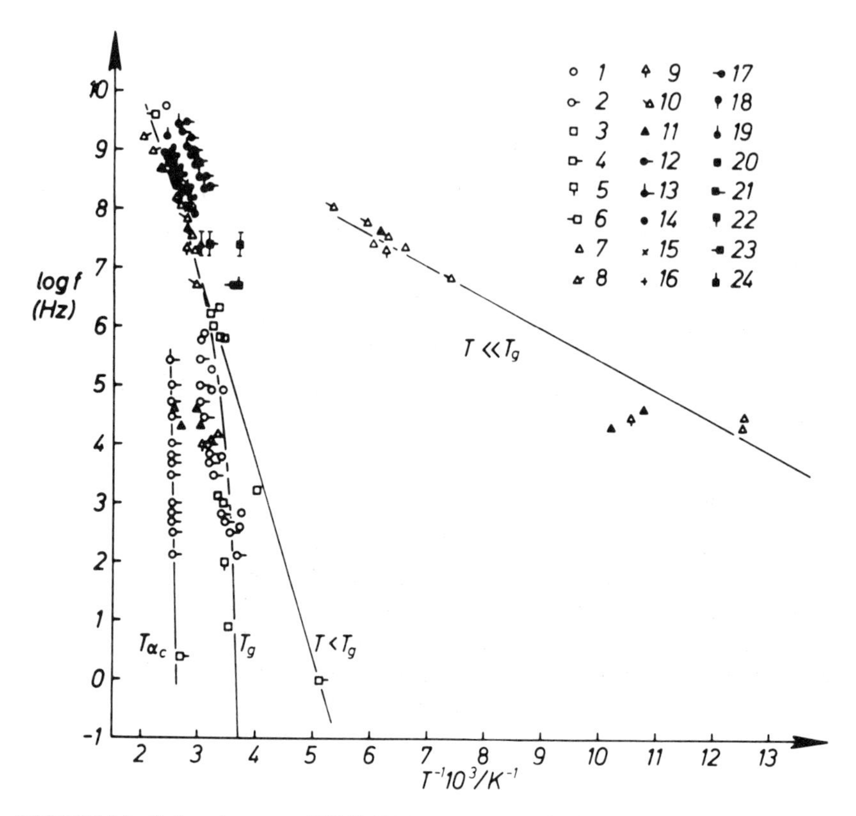

FIGURE 14 Relaxation map of PP(i). Dielectric data: (1) [17], (2) [64]; mechanical data: (3) [17], (4) [65], (5) [66], (6) [35]; NMR data: (7) [17], (8) [67], (9) [66], (10) [68], (11) [69]; ESR data: (12–18) probes I, III, XIV, IV, XXIX, IX, and XVI, resp. [45], (19) probe VII [28], (20) T_{50G} of probe III [28], (21) T_{50G} of probe VII [28,31], (22) T_{50G} of probe VII [70], (23–24) T_u of probes VII, and XXIX, resp. [12].

attains a selected value 3.2×10^8 Hz is given as a function of probe M_w. Polymer segments show this relaxation frequency at a temperature 400 ± 20 K (estimated from Figure 14). Figure 15 shows that T_s approaches this temperature with increasing probe M_w.

The polymers of Figure 14 have a degree of crystallinity of ~60%. The E_a values of radicals are practically independent of probe M_w ($E_a = 45 \pm 5$ kJ mol^{-1}). This may indicate the large size of relaxing polymer segments [46]. Vasserman and Barashkova have reported that τ_R of small probe I decreases and its E_a increases with decreasing crystallinity of PP(i) [71]. It is generally approved that the decrease of crystallinity facilitates relaxations in the amorphous phase. The decrease of τ_R manifests this phenomenon while the

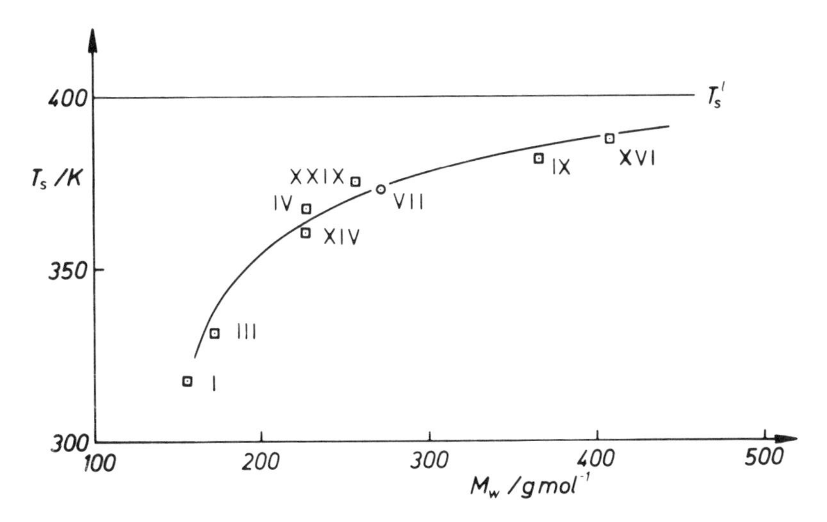

FIGURE 15 Temperature T_s, at which f_{eff} of probe radicals in PP(i) attains a selected value 3.2 $\times$ 10^8 Hz *vs.* probe M_w. (□) from data of Kovarskii *et al.* [45], (○) from data of Rabold [28], T_s^l = the loss maximum temperature of segmental relaxations at f = 3.2 $\times$ 10^8 Hz (estimated from Figure 14).

increase of E_a shows that effective activation energies of small probes do not directly describe the energy barrier of segmental motions [46].

3.7. Poly(vinyl acetate) (PVAC)

PVAC shows the three relaxations, T_g (β), $T < T_g$ (γ) and $T \ll T_g$ (δ), which are typical for amorphous polymers with pendant side groups. The γ relaxation is usually centered near 150 K (1–10 Hz) [72]. It is generally thought to involve motions of the $-OCOCH_3$ side groups [1]. The δ process occurs at low frequencies (1–10 Hz) in the temperature region below 100 K [72,73] but it is shifted to ~170 K at 2 $\times$ 10^6 Hz [74]. It arises from reorientational motion of side chain methyl groups [72]. We have compiled in Figure 16 data of dielectric [1,74], mechanical [1,35,73–75], NMR [76], and ESR [33,38, 77,78] measurements.

Figure 16 shows the general principle that f_{eff} values of probe radicals diminish with increasing probe M_w and approach f_{eff} values of label radicals.

At low temperatures ($T < 350$ K) the f_{eff} values of large probes VII and XVII practically coalesce with those of label XVII. At higher temperatures labels seem to show somewhat smaller f_{eff} values than large probes. This departure may however originate from anisotropic rotation of labels at high temperatures (subsplitting of the high field hyperfine peak) [77].

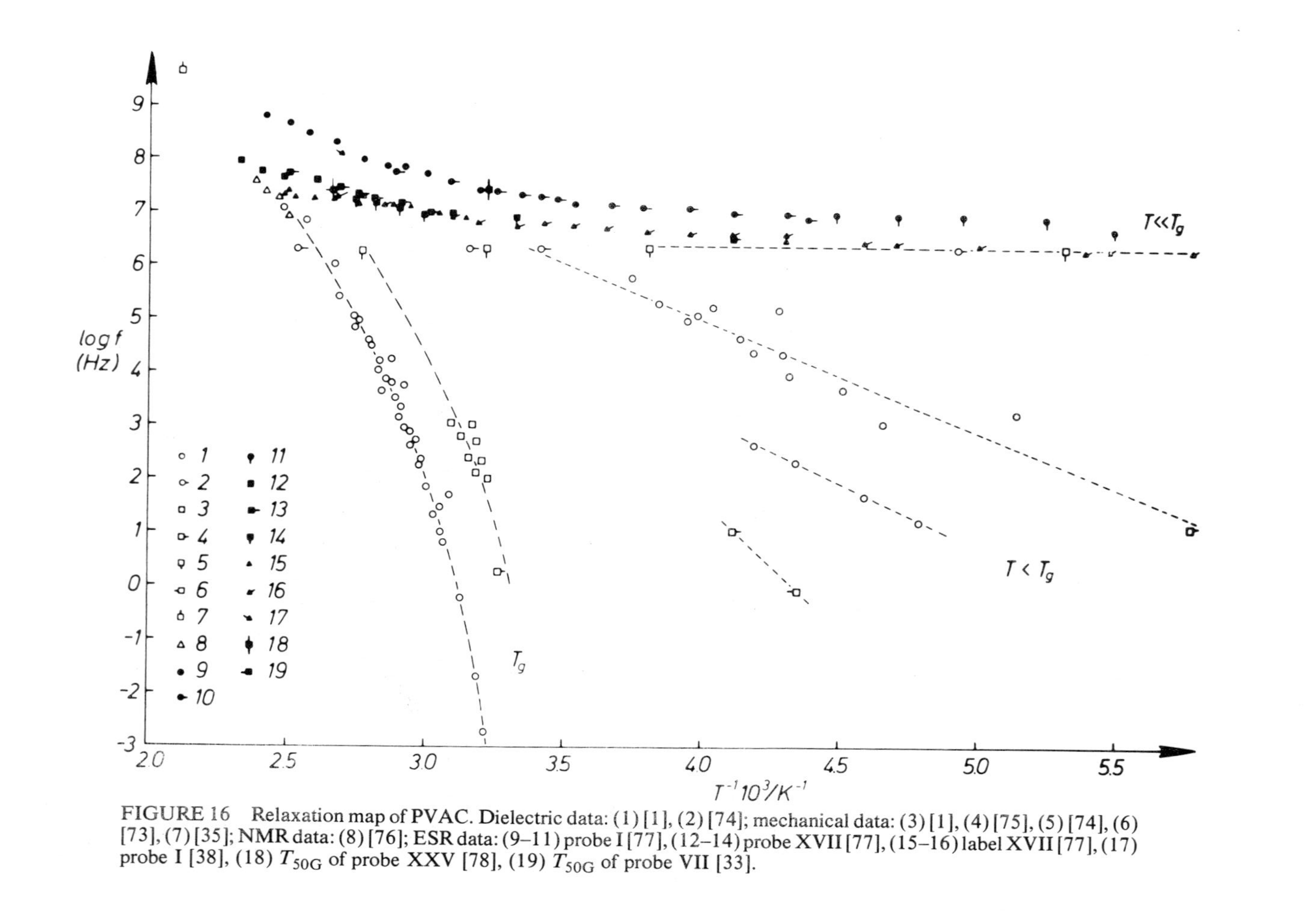

FIGURE 16 Relaxation map of PVAC. Dielectric data: (1) [1], (2) [74]; mechanical data: (3) [1], (4) [75], (5) [74], (6) [73], (7) [35]; NMR data: (8) [76]; ESR data: (9–11) probe I [77], (12–14) probe XVII [77], (15–16) label XVII [77], (17) probe I [38], (18) T_{50G} of probe XXV [78], (19) T_{50G} of probe VII [33].

From Figure 16 it can be concluded that radicals respond at high temperatures ($T > 370$ K) to the merged T_g + local mode relaxation.

While log f of the T_g relaxation attains a value ~6.2 in the temperature region of 360 K radicals respond between temperatures 310 K–360 K to a segmental γ relaxation whose activation energy 40 ± 4 kJ mol^{-1} [1] is in agreement with E_a values of probe radicals (probe I 39 kJ mol^{-1}, probe XVII 29.3 kJ mol^{-1}) [77]. E_a of radicals diminishes abruptly at a discontinuity temperature T_d ~310 K. Below this temperature log f of the γ process attains values < 6.2. Therefore radicals do not "see" these motions any more but they respond solely to the δ relaxation of methyl groups. E_a values of radical rotation (~5 kJ mol^{-1}) of this low temperature region are in agreement with the activation energies typical for motions of pendant side groups (δ relaxations) (< 4 to 12 kJ mol^{-1}) [79]. Figure 16 shows that the rotational frequencies of label and probe radical XVII coalesce with the frequency of δ relaxation while small probe I shows somewhat higher f_{eff} values.

3.8. Polyamide (PA) 610

Polyamide 610 shows three distinct relaxations: T_g (α), and two $T<T_g$ processes (β and γ) [1]. It is generally approved that all these processes originate with molecular motions in the amorphous domains. T_g relaxation is related to the breaking of hydrogen bonds [80]. The β relaxation is postulated to originate with motions of both non-hydrogen-bonded polar groups and polymer-water complex units [81]. The γ relaxation arises from the motions of the $[—CH_2—]_n$ units between amide groups and from the motions of dipolar amide groups [1].

The rotational frequencies of label radical XX in PA 610 [30] are compared with dielectric [1,82] and mechanical [1] relaxation data of PA 610 in Figure 17.

It is seen that the label radicals respond to the motions of short polymer segments (γ relaxation) in the studied temperature region.

3.9. Poly(vinyl chloride) (PVC)

Figure 18 gives the relaxation map of PVC. PVC shows three relaxations: T_g (β), $T<T_g$ (γ), and $T \ll T_g$ (δ) processes [1,83]. T_g and $T<T_g$ relaxations are the normal co-operative and local mode processes, respectively [1]. $T \ll T_g$ relaxation has been observed at low temperatures (at 18 K when $f = 7.2 \times 10^3$ Hz) and it originates possibly from wagging motion of —Cl substituents [83].

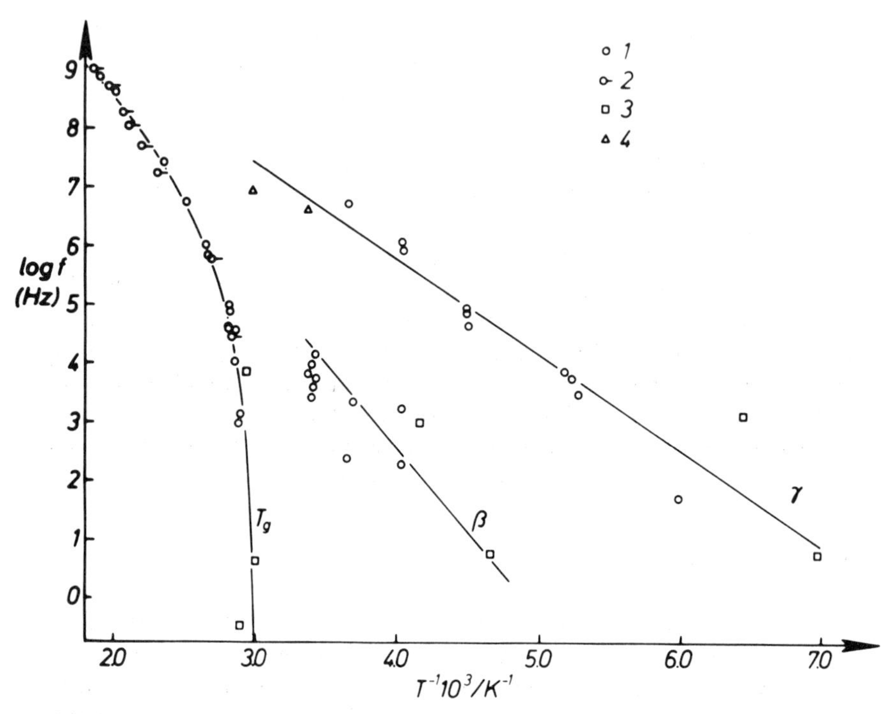

FIGURE 17 Relaxation map of PA 610. Dielectric data: (1) [1], (2) [82]; mechanical data: (3) [1]; ESR data: (4) label XX [30].

Figure 18 shows that the large probe VII responds at T_{50G} to the merged $T_g + T<T_g$ relaxation. At low temperatures ESR data deviate clearly from T_g relaxation line: rotational frequencies of radicals III and VII are at T_u 2–3 decades higher than the frequency of T_g relaxation at this temperature. While ESR spectra are insensitive to motions with log $f \lesssim 6.2$ radicals III and VII respond at T_u to $T<T_g$ (γ) relaxation. Further support for this hypothesis is obtained when the E_a value of rotations of radical VII is estimated by means of T_{50G} and T_u data. The obtained value 80 ± 10 kJ mol^{-1} is somewhat higher than E_a of γ relaxation (54–63 kJ mol^{-1}) [1] but clearly smaller than the activation energy of T_g relaxation in the same temperature region (~350 kJ mol^{-1}; estimated from Figure 18).

It is evident that at $T \ll T_u$ only wagging motions of −Cl atoms can attain so high frequencies that they can contribute to ESR spectra of nitroxides. Accordingly probe XVIII seems to respond at 298 K to this δ relaxation.

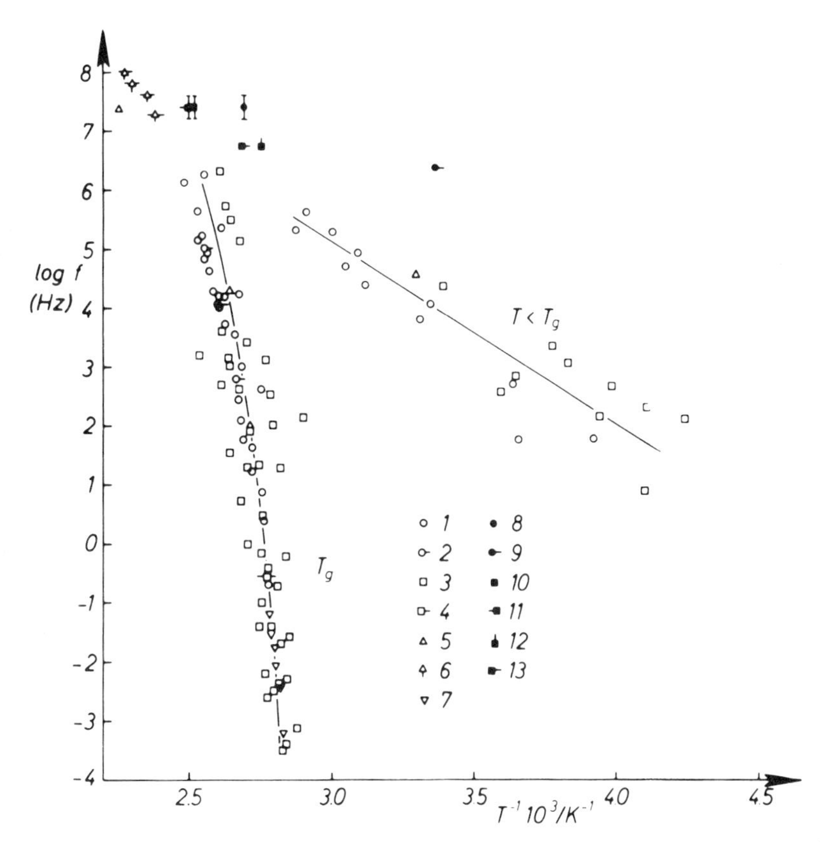

FIGURE 18 Relaxation map of PVC. Dielectric data: (1) [1], (2) [84]; mechanical data: (3) [1], (4) [84]; NMR data: (5) [49], (6) [84]; thermal data: (7) [84]; ESR data: (8) T_{50G} of probe III [28], (9) probe XVIII [85], (10) T_{50G} of probe VII [28,31], (11) T_{50G} of probe VII [33], (12) T_u of probe III [12], (13) T_u of probe VII [12].

3.10. Polystyrene (PS)

Atactic polystyrene shows the T_g relaxation (α) and three local processes (β, γ, and δ) [1]. It has been proposed that β and γ processes may originate from local motions of defects along the polymer chain [1]. The δ relaxation is attributed to the motion of the phenyl groups [73]. Literature data describing these relaxations are given in Figure 19 with ESR data of different label and probe radicals. ESR data show a discontinuity temperature T_d at ~415 K. Below this temperature E_a values of radical motions are extremely low: probe I, $E_a = 12\,\text{kJ mol}^{-1}$ [37], and label XXX, $E_a = 6.3\,\text{kJ mol}^{-1}$ [29]. These values

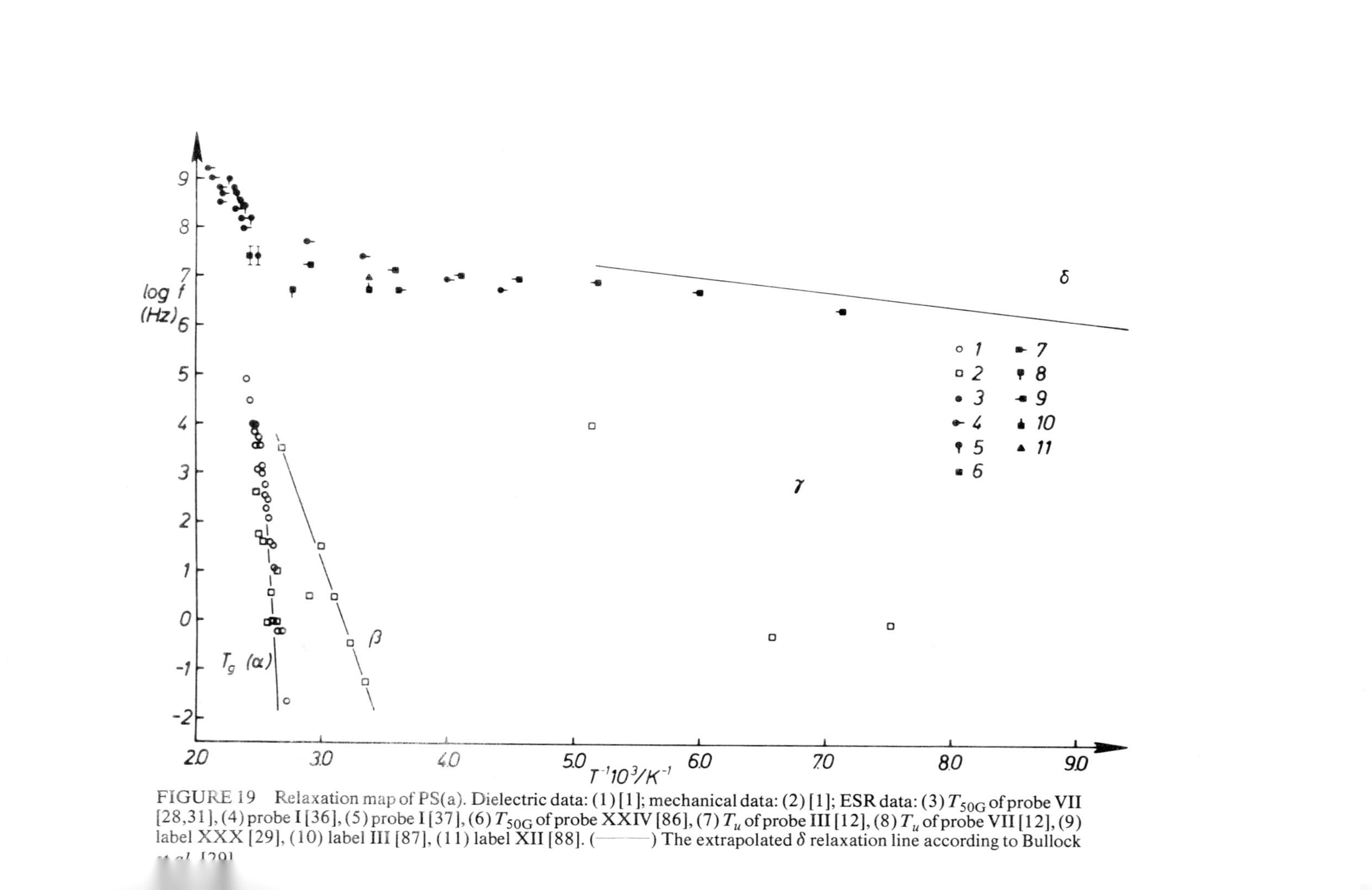

FIGURE 19 Relaxation map of PS(a). Dielectric data: (1) [1]; mechanical data: (2) [1]; ESR data: (3) T_{50G} of probe VII [28,31], (4) probe I [36], (5) probe I [37], (6) T_{50G} of probe XXIV [86], (7) T_u of probe III [12], (8) T_u of probe VII [12], (9) label XXX [29], (10) label III [87], (11) label XII [88]. (———) The extrapolated δ relaxation line according to Bullock *et al.* [29]

are in agreement with the activation energy of δ relaxation of PS (6.7 kJ mol^{-1}) [29].

It is evident that T_d (which in the case of PS is $\sim T_{50G}$) is related to the glass transition. It may manifest the fact that the frequency of T_g relaxation is attaining the level where it can modulate the linewidths of ESR spectra. Figure 20 shows, however, that T_d of probe radicals is clearly smaller than T_g at the corresponding relaxation frequency. This difference may be explained by means of the size differences between the radical and the activation volume of relaxing polymer segments [89].

4. CONCLUSIONS

The above discussion shows clearly that spin labels and probes are delicate indicators of the dynamic state of their environment: At high temperatures radicals respond to the merged $T_g + T<T_g$ relaxation; if the frequency of the T_g relaxation is retarded below $\sim 2 \times 10^6$ Hz radicals are responding to segmental motions (if any are present) and when these are retarded below 2×10^6 Hz radicals may still respond to δ relaxation. There seems to be no single *a priori* relaxation process to which radicals tend to respond: all molecular motions which lead to reorientation of the g- and A-tensor system of radical in relation to the external magnetic field can contribute to the obtained f_{eff} value. Because the rotations of radicals modulate the linewidths of ESR spectra in the τ_R range 10^{-7} s–5×10^{-11} s the linear ESR measurements give information about motions in the frequency range 1.6×10^6 Hz–3×10^9 Hz. The ESR method is therefore a typical high frequency measurement method while the radicals do not "see" the motions which are below this frequency range.

Examination of the data which was compiled above leads us to construct a "typical" relaxation diagram of nitroxides in thermoplastics (Figure 20).

The diagram of Figure 20 is one containing the broad features to be found in a number of relaxation maps while ignoring details specific to a certain polymer.

We divide the temperature axis into four different regions: (I) the high temperature, (II) the medium temperature, (III) the low temperature, and (IV) the very low temperature region.

In the very low temperature region all relaxations have frequencies which are smaller than $\sim 10^6$ Hz. Therefore radicals do not "see" these motions and exhibit accordingly the frozen (powder) spectrum. At the limiting temperature between regions III and IV the frequency of the least hindered relaxation (δ process) attains the value $>10^6$ Hz. Therefore the ESR spectrum begins to depart from its rigid limit value which is seen as a decrease of the peak-to-peak separation. In the temperature region III radicals respond solely to δ

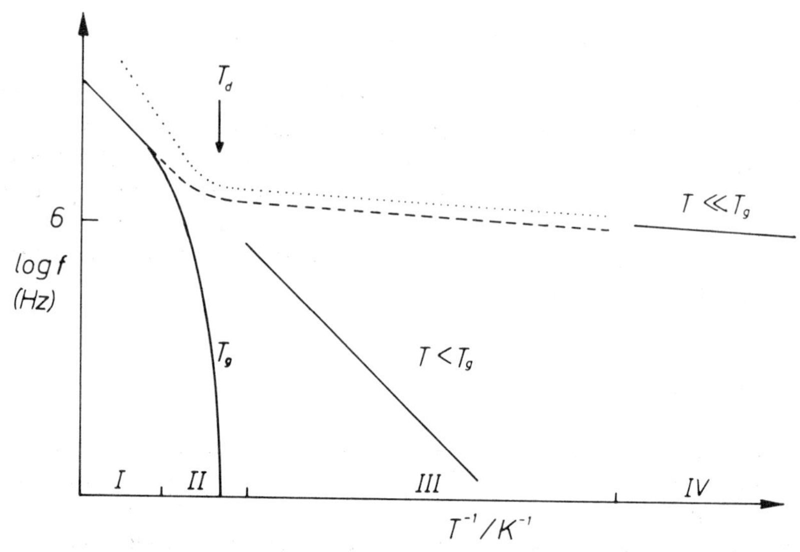

FIGURE 20 A typical relaxation diagram of nitroxide labels and probes in amorphous phase of thermoplastics; (· · · ·) small probes, (- - -) large, flexible probes and labels. T_d = the discontinuity temperature. T_g = the glass transition, $T<T_g$ = the local segmental relaxation, $T \ll T_g$ = the local side group relaxation. I = the high temperature region, II = the medium temperature region, III = the low temperature region, and IV = the very low temperature region. The numerical value log $f = 6$ refers to polymers with T_g values >300 K.

relaxation while the frequencies of other relaxations are still so slow that they do not have an effect on the linewidths of ESR spectra. Both probes and labels show E_a values which are in agreement with the activation energies of δ processes. The effective frequency data of labels and of large, flexible probes coalesce with each other which also indicates the local mode nature of radical–polymer interaction.

The correlation between the absolute values of $T \ll T_g$ and radical relaxations seems to be related to the stiffness of polymer chains. Large probes show in flexible polymers in the temperature region III clearly smaller f_{eff} values than those of δ relaxation while in more stiff polymers (with higher T_g values) rotations of large probes and labels are in accordance with δ relaxation. Small probes show in flexible polymers f_{eff} values which are in accordance with δ relaxation data while in more stiff polymers they may have higher rotational frequencies as those of δ relaxation.

It is interesting to refer in this connection to the effect of additives to δ relaxation. It is well known that already small amounts of additives (e.g., plasticizers) increase strongly the intensity of δ relaxation [72]. Therefore it is tempting to speculate that also nitroxides facilitate δ relaxation in their nearest environment while dynamic radical data and δ relaxation data seem to be correlated generally to each other. Therefore the application of nitroxide labeled plasticizers seems to afford a new method to study the effects of additives upon δ relaxation.

The medium temperature region (II) is characterized by the discontinuity temperature T_d where the frequency of segmental $T<T_g$ (γ) relaxation attains the level of the ESR method. The E_a values of radicals change abruptly from 5–10 kJ mol^{-1} to values which are characteristic to the γ relaxation (the order of magnitude of 40 kJ mol^{-1}). Small probes show high E_a and pre-exponential factor values which we have explained by means of temperature dependence of E_a [46]. However, these parameters decrease with increasing probe M_w approaching values which are characteristic for relaxations of polymer chains.

The frequency of T_g relaxation attains also in the region II the level of ESR measurements. Therefore radicals can show complex superposition ESR spectra [28] and/or different radicals may respond to different relaxations [40]. Evidently accurate spectral simulations are needed to analyse ESR data of this temperature region profoundly.

In the high temperature region (I) where T_g and local mode relaxations are merged the ESR data can be interpreted quantitatively [46]. The radical rotations obey an Arrhenius equation. The rotational frequencies of large, flexible probes and of labels seem to coalesce with the relaxation frequencies of polymer chains although the bulky labels may show somewhat lower frequencies than chains in polymers with very flexible chain structure [90].

When the size differences of different probes are taken into account the f_{eff} data of probes can be reduced to a single master curve which coalesces with the merged T_g + local mode relaxation line [46].

Partially crystalline polymers show often complex behavior which can be attributed to complex relaxation patterns [91] and to interactions between the crystalline phase and radicals [30,42,60].

The knowledge concerning interactions of label and probe radicals with polymer molecules is still far from complete.

The indirect nature of label and probe methods causes interpretation difficulties which can be settled only by systematic studies of structural effects upon the dynamic ESR data of radicals which are tumbling in condensed polymeric systems.

Such studies which are in progress in several laboratories will evidently in the near future give still more quantitative character for these new relaxation measurement methods.

ACKNOWLEDGEMENTS

We would like to thank Professor D. Braun for his support throughout the development of this work.

One of us (P. T.) is indebted to the Alexander von Humboldt Foundation for a research fellowship.

REFERENCES

1. N.G. McCrum, B.E. Read, and G. Williams, "*Anelastic and Dielectric Effects in Polymeric Solids*," Wiley, Bristol, 1967.
2. P.I. Vincent, *Polymer*, **15**, 111 (1974).
3. A. Hiltner and E. Baer, *Polymer*, **15**, 805 (1974).
4. E. Sacher, *J. Macromol. Sci.-Phys.*, **B11**, 403 (1975).
5. A.R. Berens and H.B. Hopfenberg, *Polymer*, **19**, 489 (1978).
6. P. Törmälä, G. Weber, and J.J. Lindberg, *Rheol. Acta*, **17**, 201 (1978).
7. A.B. Baschirow, J.W. Selenew, and S.K. Achundow, *Plaste Kautsch.*, **23**, 104 (1976).
8. N. Bach Van and C. Noel, *J. Polym. Sci., Polym. Chem. Ed.*, **14**, 1627 (1976).
9. M.C. Lang, C. Noel, and A.P. Legrand, *J. Polym. Sci., Polym. Phys. Ed.*, **15**, 1319 (1977).
10. J.B. Enns and R. Simha, *J. Macromol. Sci.-Phys.*, **B13**, 11 (1977).
11. J.B. Enns and R. Simha, *ibid.*, p. 25.
12. A.L. Buchachenko, A.L. Kovarskii, and A.M. Vasserman, in "*Advances in Polymer Science*," (Z.A. Rogovin, ed.), Wiley, New York, 1974, p. 26.
13. P. Törmälä and J.J. Lindberg, in "*Structural Studies of Macromolecules by Spectroscopic Methods*," (K.J. Ivin, ed.), Wiley, London, 1976, p. 255.
14. A.T. Bullock and G.G. Cameron, *ibid.*, p. 273.
15. A.M. North, *Int. Rev. Sci. Phys. Chem. Ser. Two*, **8**, 1 (1975).
16. D.W. McCall, *J. Elastom. Plast.*, **8**, 60 (1976).
17. W.P. Slichter, *J. Polym. Sci., C*, **14**, 33 (1966).
18. G. Allen and A. Maconnachie, *Brit. Polym. J.*, **9**, 184 (1977).
19. G. Weber, *Progr. Coll. & Polymer Sci.*, in press.
20. J.A. Sauer, G.C. Richardson, and D.R. Morrow, *J. Macromol. Sci.-Revs. Macromol. Chem.*, **C9**, 149 (1973).
21. R.F. Boyer, *Polymer*, **17**, 997 (1976).
22. E.G. Rozantsev, "*Free Nitroxyl Radicals*," Plenum, New York, 1970.
23. P. Törmälä, H. Lättilä, and J.J. Lindberg, *Polymer*, **14**, 481 (1973).
24. D. Kivelson, *J. Chem. Phys.*, **33**, 1094 (1960).
25. J.H. Freed, in "*Spin Labeling: Theory and Applications*" (L.J. Berliner, ed.), Academic Press, New York, 1976, p. 53.
26. R.C. McCalley, E.J. Shimshick, and H.M. McConnell, *Chem. Phys. Lett.*, **13**, 115 (1972).
27. A.N. Kuznetsov, A.M. Vasserman, A.U. Volkov, and N.N. Korst, *Chem. Phys. Lett.*, **12**, 103 (1971).
28. G.P. Rabold, *J. Polym. Sci., A-1*, **7**, 1203 (1969).
29. A.T. Bullock, G.G. Cameron, and P.M. Smith, *J. Polym. Sci., Polym. Phys. Ed.*, **11**, 1263 (1973).
30. P. Törmälä, Ph.D. Dissertation, Department of Wood and Polymer Chemistry, Helsinki University, Helsinki, 1973.
31. P.L. Kumler and R.F. Boyer, *Macromolecules*, **9**, 903 (1976).
32. P. Törmälä and G. Weber, *Polymer*, **19**, (1978).
33. D. Braun, P. Törmälä, and G. Weber, *Polymer*, **19**, 1026 (1978).
34. B. Stoll, W. Pechhold, and S. Blasenbrey, *Kolloid-Z.*, **250**, 1111 (1972).
35. G.D. Patterson, *J. Polym. Sci., Polym. Phys. Ed.*, **15**, 455 (1977).
36. V.B. Stryukov and G.V. Korolev, *Vysokomol. Soedin. A*, **11**, 419 (1969).
37. V.B. Stryukov, T.V. Sosnina, and A.M. Kraitsberg, *Vysokomol. Soedin. A*, **15**, 1397 (1973).
38. A.T. Alexandrova, A.M. Vasserman, A.L. Kovarskii, and A.A. Tager, *Vysokomol. Soedin. B*, **18**, 322 (1976).

39. R. Kosfeld and U. Mylins, *Kolloid-Z.*, **250**, 1088 (1972).
40. P.L. Kumler and R.F. Boyer, *Polym. Prepr., Am. Chem. Soc., Div. Polym. Chem.*, in press.
41. P. Törmälä, *J. Appl. Polym. Sci.*, in press.
42. M.C. Lang, C. Noel, and A.P. Legrand, *J. Polym. Sci., Polym. Phys. Ed.*, **15**, 1329 (1977).
43. A.M. Shapiro, V.B. Stryukov, B.A. Rozenberg, G.L. Grigoryan, and E.G. Rozantsev, *Vysokomol. Soedin. B*, **17**, 265 (1975).
44. P. Törmälä and J. Tulikoura, *Polymer*, **15**, 248 (1974).
45. A.L. Kovarskii, A.M. Vasserman, and A.L. Buchachenko, *Vysokomol. Soedin. A*, **13**, 1647 (1971).
46. G. Weber and P. Törmälä, *Colloid Polym. Sci.*, in press.
47. W. Pechhold and S. Blasenbrey, *Kautsch. Gummi, Kunstst.*, **25**, 195 (1972).
48. H.S. Gutowsky, A. Saika, M. Takeda, and D.E. Woessner, *J. Chem. Phys.*, **27**, 534 (1957).
49. D.W. McCall and D.R. Falcone, *JCS, Faraday Trans. 2*, **66**, 262 (1970).
50. R.F. Boyer, *Plastics Polym.*, **41**, 15 (1973).
51. W. Reddish and J.T. Barrie, *IUPAC Symposium, Wiesbaden*, 1959, Lecture I A 3.
52. S. Hartmann, *Diploma-work*, Stuttgart University, Stuttgart, 1967.
53. T. Peterlin-Neumaier and T. Springer, *J. Polym. Sci., Polym. Phys. Ed.*, **14**, 1351 (1976).
54. U. Haeberlen, *Kolloid-Z.*, **225**, 15 (1968).
55. K. Bergmann and K. Nawotki, *Kolloid-Z.*, **219**, 132 (1967).
56. Y. Inoue, A. Nishioka, and R. Chujo, *Makromol. Chem.*, **168**, 163 (1973).
57. V.B. Stryukov and E.G. Rozantsev, *Vysokomol. Soedin. A*, **10**, 626 (1968).
58. T.V. Sosnina, A.M. Kraitsberg, B.R. Smirnov, V.B. Stryukov, and G.V. Korolev, *Vysokomol. Soedin. B*, **16**, 20 (1974).
59. A.B. Shapiro, K. Baimagambetov, V.A. Radzig, and E.G. Rozantsev, *Vysokomol. Soedin. B*, **15**, 300 (1973).
60. A.T. Bullock, G.G. Cameron, and P.M. Smith, *Eur. Polym. J.*, **11**, 617 (1975).
61. A.T. Bullock, G.G. Cameron, and P.M. Smith, *Macromolecules*, **9**, 650 (1976).
62. A.J. Bur, in *High Polymers*, **25**, (L.A. Wall, ed.), Wiley, New York, 1972, p. 475.
63. M.G. Brereton, G.R. Davies, A. Rushworth, and J. Spence, *J. Polym. Sci., Polym. Phys. Ed.*, **15**, 583 (1977).
64. H. Krämer and K.-E. Helf, *Kolloid-Z.*, **180**, 114 (1962).
65. H.A. Flocke, *Kolloid-Z.*, **180**, 118 (1962).
66. J.G. Powles and P. Mansfield, *Polymer*, **3**, 339 (1962).
67. T. Kawai, I. Ioshimi, and A. Mirai, *J. Phys. Soc. Japan*, **16**, 2356 (1966).
68. U. Kienzle, F. Noack, and J. Von Schultz, *Kolloid-Z.*, **236**, 129 (1970).
69. V.J. McBrierty, D.C. Douglass, and D.R. Falcone, *JCS, Faraday Trans. 2*, **68**, 1051 (1972).
70. S.E. Keinath, P.L. Kumler, and R.F. Boyer, *Polym. Prepr., Am. Chem. Soc., Div. Polym. Chem.*, **18**, 456 (1977).
71. A.M. Vasserman, and I.I. Barashkova, *Vysokomol. Soedin. B*, **19**, 820 (1977).
72. J.A. Sauer and R.G. Saba, *J. Macromol. Sci.-Chem.*, **A3**, 1217 (1961).
73. V. Frosini and A.E. Woodward, *J. Polym. Sci.*, *A-2*, **7**, 525 (1969).
74. H. Turn and K. Wolf, *Kolloid-Z.*, **148**, 16 (1956).
75. K. Schnieder and K. Wolf, *Kolloid-Z.*, **134**, 149 (1953).
76. G.P. Mikhailov and V.A. Shevelev, *Vysokomol. Soedin. A*, **8**, 1542 (1966).
77. A.M. Vasserman, T.A. Alexandrova, and A.L. Buchachenko, *Eur. Polym. J.*, **12**, 691 (1976).
78. A.T. Bullock, G.G. Cameron, C.B. Howard, and N.K. Reddy, *Polymer*, **19**, 352 (1978).
79. J.A. Sauer, *J. Polym. Sci.*, *C*, **32**, 69 (1971).
80. R.D. Andrews, *J. Polym. Sci.*, *C*, **14**, 261 (1966).
81. Y.S. Papir, S. Kapur, C.E. Rogers, and E. Baer, *J. Polym. Sci.*, *A-2*, **10**, 1305 (1972).
82. R.H. Boyd, and C.H. Porter, *J. Polym. Sci.*, *A-2*, **10**, 647 (1972).
83. J.M. Crissman, J.A. Sauer, and A.E. Woodward, *J. Polym. Sci.*, *A*, **2**, 5075 (1964).
84. E. Donth and W. Riechert, *Plaste Kautsch.*, **24**, 642 (1977).
85. P. Törmälä, *Angew. Makromol. Chem.*, **37**, 135 (1974).

86. P.L. Kumler, S.E. Keinath, and R.F. Boyer, *Polym. Prepr., Am. Chem. Soc., Div. Polym. Chem.*, **17**, 28 (1976).
87. S.L. Regen, *J. Am. Chem. Soc.*, **97**, 3108 (1975).
88. Z. Veksli and W.G. Miller, *Macromolecules*, **10**, 686 (1977).
89. R.F. Boyer and P.L. Kumler, *Macromolecules*, **10**, 461 (1977).
90. P. Törmälä, *Polymer*, **15**, 124 (1974).
91. R.F. Boyer, *J. Polym. Sci., C*, **50**, 189 (1975).

DISCUSSION

L. J. Berliner (Ohio State University, Columbus, Ohio): If one uses different spin labels, e.g., differing in size or geometry, at the same "site", and if all the results are self-consistent with one another, then the observation is *not* an artifact or a perturbation by the label.

G. G. Cameron (University of Aberdeen, Old Aberdeen, Scotland): Could you please clarify your comment on the idea that there is no sharp distinction between spin labels and spin probes?

P. Törmälä: The examination of Arrhenius plots of different probes and labels, shows that the rotational frequency of probe radicals at a constant temperature, decreases smoothly with increasing molecular weight of the probes and rapidly approaches the rotational frequency of label radicals.

P. L. Kumler (State University of New York, Fredonia, New York): Would you care to comment on the lack of variation in T_{50G} for poly(vinylidene fluoride) as the probe size is varied?

P. Törmälä: The lack of variation of T_{50G} for PVF_2 may indicate that the motional unit is very small and that the limiting value of T_{50G} has already been achieved in the case of even the smallest probe.

SPIN LABEL STUDIES

Spin-Labelling Studies of the Dynamics of Synthetic Macromolecules in Solution

ANTHONY T. BULLOCK

Department of Chemistry, University of Aberdeen, Meston Walk, Old Aberdeen AB9 2UE, Scotland

INTRODUCTION

Apart from their intrinsic interest, studies of the dynamics of linear polymers in solution can aid in the rationalization of certain aspects of the kinetics of polymerizations and, under certain conditions, to "gas phase" values of barriers to segmental reorientation of the chain. Techniques which give information on the dynamics of solutions of macromolecules include nuclear magnetic resonance relaxation measurements[1] (especially ^{13}C and D), dielectric relaxation,[2] fluorescence depolarization,[3] ultrasonic relaxation,[4] and Rayleigh scattering.[5] The most recent addition to this armoury of techniques is that of spin-labelling.[6] The manner in which this technique may complement and add to the information obtained by the other techniques is shown in Table 1. In general, all techniques are capable of measuring frequencies of segmental reorientation although dielectric studies are restricted to polymers possessing an electric dipole moment perpendicular to the main chain axis and are preferably made in non-polar solvents. Only

TABLE 1
Relaxation techniques

Method	Segmental relaxation	Chain-end relaxation
n.m.r. relaxation	+	−
dielectric relaxation	+	−
ultrasonic relaxation	+	−
fluorescence depolarization	+	+
Rayleigh scattering	+	−
spin-labelling	+	+

fluorescence depolarization studies are capable of giving useful information on both segmental and chain-end rotation[3] but they have the disadvantage that the fluorescent labels are usually very bulky and may well perturb the macromolecule under consideration. It will be seen later that the advantages the spin-labelling technique possesses are, firstly, that both segmental and chain-end mobility may be measured. Secondly, the measurements and subsequent analysis are rapid and straightforward and, finally, the size of the spin label is readily altered so that its perturbing effect on the polymer chain (if any) may be assessed.

Theory and Basis of Spin-Labelling

Nitroxides are the radicals of choice for spin-labelling studies. The reasons for this are three-fold. Firstly, they are usually stable, secondly they may be prepared carrying a variety of functional groups suitable for coupling to the polymer and, finally, they possess well-defined anisotropic hyperfine coupling and g-tensors. To illustrate qualitatively how they can give information about polymer dynamics, figure 1 shows the basic structure and axis system for the nitroxide group (R_1 will usually be the polymer chain).

Omitting nuclear Zeeman terms, the Hamiltonian for such a group may be written

$$\mathcal{H} = \beta\, \mathbf{H} \cdot \mathbf{g} \cdot \mathbf{S} + \sum_i \mathbf{S} \cdot \mathbf{T}_i \cdot \mathbf{I}_i \tag{1}$$

where $\mathbf{g}$ and $\mathbf{T}_i$ are the g-tensor and the hyperfine coupling tensor to nucleus i. Usually the spectrum is dominated by coupling to ^{14}N and equation 1 becomes

$$\mathcal{H} = \beta\, \mathbf{H} \cdot \mathbf{g} \cdot \mathbf{S} + \mathbf{S} \cdot \mathbf{T} \cdot \mathbf{I} \tag{2}$$

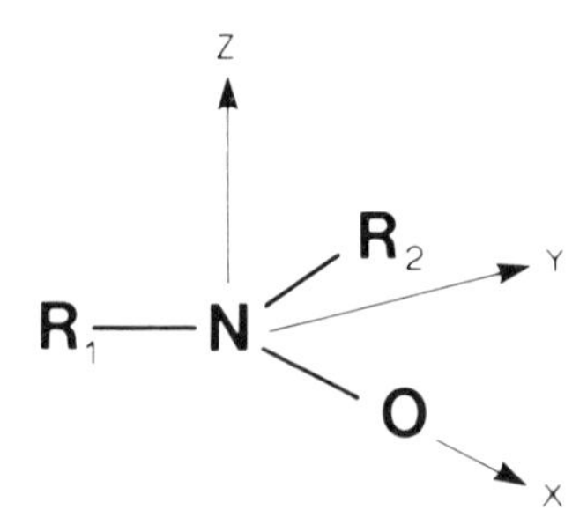

FIGURE 1 The axis system for the nitroxide group (R_1—usually the polymer chain).

The effect of rapid tumbling is to remove the anisotropy implicit in (1) and (2) and to give a spectrum whose centre is defined by

$$g_{iso} = \tfrac{1}{3}(g_{xx} + g_{yy} + g_{zz}) \qquad (3)$$

and splitting given by

$$T_{iso} = a = \tfrac{1}{3}(T_{xx} + T_{yy} + T_{zz}) \qquad (4)$$

The tumbling of the radical has, however, a secondary effect in that it modulates the field experienced by the unpaired electron. This provides a relaxation mechanism for both the longitudinal and transverse components of the magnetization of the system. Detailed analysis leads to the expression[7]

$$T_2^{-1}(m_1, m_2 \ldots m_n) = A + \sum_{i=1}^{n} B_i m_i + \sum_{i=1}^{n} C_i m_i^2 + \sum_{i \neq j=1}^{n} E_{ij} m_i m_j \qquad (5)$$

Explicitly, for the simplest case where only the anisotropy of g and the coupling to the ^{14}N nucleus need be considered,

$$T_2^{-1}(m_I) = \left[\frac{b^2}{20}(3 + 7u) + \frac{1}{15}(\Delta\gamma H_0)^2\left(\frac{4}{3} + u\right) + \frac{1}{8}b^2 m_I^2\left(1 - \frac{u}{5}\right) - \frac{1}{5}b\Delta\gamma H_0 m_I\left(\frac{4}{3} + u\right)\right]\tau_c + X \qquad (6)$$

For Lorentzian lines, the line-width parameter T_2 is related to the peak-to-peak width of the derivative spectrum $\Delta\nu$(Hz) by $T_2 = (\pi\sqrt{3}\,\Delta\nu)^{-1}$. Other parameters in equation (6) are defined as follows:

$$b = (4\pi/3)\,[T_{zz} - \tfrac{1}{2}(T_{xx} + T_{yy})]$$

$$\Delta\gamma = -\frac{|\beta|}{\hbar}[g_{zz} - \tfrac{1}{2}(g_{xx} + g_{yy})]$$

and

$$u = 1/(1 + \omega_0^2 \tau_c^2).$$

T_{ii} and g_{jj} are the principal values of the hyperfine coupling and g tensors respectively (the former in Hz). H_0 and ω_0 are the applied magnetic field and corresponding Larmor angular frequency respectively and τ_c is the rotational correlation time. The spin-lattice or nonsecular broadening, represented by u, is usually negligible for spin-labelled polymers in solution since τ_c is normally greater than 10^{-10} s and the Larmor frequency is *ca.* $2\pi \times 9.3 \times 10^9$ rads s^{-1}. The remaining term requiring definition in equation (6) is X. This is simply the broadening by mechanisms independent of the component of the nuclear spin quantum number m_I. Neglecting nonsecular terms, equation (6) becomes

$$T_2(m_I)^{-1} = \left[\frac{3b^2}{20} + \frac{4}{45}(\Delta\gamma H_0)^2 + \frac{b^2 m_I^2}{8} - \frac{4}{15} b\Delta\gamma H_0 m_I\right]\tau_c + X \qquad (7)$$

which is readily rearranged to eliminate X, thus

$$T_2(O)/T_2(m_I) = 1 - \frac{4}{15}\tau_c b\Delta\gamma H_0 T_2(O) m_I + \frac{1}{8}\tau_c b^2 T_2(O) m_I^2 \qquad (8)$$

Using $R_\pm$ to represent $T_2(O)/T_2(\pm 1)$

$$R_+ + R_- - 2 = (\tfrac{1}{4})\,\tau_c b^2\, T_2(O) \qquad (9)$$

It is usual to obtain accurate values of $R_\pm$ by measuring peak-to-peak intensities Y of the relevant lines, thus

$$R_\pm = T_2(O)/T_2(\pm 1) = [Y(O)/Y(\pm 1)]^{1/2} \qquad (10)$$

Before describing the method used to determine b, two assumptions have been made and must be clearly recognized since they are often ignored in the literature. The first is that all equations from (5) onwards are based on a model of isotropic rotational diffusion. Freed has removed this restriction[8,9] but in practice it is not always easy to use his analysis. Secondly, in deriving equation (8) it was assumed that X represented a source of homogeneous broadening. As will be seen later, there is a source of inhomogeneous broadening in most spin-labelled polymers. For example, if, in figure 1, R_1 represents the polymer chain and R_2 the t-butyl group then unresolved couplings to the protons in the latter, although small, can make a substantial contribution to the observed line widths due to the fact that Σm_H takes values from $-9/2$ to $+9/2$. The problem of inhomogeneous broadening will be considered in more detail in the next section.

The anisotropy parameter b may be determined by measuring a^N, the isotropic coupling constant together with the extrema separation in the powder spectrum of the solid labelled polymer. Provided that the randomly oriented radicals in the powder are at their rigid limit, i.e., $\tau_c \rightarrow \infty$, it may be shown that this separation is equal to $2T_{zz}$ (figure 2). Methods have been devised for determining T_{zz} from the powder spectrum when the rigid limit has not been

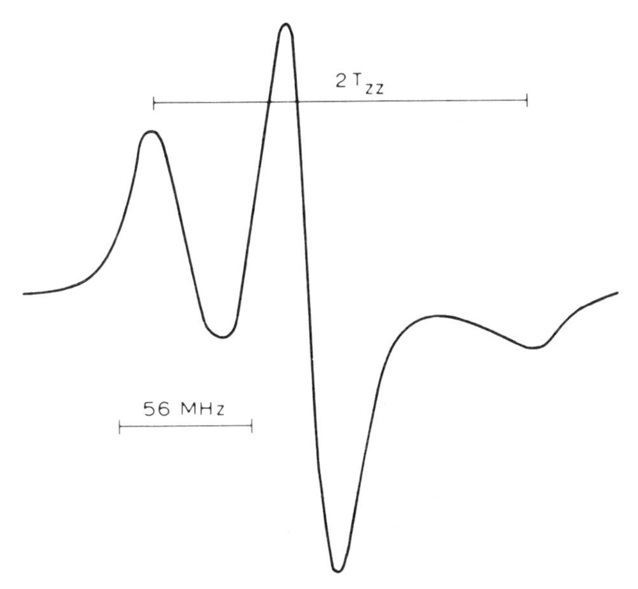

FIGURE 2 Powder spectrum of polystyrene labelled with *p*-t-butyl nitroxide. Reproduced from A.T. Bullock, G.G. Cameron, and P.M. Smith, *J. Polym. Sci., Polym. Phys. Ed.*, **11**, 1263 (1973) by permission of John Wiley & Sons, Inc.

reached.[6,10] From the definition of b together with equation (4) it readily follows that

$$b = 2\pi\,(T_{zz} - a^N) \tag{11}$$

Corrections for Inhomogeneous Broadening

Figure 3 illustrates the concept of an inhomogeneously broadened line. Briefly, the observed line width is partly determined by a static distribution of local fields and partly by the widths of the individual "spin packets". This latter width is described by the parameter T_2 used in the equations of the previous section. There are several techniques available for determining the spin packet width, namely power saturation studies,[11,12] computer simulations,[6,13,14] and spin-echo techniques.[15] Recently, the natural line widths have been increased by the admission of oxygen[16] until the observed width was in effect the spin packet width. This is essentially equivalent to increasing X in equations (6) and (7). On the whole, it seems that computer simulations are likely to prove the most reliable technique since not only can they take into

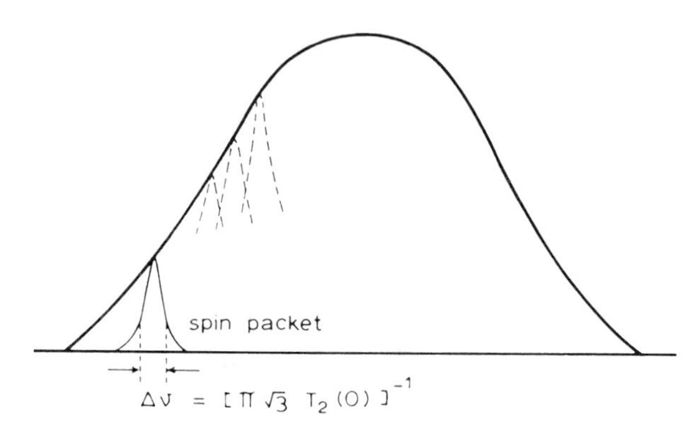

FIGURE 3 An inhomogeneously broadened line.

account unresolved couplings (provided that a reasonable estimate of the appropriate coupling constants can be made) but overlap of partially-resolved lines is readily accommodated.

As an example of the method, we will consider segmentally-labelled polystyrene in which one in *ca* 1,000 monomer units has the structure

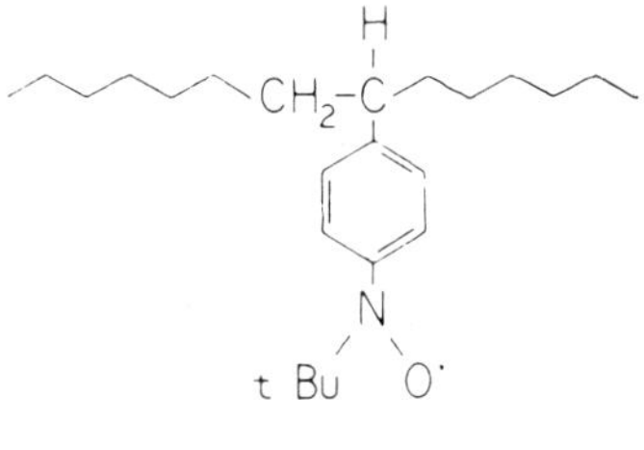

I

The reason for the low density of labelling is to avoid intrachain spin exchange; the method of preparation is described in reference 17. A typical spectrum is shown in figure 4 from which it may be seen that couplings to the ^{14}N nucleus and the aromatic protons are resolved whilst those to the t-butyl protons are not.

The computer program used has facilities for plotting the simulated spectrum on a graph plotter and also prints out the coordinates of all the turning points in the first-derivative spectrum. Simulations are made for a range of all the relevant coupling constants and the "true" peak-to-peak line width $\Delta\nu$(Hz) until a good match is obtained with an experimental spectrum chosen at random. Such a match is shown in figure 5. The simulations are then repeated

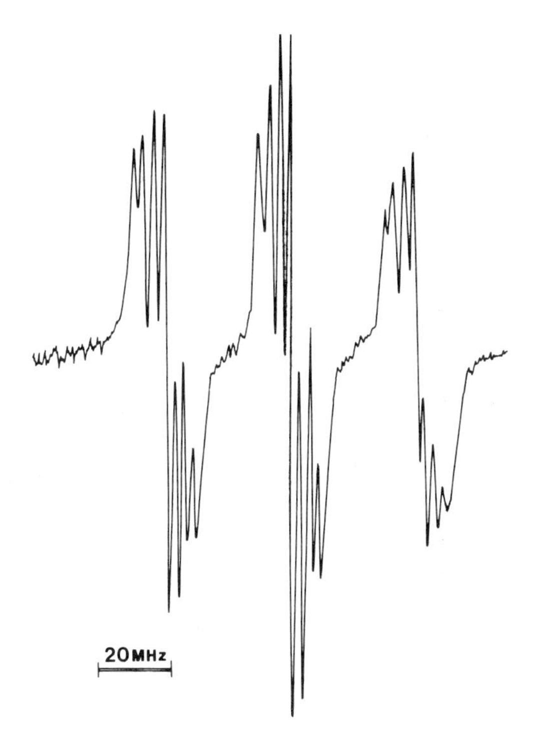

FIGURE 4 E.s.r. spectrum of a polystyrene fraction. ($\bar{M}_n$ = 1950) labelled with *p*-t-butyl nitroxide groups (polymer concentration 1% in toluene at room temperature). Reprinted with permission from ref. 18. Copyright by the American Chemical Society.

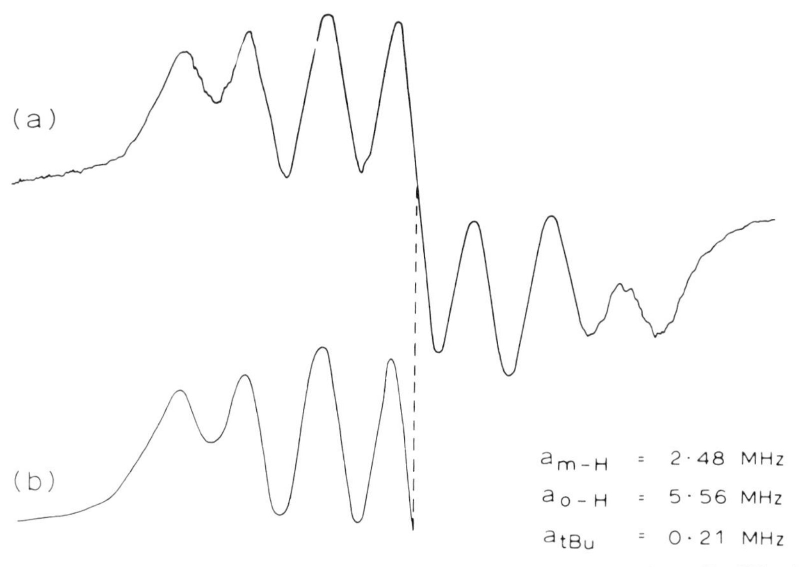

FIGURE 5 Recorded (a) and simulated (b) spectra of the central multiplet ($m_I = 0$) of Fig. 4. Reprinted with permission from ref. 18. Copyright by the American Chemical Society.

over a suitable range of values of "true" or input, line widths and calibration plots are made of observed *versus* true line widths and observed values of $R_{\pm}$ *versus* true $T_2(O)/T_2(\pm 1)$ ratios. Examples of such plots for *para*- labelled polystyrene are shown in figures 6 and 7 respectively.

Segmentally- and End-Labelled Polystyrene. A Case History

A priori it is expected that the dynamics of a macromolecule in solution will be determined by the following four factors: (i) intramolecular steric effects; (ii) molecular weight; (iii) viscous drag exerted by the solvent; and (iv) intermolecular interactions dependent upon polymer concentration.

All four factors have been studied[18,19] but we will omit a discussion of (iv) noting only that observed rotational correlation times for segmentally-labelled polystyrene showed no dependence on polymer concentration until this reached a value of *ca* 10% by weight.[19]

Seven narrow fraction GPC calibration standards of polystyrene were all lightly labelled in the *para* position of the phenyl ring with the t-butyl nitroxide label (I). The number average molecular weights were 1950, 3100, 9700,

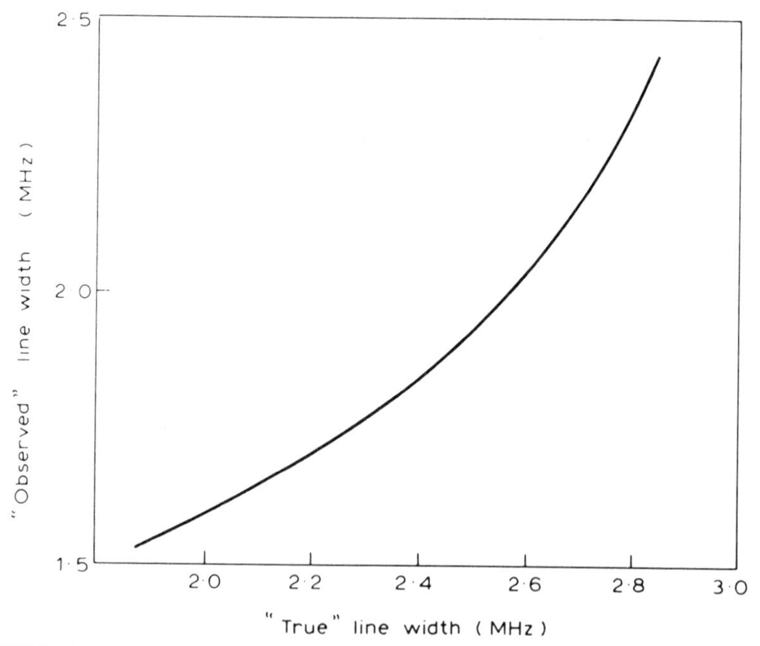

FIGURE 6 Observed *vs.* true peak-to-peak line width calibration plot. Reprinted with permission from ref. 18. Copyright by the American Chemical Society.

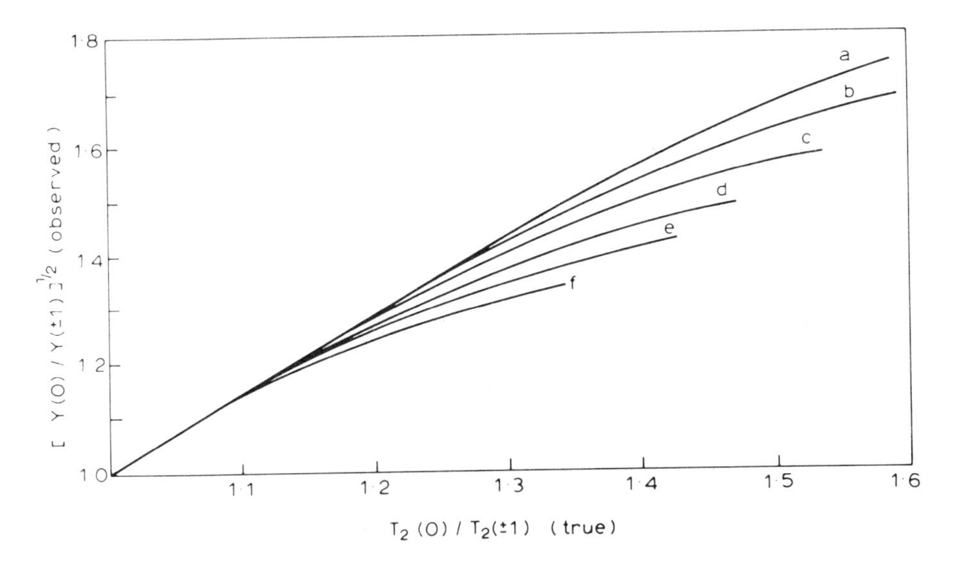

FIGURE 7 Observed $[Y(O)/Y(\pm 1)]^{1/2}$ *vs.* true $T_2(O)/T_2(\pm 1)$ plots for the following values of $\Delta\nu$ $(=[\sqrt{3}\pi\, T_2(O)]^{-1})$ in MHz: (a) 1.90, (b) 2.00, (c) 2.10, (d) 2.20, (e) 2.30, and (f) 2.40. Reprinted with permission from ref. 18. Copyright by the American Chemical Society.

19,650, 49,000, 96,200 and 193,000. These will be designated PS1 to PS7 respectively and figure 8 shows the dependence of τ_c on molecular weight for 1% solutions in toluene at 294.2, 312.6 and 345.2 K. All plots show an independence of τ_c on chain length at high molecular weights which is characteristic of a segmental, or local mode, relaxation associated with a correlation time τ_{lm}. As the molecular weight decreases then it is expected, and found, that "end-over-end" rotation of the whole macromolecule will begin to contribute to τ_c. Characterizing this mode by the correlation time τ_{eoe}, the observed correlation time τ_c will be given by

$$1/\tau_c = (1/\tau_{eoe}) + (1/\tau_{lm}) \qquad (12)$$

For reasons to be discussed shortly, correlation times were measured over a range of temperatures for the seven labelled polystyrenes in two good solvents, in the thermodynamic sense, namely toluene and α-chloronaphthalene. Measurements were also made in one poor solvent, cyclohexane. At a given temperature, plots of τ_c *versus* M_n all followed the general behavior shown in figure 8. It was thus possible to obtain values of τ_{eoe} from the observed τ_c and the plateau value τ_{lm}. It is possible to make theoretical estimates of τ_{eoe}. For example, for a non-draining coil[19,20]

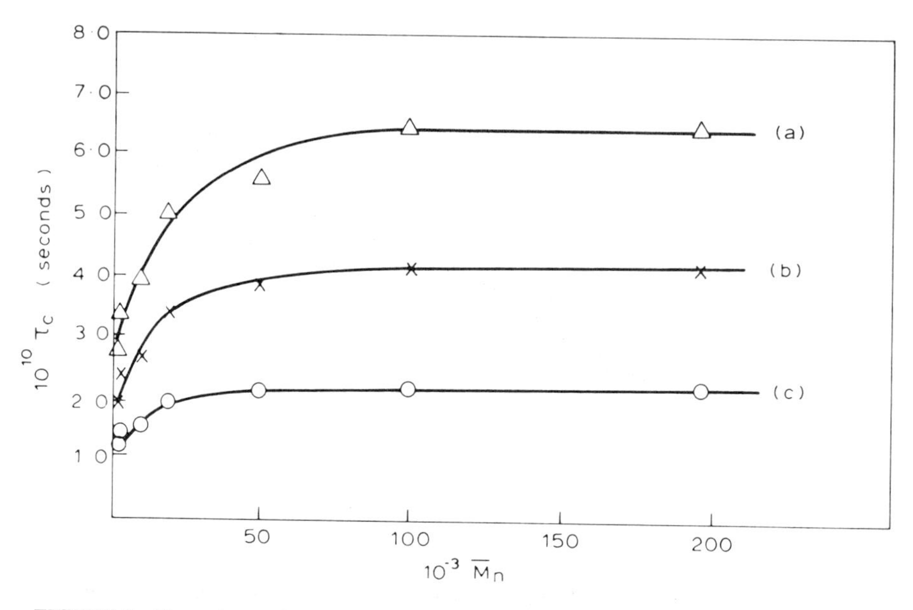

FIGURE 8 Dependence of τ_c upon molecular weight at (a) 294.2, (b) 312.6, and (c) 345.2 K for 1% solutions of labelled polystyrene in toluene. Reprinted with permission from ref. 18. Copyright by the American Chemical Society.

$$\tau_{eoe} = 0.42\ M[\eta]\eta_0/3RT \tag{13}$$

where M is the molecular weight, $[\eta]$ the limiting viscosity number and η_0 the viscosity of the solvent. Two tests of this equation are possible. Firstly, provided that $[\eta]$ is independent of temperature, as it is for toluene, then equation (13) shows that a plot of $\log_{10}(T\,\tau_{eoe})$ *versus* $1/T$ should give an activation energy equal to that for viscous flow of the pure solvent. Such a plot is shown in figure 9 from which $E_a = 8.8 \pm 1.7$ kJ mol^{-1}. This is in good agreement with that for viscous flow in toluene, E_η, which is[21] 9.00 kJ mol^{-1}. Secondly, values of τ_{eoe} calculated from equation (13) may be compared with values derived from experimental measurements. Table 2 shows the excellent agreement between experimental and theoretical estimates of τ_{eoe} for PS1 at 55°C in 1% solutions. For PS2 and higher fractions, however, the agreement is not good. As has been noted elsewhere,[19] this is almost certainly due to the inclusion in equation (12) of only the shortest and the longest normal modes.

Before continuing this discussion it is important to establish that the motion of the nitroxide label accurately reflects the dynamics of the chain to which it is bonded and also that this motion is indeed isotropic, as assumed in deriving τ_c from the corrected line widths. Two modes which would invalidate both these

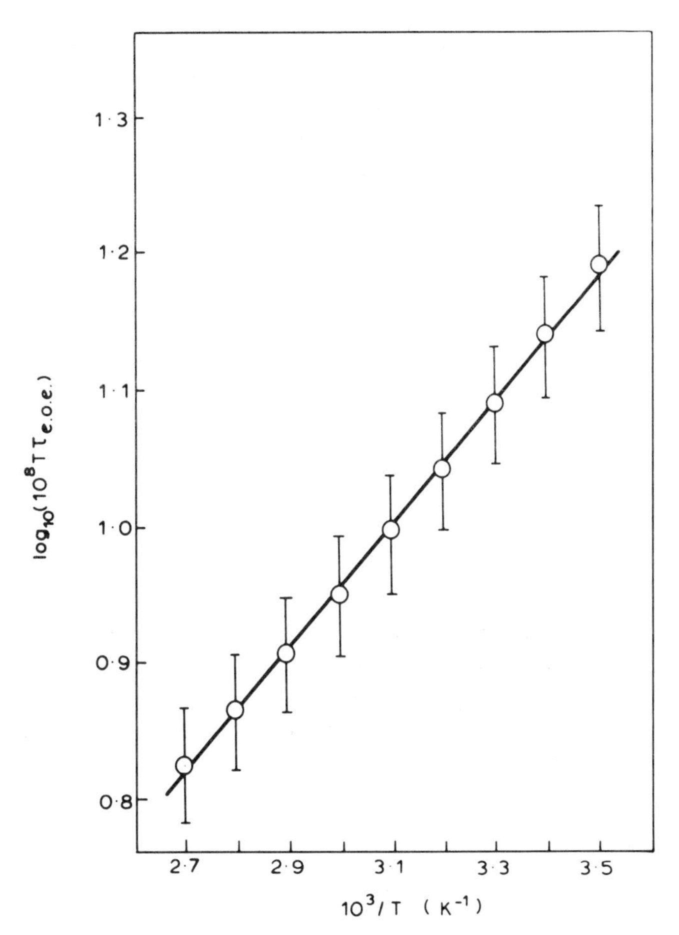

FIGURE 9 $\text{Log}_{10}(10^8\, T\, \tau_{eoe})$ *vs.* $10^3/T$ for segmentally labelled polystyrene in toluene ($\bar{M}_n =$ 1950): $E_a = 8.8 \pm 1.7$ kJ mol^{-1}. Reprinted with permission from ref. 18. Copyright by the American Chemical Society.

TABLE 2

Correlation times τ_{eoe} for end-over-end rotation of PS1 at 55°C in 1% solutions

Solvent	τ_{eoe} from equa. 13/ $s \times 10^{-10}$	τ_{eoe} experimental/ $s \times 10^{-10}$
toluene	2.0	2.4
cyclohexane	2.8	2.6
α-chloronaphthalene	7.5	6.0

assumptions are rapid rotation about (a) the *para*–carbon–nitrogen bond in I and (b) the main-chain carbon–phenyl carbon bond. If either of these modes were operative then changing the size of the pendant group bonded to the nitrogen atom would alter the apparent activation energy for the rotational diffusion process determining τ_c. A heterodisperse sample of polystyrene (mol. wt. *ca* 115,000) has been lightly labelled with II and then I.[6]

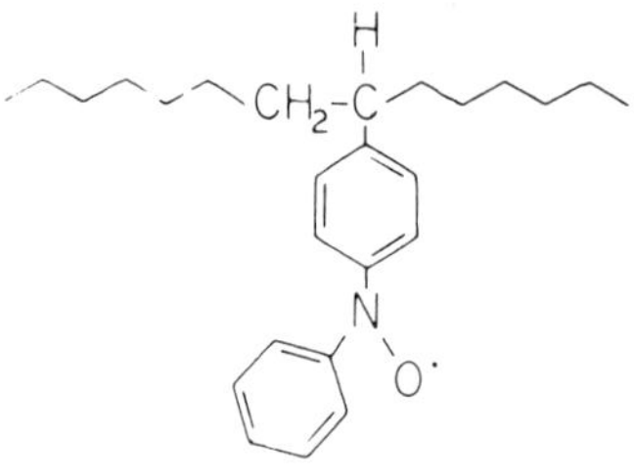

II

Figure 10 shows a composite Arrhenius plot for the two polymers together with a point derived from some ^{1}H spin-lattice time measurements.[22] Clearly, the close agreement between the correlation times for I and II over a range of temperatures supports the assumptions that the motion is isotropic and is characteristic of the polystyrene chain. Further support comes from some measurements of τ_c for *meta*-labelled polystyrene[23] (III).

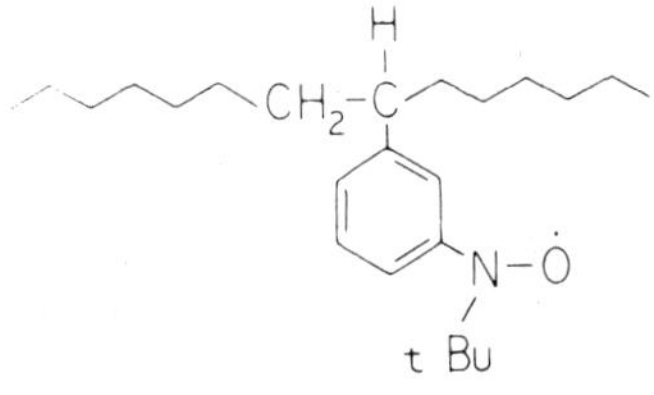

III

At 52°C, 1% toluene solutions gave correlation times of $(3.2 \pm 0.3) \times 10^{-10}$ s and $(3.9 \pm 0.4) \times 10^{-10}$ s for the *para*- and *meta*-labelled polymers respectively. The good agreement between these two values is further evidence that the motion of the spin label is accurately reflecting that of the segment of polymer chain to which it is bonded. Having established this, it is now possible to address the problem of assessing the relative contributions the viscous drag of the solvent and intramolecular contributions make in determining τ_{lm}.

Helfand has discussed the problem of Brownian diffusion over a potential

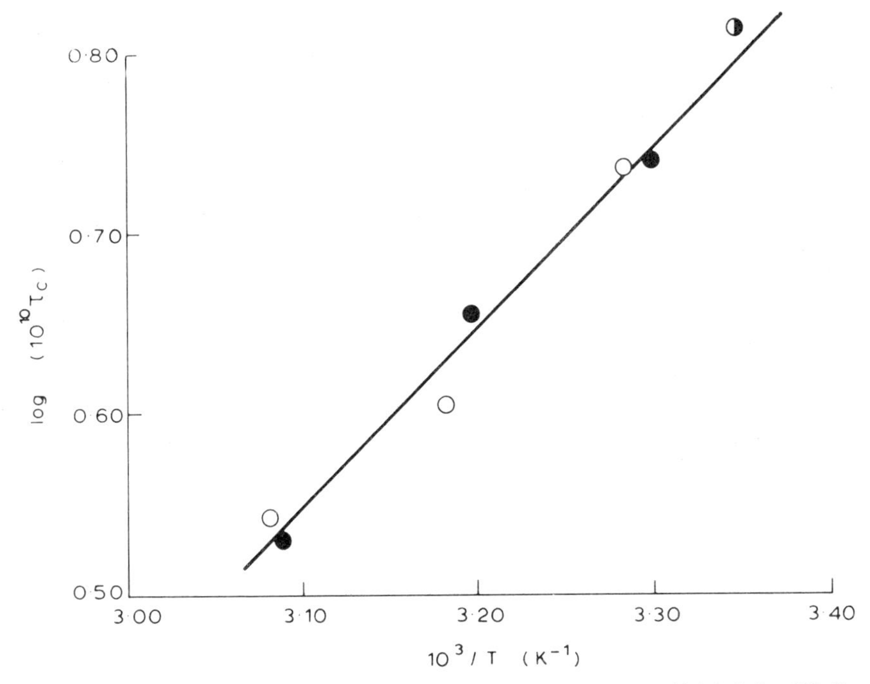

FIGURE 10 Composite Arrhenius plot for heterodisperse polystyrene with labels I and II: ○, label I; ●, label II; ◑, n.m.r. from ref. 22. Reprinted with permission from ref. 18. Copyright by the American Chemical Society.

barrier.[24] With suitable modifications it is easy to show that, in the limit of high viscous damping,

$$\tau_c = \tau_0 \exp[(E^* + E_\eta)/RT] \tag{14}$$

where E^* is the "gas phase" or intramolecular barrier to rotation and E_η is the activation energy for viscous flow of the solvent. Equation (14) may be compared with the empirical equation found to fit all our results so far, namely

$$\tau_c = \tau_0 \exp(E_{\text{tot}}/RT) \tag{15}$$

whence

$$E^* = E_{\text{tot}} - E_\eta \tag{16}$$

For toluene, α-chloronaphthalene and cyclohexane, $E_{\text{tot}} = 18.0 \pm 0.8$, 26.4 ± 1.3 and 26.1 ± 0.8 kJ mol^{-1} and $E_\eta - 9.00$[21], 17.78[25,26] and 12.55[21] kJ mol^{-1}. Thus $E^* = 9.00 \pm 0.8$ (toluene), 8.6 ± 1.3 (α-chloronaphthalene) and $13.6 \pm$

0.8 (cyclohexane). For the good solvents in which the polymer chains are well-extended the values of E^* are equal to within experimental error. This supports the use of the local mode model and, in particular, the adequacy of Helfand's treatment in combining the effects of intramolecular interactions with the viscous drag of the solvent. The most reasonable interpretation of the high values of E^* for polystyrene in cyclohexane (*ca* 50% higher than in the other two solvents) is that it reflects the increase in intrachain steric interactions consequent upon the more tightly-coiled configuration of the polymer in this solvent.

Having delineated in some detail the segmental motions of polystyrene chains in solution it is of interest to examine the rotational mobility of the chain ends. A method exists for labelling polymers with nitroxide groups specifically at the chain ends and is applicable to all polymers which can be prepared carbanionically.[27] In brief, it consists of terminating, or "killing" the living polymer by adding 2-methyl-2-nitrosopropane (MNP). The reaction scheme for polystyrene is

$$\sim CH_2\text{-}CH(C_6H_5)^{\ominus} + t\,BuNO\ (MNP) \xrightarrow{THF} \sim CH_2\text{-}CH(C_6H_5)\text{-}N(t\,Bu)\text{-}O^{\ominus}$$

$$\xrightarrow[Ag_2O]{MeOH} \sim CH_2\text{-}C\overset{*}{H}(C_6H_5)\text{-}N(t\,Bu)\text{-}O^{\cdot}$$

IV

The resultant polymer was purified by repeated reprecipitations and had a molecular weight of *ca* 2.3×10^5. The spectrum of a dilute solution of the polymer in toluene is shown in figure 11. It was found to be characterized by three features,[28] namely (a) couplings to ^{14}N of 41.83 ± 0.25 MHz and to ^1H (marked * in IV) of 8.70 ± 0.20 MHz, (b) line-width variations dependent on both m_H and m_N, the relevant nuclear spin quantum numbers, and (c) a negligible dependence of a^H on temperature.

For end-labelled polystyrene, there are six equations implicit in equation (5). Although the principal values and axes of the coupling tensor to ^1H are not

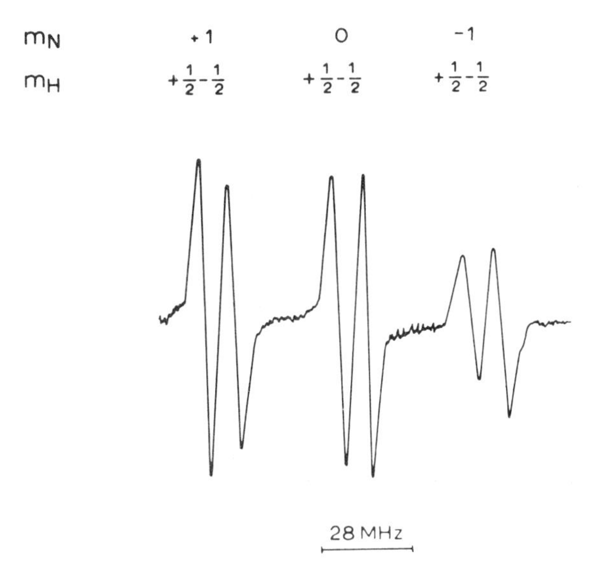

FIGURE 11 E.s.r. spectrum of end-labelled polystyrene in solution (3%). The signs of m_H and m_N are discussed in the text. Reproduced from ref. 28 by permission of The Chemical Society.

known, it is possible to eliminate A, B_H, C_H and $E_{N,H}$. Labelling the lines (m_H, m_N) we obtain

$$T_2^{-1}\ (+½, +1) + T_2^{-1}\ (+½,-1) - 2T_2^{-1}\ (+½, 0) = 2C_N \qquad (17)$$

whence

$$R_+ + R_- - 2 = 2C_N T_2\ (+½, 0) \qquad (18)$$

where $R_\pm$ are again the line-width ratios

$$T_2\ (+½, 0)/T_2\ (+½, \pm 1) \text{ and } C_N = (b^2/8)\tau_c.$$

It has been assumed that the isotropic coupling constants to ^{1}H and ^{14}N were both positive. There is unequivocal experimental[29] and theoretical[30] support for this assumption with regard to a^N. A detailed analysis shows that the sign of a^H, and hence m_H for a given line, is unimportant for our purpose. Provided that the sign chosen is used consistently, equation (18) obtains for both positive and negative values of a^H and m_H. Both sets of lines ($m_H = \pm½$) were used in calculating τ_c.[28]

The results were again analysed in terms of equations (14) and (15). Table 3

TABLE 3
Relaxation parameters for spin-labelled polystyrene in various solvents

Labelling site/solvent	E_{tot}/kJ mol^{-1}	$-\log_{10}\tau_0$	$10^{10}\tau$/s(298K)
end/toluene	14.8 ± 0.8	12.2 ± 0.1	2.4 ± 0.2
side-chain/toluene	18.0 ± 0.8	12.4 ± 0.1	5.9 ± 0.5
end/α-chloronaphthalene	23.0 ± 0.2	12.95 ± 0.05	11.9 ± 1.5*
side-chain/α-chloronaphthalene	26.4 ± 1.3	12.98 ± 0.06	35.4 ± 3.5*
end/cyclohexane	21.8 ± 1.4	13.1 ± 0.1	4.9 ± 0.1
side-chain/cyclohexane	26.1 ± 0.8	13.6 ± 0.1	10.0 ± 0.3*

*extrapolated values for comparison purposes

TABLE 4
Intramolecular barriers to rotation in labelled polystyrene E^*/kJ mol^{-1}

Solvent	side-chain	end-labelled
toluene	9.0 ± 0.6	5.8 ± 0.8
α-chloronaphthalene	8.6 ± 1.3	5.2 ± 0.2
cyclohexane	13.6 ± 0.8	9.2 ± 0.1

compares the relaxation parameters τ_0 and E_{tot} for the end- and segmentally-labelled polymers in three solvents while table 4 compares the different values of the intramolecular barriers, E^*. The barriers for a given site in the good solvents are equal, again supporting our use of a local mode model and the adequacy of Helfand's treatment. Furthermore, it seems quite reasonable that values of E^* for chain-end rotation are lower than those for segmental rotation in the thermodynamically good solvents. Again, in cyclohexane there is a markedly higher barrier reflecting the tightly-coiled configuration of the polymer in this solvent. This is in accord with the fact that k_t, the radical-radical termination rate coefficient, decreases in poor solvents.[31]

Table 3 shows that in the good solvents the pre-exponential factors, expressed as $\log_{10} \tau_0$, are equal for side-chain and end-labelled polymers in a given solvent. However, in cyclohexane there is a marked difference in $\log \tau_0$ for the two sites. The sign of this difference clearly shows a larger entropy of activation for rotation of the side-chain labelled segments in the tightly-coiled chain and implies that the ends of the macromolecule are to some extent excluded from the coil.

As in the case of the segmentally labelled polymer, it is necessary to demonstrate that the motion of the nitroxide label IV accurately reflects that of the polymer chain end. The only possible reason why it might not is rapid rotation about the label-polymer bond. Fortunately, internal spectroscopic evidence rules out this possibility. The coupling constant of the β-hydrogen is, in a nitroxide radical, given by[32]

$$a^{\mathrm{H}}/mT = Q^{\mathrm{H}}_{\mathrm{NCH}} \cdot \rho_{\mathrm{N}} \langle\cos^2 \theta\rangle \tag{19}$$

where a^{H} is in field units, θ is the angle which the N–C–H plane makes with the symmetry axis of the $2p_\pi$ orbital on the nitrogen atom and $Q^{\mathrm{H}}_{\mathrm{NCH}} \cdot \rho_{\mathrm{N}} \sim 2.6\ mT$* (73MHz $\equiv 4.6 \times 10^8$ rads s^{-1}). If the rotational frequency about the C–N bond becomes comparable to $Q^{\mathrm{H}}_{\mathrm{NCH}} \cdot \rho_{\mathrm{N}}$ (in radial frequency units) then $\langle\cos^2 \theta\rangle = ½$ and $a^{\mathrm{H}} \sim 1.3mT$. Since $a^{\mathrm{H}} \sim 0.3mT$ and shows no marked dependence on temperature, we may confidently conclude that the rotational frequency about the C–N bond is less than 4.6×10^8 rads s^{-1}, corresponding to a correlation time of $> 2 \times 10^{-9}$ s. Measured correlation times were always less than 5×10^{-10} s and thus the rotational mobility of the label is strictly governed by the mobility of the end of the polymer chain.

CONCLUSIONS

In this chapter we have attempted to show how the spin-labelling technique can lead to a detailed understanding of the dynamics of macromolecules in solution. Emphasis has been laid on the need to allow for inhomogeneous broadening and to check for isotropic rotational diffusion. It has been shown that, partly through the synthetic flexibility of nitroxides and partly through internal spectroscopic evidence, it is possible to ensure that the dynamics of the label accurately reflect those of the macromolecule. In suitable cases "gas phase" barriers to rotation of segments and chain ends may be obtained and correlated with certain kinetic parameters relevant to free radical polymerizations. In a more fundamental sense it has been found possible to quantitatively separate two effects which determine segmental and chain-end mobilities, namely, intramolecular steric effects and the viscous drag of the solvent.

---

*1 millitesla ≡ 10 Gauss

REFERENCES

1. A. Allerhand and R.K. Hailstone, *J. Chem. Phys.*, **56**, 3718 (1972).
2. H. Block and A.M. North, *Adv. Mol. Relaxation Processes*, **1**, 309 (1970).
3. A.M. North and I. Soutar, *J.C.S. Faraday I*, **68**, 1101 (1972).
4. H.J. Bauer, H. Hässler and M. Immendörfer, *Disc. Faraday Soc.*, **49**, 238 (1970).
5. R. Pecora, *Disc. Faraday Soc.*, **49**, 222 (1970).

6. A.T. Bullock and G.G. Cameron, ch. 15 of "*Structural Studies of Macromolecules by Spectroscopic Methods*", ed. K.J. Ivin, J. Wiley and Sons, London, New York, Sydney, Toronto, 1976, p. 273.
7. A. Hudson and G.R. Luckhurst, *Chem. Rev.*, **69**, 191 (1969).
8. J.H. Freed, *J. Chem. Phys.*, **41**, 2077 (1964).
9. S.A. Goldman, G.V. Bruno, C.F. Polnaszek and J.H. Freed, *J. Chem. Phys.*, **56**, 716 (1972).
10. A.T. Bullock, G.G. Cameron and V. Krajewski, *J. Phys. Chem.*, **80**, 1792 (1976).
11. T.G. Castner, *Phys. Rev.*, **115**, 1506 (1959).
12. A.T. Bullock and L.H. Sutcliffe, *Trans. Faraday Soc.*, **60**, 2112 (1964).
13. G. Poggi and C.S. Johnson, Jr., *J. Magn. Reson.*, **3**, 436 (1970).
14. A.T. Bullock, G.G. Cameron and P.M. Smith, *J. Phys. Chem.*, **77**, 1635 (1973).
15. I.M. Brown, *J. Chem. Phys.*, **58**, 4242 (1973).
16. A.N. Kuznetsov, A.Y. Kolkov, V.A. Livshits and A.T. Mirsoian, *Chem. Phys. Lett.*, **26**, 369 (1974).
17. A.T. Bullock, G.G. Cameron and P.M. Smith, *Polymer*, **13**, 89 (1972).
18. A.T. Bullock, G.G. Cameron and P.M. Smith, *J. Phys. Chem.*, **77**, 1635 (1973).
19. A.T. Bullock, G.G. Cameron and P.M. Smith, *J. Chem. Soc., Faraday II*, **70**, 1202 (1974).
20. B.H. Zimm, *J. Chem. Phys.*, **24**, 269 (1956).
21. *Selected Values of Physical and Thermodynamic Properties of Hydrocarbons and Related Compounds.* (American Petrol. Inst. Research Project No. 44, 1953).
22. D.W. McCall and F.A. Bovey, *J. Polymer Sci.*, **45**, 530 (1960).
23. A.T. Bullock, G.G. Cameron and P.M. Smith, *Polymer*, **14**, 525 (1973).
24. E. Helfand, *J. Chem. Phys.*, **54**, 4651 (1971).
25. A. Bhanumathi, *Indian J. Pure Appl. Phys.*, **1**, 79 (1963).
26. W.M. Heston, Jr., E.J. Henelly and C.P. Smyth, *J. Amer. Chem. Soc.*, **72**, 2071 (1950).
27. A.R. Forrester and S.P. Hepburn, *J. Chem. Soc. (C)*, **701** (1971).
28. A.T. Bullock, G.G. Cameron and N.K. Reddy, *J. Chem. Soc., Faraday I*, **74**, 727 (1978).
29. T.J. Stone, T. Buckman, P.L. Nordio and H.M. McConnell, *Proc. Natl. Acad. Sci., U.S.*, **54**, 1010 (1965).
30. G. Giacometti and P.L. Nordio, *Mol. Phys.*, **6**, 301 (1963).
31. G.G. Cameron and J. Cameron, *Polymer*, **14**, 107 (1973).
32. E.G. Janzen, "*Topics in Stereochemistry*", ed. N.L. Allinger and E.L. Eliel (Wiley-Interscience, New York, London, Sydney and Toronto, 1971) vol. 6, p. 189.

DISCUSSION

A. Yelon (Ecole Polytechnique, Montreal, Canada): How do you separate E^* and E_η from E_{total}?

A. T. Bullock: As indicated in the text, E_{total} is simply the observed activation energy obtained by plotting $\log_{10} \tau_c$ versus $1/T$ (equa. 15). The Helfand treatment, however, gives $\tau_c = \tau_0 \exp[(E^* + E_\eta)/RT]$ (equa. 14). Thus $E_{\text{total}} = E^* + E_\eta$. The activation energy for viscous flow of the solvent, E_η, is readily obtained and thus, so is E^*.

J. Sohma (Hokkaido University, Sapporo, Japan): Do you think that E^* is the quantity compared with the rotational barrier in the gas phase?

A. T. Bullock: Yes, although I would prefer to put 'gas phase' in inverted commas.

J. Sohma: How do you know the potential barrier for rotation in a polymer molecule in the gas phase?

A. T. Bullock: Using Helfand's treatment, and knowing E_η (equa. 14 and 15) we derive the barrier to rotation separately from the viscous drag of the solvent. It is in this sense that we use the words 'gas phase'.

J. Sohma: Do you have any other values for the rotational barrier in the gas phase, which were determined by other experimental methods?

A. T. Bullock: No, of course not. We simply cannot get polymer molecules into the gas phase. However, I believe that Helfand's theory has been used in some ^{13}C T_1 results for polystyrene in various solvents with essentially the same results reported here. Further, Cameron has pointed out elsewhere, that Helfand's treatment applied to our spin-labelled polyethylene data yields a barrier to rotation very close to the true gas phase value for ethane.

A. Gupta (Jet Propulsion Laboratory, California Institute of Technology, Pasadena, California): What is the effect of the nature of the solvent on intermolecular interaction in your system?

A. T. Bullock: We normally take care to avoid polymer–polymer interactions by using dilute solutions (~1%) so I am unable to comment on your question. However, it is possible to study τ_c at a given temperature as a function of polymer concentration (see below).

A. Gupta: Is this a good method for studying chain entanglement or intermolecular interactions?

A. T. Bullock: Yes, although our results indicate that a theory other than the entanglement theory may be more consistent in concentrated polymer solutions (ref. 19 of text). Briefly, τ_c is independent of concentration until a certain critical value, C_c, is reached. It then increases sharply with concentration. C_c depends on molecular weight. For polystyrene in toluene, C_c = 20% for PS 2 ($\bar{M}_n$ = 3,100) and 16% for PS 7 ($\bar{M}_n$ = 193,000).

L. J. Berliner (Ohio State University, Columbus, Ohio): What about local microstructural effects of two "identical" solvents?

A. T. Bullock: The fact that the same value of E^* is obtained with solvents having widely different viscosities and E_η's, but the same solvent power, clearly shows the absence of abnormal microstructural effects. The question is therefore rather hypothetical.

W. G. Miller (University of Minnesota, Minneapolis, Minnesota): The Helfand treatment is a local mode mechanism whereby only a few bonds are undergoing motion, with the rest of the chain fixed. Have you considered whether or not such a small number of bond rotations is sufficient to give isotropic motion to the nitroxide? Also, considering the Helfand mechanism, do you consider the effect in going from a good to a poor solvent, one of changing the rotamer population, or what?

A. T. Bullock: To the first part I would point out that molecular models show that it is only necessary for a few (say 3 or 4) monomer units either side of the label to undergo rotation in order that the labelled unit rotates isotropically. The second query has already been answered, namely, in going from a good to a poor solvent, the polymer "tightens up", i.e., $(\bar{r}^2)^{1/2}$, the root mean squared end to end chain length decreases and intramolecular steric interactions increase. This explains the increase in E^* in going from toluene to cyclohexane.

P. Törmälä (Tampere University of Technology, Tampere, Finland): It is understandable to think that the rotations of a nitroxide label which is bonded to a hypothetical fixed carbon chain should be cylindrically symmetrical. But, both American and Russian researchers have shown that this motion changes rapidly to isotropic motion when the number of carbon atoms in the chain is increased and rotation about carbon–carbon bonds is assumed.

P. Törmälä: Do you think that you will get the same values of activation energy for segmental motion if you use, instead of E_η, the apparent activation energies of rotation for the corresponding probe radicals in the solvents studied?

A. T. Bullock: Yes. It has been known for some time that the rotational diffusion motion of small molecules in many solvents has the same activation energy as that for viscous flow of the solvent being used. This has been demonstrated by NMR and dielectric relaxation studies as well as by ESR linewidth measurements. However, the experimental values of the rotational correlation times are usually smaller than that predicted by the Stokes-Einstein equation, often by an order of magnitude. I repeat, however, that $E(\tau_c) = E_\eta$.

Comparison of Molecular Motions Detected by Spin Label and ^{13}C NMR Experiments of PMMA in Benzene Solutions

JUNKICHI SOHMA and KENTARO MURAKAMI

Faculty of Engineering, Hokkaido University, Sapporo 060, Japan

INTRODUCTION

The spin labeling technique is very useful to detect modes of rapid motion in polymer systems.[1–6] This method always introduces a foreign molecule, that is, a labeling radical, into a system. The existence of such foreign molecules may perturb the system, and therefore, the result obtained by the spin labeling method might not be genuine, but different from that of the unlabeled system. One may say that the concentration of the labeling radical is too small to cause disturbance to the system and the effect is negligible, but this is incorrect. It is true that a small concentration produces merely negligible effects on the bulk nature. However, ESR, which observes a labeled radical in the system, provides us with information on the radicals and their particular sites. Thus, the environment around the radical, which might be perturbed by the labeling, is crucial in ESR.

The correlation times determined by using spin labels of different sizes or by an NMR method were compared,[2b] but the comparisons were qualitative and not detailed. It seems interesting and important to compare quantitative results obtained by the spin label experiments on a particular system with those derived from other experiments on an undisturbed system. It is difficult to compare a rotational correlation time obtained from a spin label experiment with other data taken from an unlabeled system, because the rapid mode of molecular motion, say 10^{-10} sec, in a solid system is hardly accessible by methods other than spin label experiments. It is known that correlation times determined by ^{13}C NMR relaxation cover a similar range of time, but an application of this technique is limited to a liquid phase. Two experiments,

ESR of a spin labeled system in a liquid and ^{13}C NMR of an unlabeled system, provide an experimental basis to decide whether or not the perturbation of the label to the system is negligible.

EXPERIMENTAL

The sample used is poly(methyl methacrylate), which is spin labeled at the ester end, as shown by the formula below: The method of preparation has been described previously.[6] The ratio of labeled to unlabeled side chains is about 1:100. Normal, non-labeled PMMA having the same degree of polymerization and microtacticity, was also prepared in our laboratory (MW *ca.* 8,000).

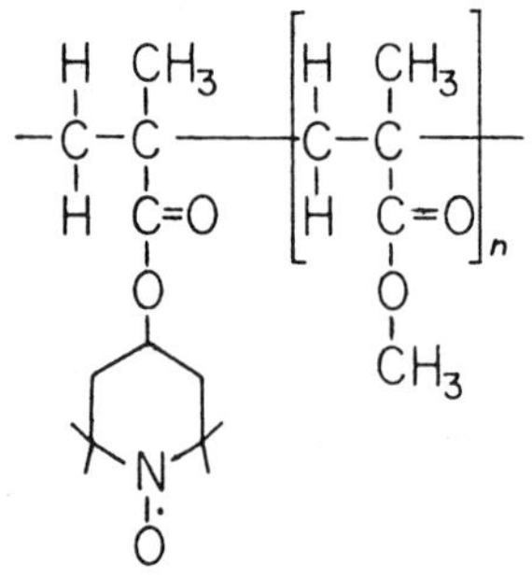

RESULTS AND DISCUSSION

ESR of the labeled PMMA

The ESR spectrum observed at 78°C from a benzene solution (0.015 g/cc) of the spin labeled PMMA is shown in Fig. 1. Each of the nitrogen triplet lines is clearly resolved and are similar to those reported by Bullock for spin labeled PMMA in toluene.[7] The spectra shown in Fig. 2 are for the $M_I = 0$ band of the triplet. Fig. 2 also illustrates the temperature variation of the line-shapes; the resolution of the spectrum increases and the line-width narrows with increasing temperature. The correlation time at each temperature can be estimated by using Kivelson's equation,[8] Eq. (1).

$$T_2(M_I)^{-1} = \tau_c\{(6 + 5M_I^2)\frac{b^2}{40} + \frac{4}{45}(\Delta\gamma \cdot H)^2 - \frac{4}{15}\, b \cdot \Delta\gamma H M_I\} + X \qquad (1)$$

where

$$b = \frac{4\pi}{3}(A_\parallel - A_\perp) \qquad (2)$$

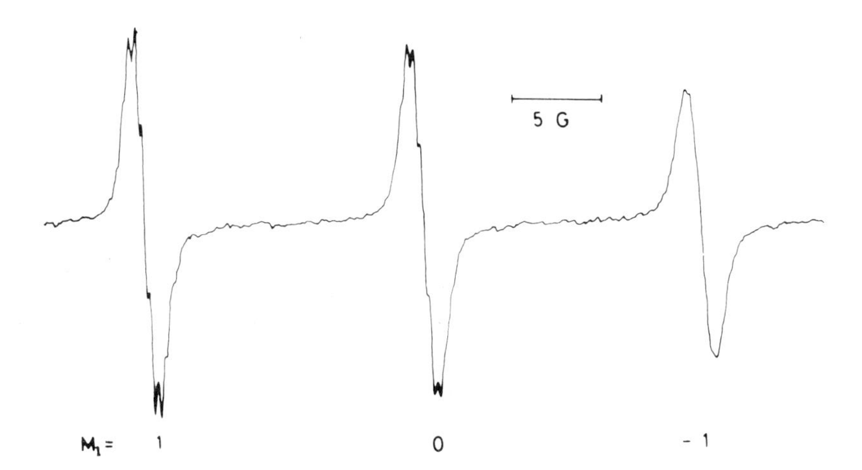

FIGURE 1 The ESR spectrum of spin labeled PMMA in C_6H_6 (conc. = 0.015 g/cc) at 78°C. Reprinted with permission from K. Murakami and J. Sohma, *J. Phys. Chem.*, **82**, 2825 (1978). Copyright by the American Chemical Society.

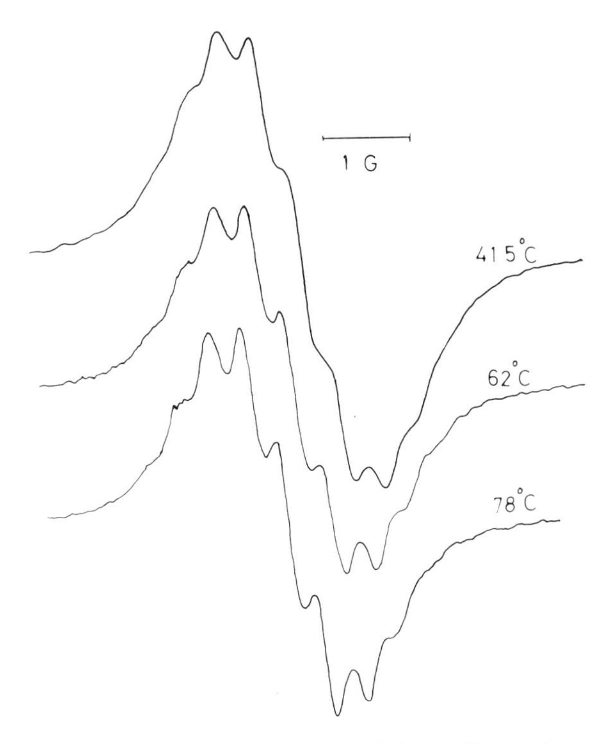

FIGURE 2 The magnified spectra of the central peak (m_I = 0) at various temperatures. Reprinted with permission from K. Murakami and J. Sohma, *J. Phys. Chem.*, **82**, 2825 (1978). Copyright by the American Chemical Society.

$$\Delta\gamma = \frac{\beta}{\hbar} \{g_{zz} - ½(g_{xx} + g_{yy})\} \tag{3}$$

For nitroxides these constants, b and $\Delta\gamma$, which are the differences between the principal values of the hf and g tensors, are known and therefore T_2 is the quantity, which can be estimated from the derivation of the correlation time, τ_c. The line-widths are determined by computer simulations and a calibration curve. The patterns shown in Fig. 3 are simulation curves using these ESR parameters and variable line-widths as adjustable parameters. The ratio, K_1/K_2, changes very sensitively with line-width. When this ratio is plotted against line-width, we get the calibration curve shown in Fig. 4. This curve is the plot of the K_1/K_2 ratios in the simulated patterns versus line-width. One can easily determine the line-width in gauss from the observed ratio by using this calibration curve. By this procedure the line-width at each temperature was determined. The line-width estimated in gauss was converted to T_2 in seconds assuming Lorentzian line-shape. The correlation time τ_c at each temperature is evaluated by inserting the values of T_2, b, $\Delta\gamma$, and H into Eq. (1); the Arrhenius plot of τ_c is shown in Fig. 5. Arrhenius plots of τ_c's are represented

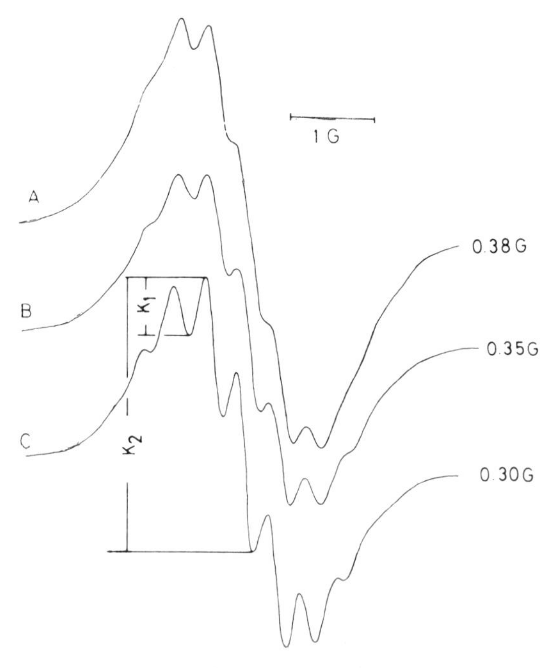

FIGURE 3 The simulated spectra with various linewidths, and the parameters K_1 and K_2.

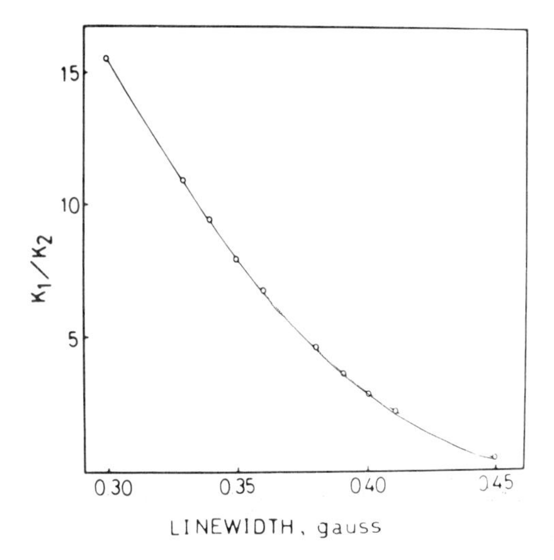

FIGURE 4 The calibration curve of the parameter K_1/K_2 *vs.* linewidths used in the simulations. Reprinted with permission from K. Murakami and J. Sohma, *J. Phys. Chem.*, **82**, 2825 (1978). Copyright by the American Chemical Society.

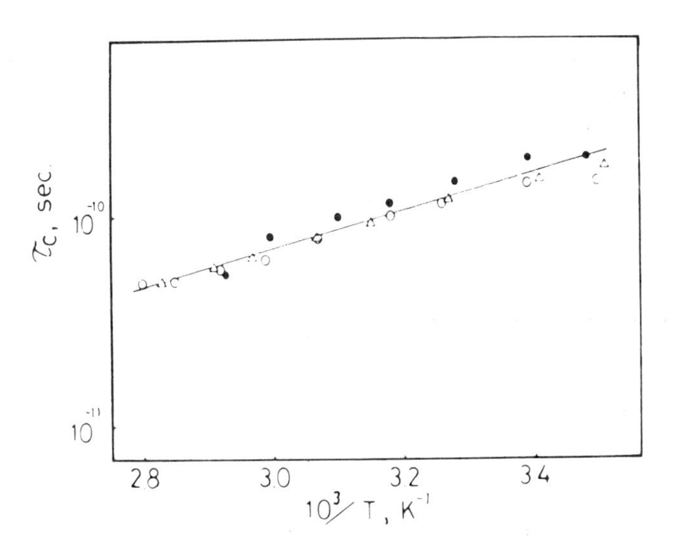

FIGURE 5 Arrhenius plot of the correlation times of spin labeled PMMA of varying concentrations: ○, 6×10^{-4} g/cc; △, 3×10^{-3} g/cc; ●, 1.5×10^{-2} g/cc. Reprinted with permission from K. Murakami and J. Sohma, *J. Phys. Chem.*, **82**, 2825 (1978). Copyright by the American Chemical Society.

T_1's in milliseconds

Temp. (°C)	$-CH_2-$	$-OCH_3$	$-\overset{|}{\underset{|}{C}}-$	$\alpha-CH_3$
28	61	652	747	110
50	81	876	1296	267
65	189	1072	1450	355

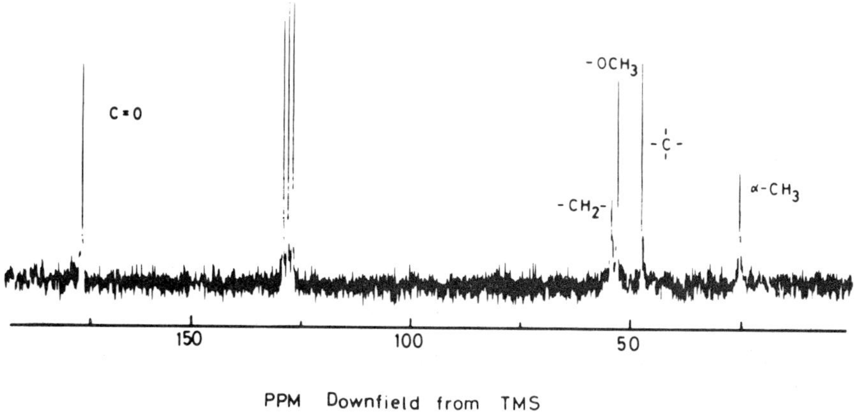

FIGURE 6 The proton-decoupled ^{13}C FT NMR spectrum of non-labeled isotactic PMMA in C_6D_6 (conc. = 0.1 g/cc) at 30°C after 800 accumulations. T_1 values are also given.

with a single line; however, the τ_c's of the most concentrated solution are longer than the others, although the activation energy is about the same (4.4 kcal/mol) for all samples.

^{13}C NMR of the non-labeled PMMA

Fig. 6 shows a ^{13}C NMR spectrum of the unlabeled PMMA in C_6D_6 solution and the peaks are assigned as shown. The spin-lattice relaxation times were determined by the partially-relaxed FT method. The T_1 value of each chemical shift in PMMA is also given in Fig. 6. The T_1's are primarily determined by the dipolar interaction with the attached protons in this system; the correlation times are estimated from the T_1's through the following equation.[9]

$$T_1^{-1} = N\gamma_C^2\gamma_H^2\hbar^2 r^{-6}\tau_c \tag{4}$$

γ_C and γ_H are the gyromagnetic ratios for carbon and a proton, respectively; r is the C–H bond distance and is taken to be 1.09Å. The molecular motion of the side chain is separated into two contributions, the internal rotation of the side chain and the motions of the other degrees of freedom. The latter mode of motion includes the overall tumbling of the bulk molecule and reorientation of the back-bone chain. That is,

$$(\tau_c)^{-1}_{\rm obs} = (\tau_c)^{-1}_{\rm R} + (\tau_c)^{-1}_{\rm I} \qquad (5)$$

where $(\tau_c)_{\rm obs}$ stands for the observed τ_c of the side chain, $(\tau_c)_{\rm I}$ stands for the internal rotation, and $(\tau_c)_{\rm R}$ represents whole molecular tumbling and back-bone reorientation.[10] This relation between correlation times holds in both NMR and ESR. It is reasonable to take the experimental NMR value of the correlation time of the methylene carbon as $(\tau_c)_{\rm R}$. By inserting this value for $(\tau_c)_{\rm R}$ and $(\tau_c)_{\rm obs}$ in Eq. (5) the correlation time of the methoxy group in the side chain is estimated to be 3.8×10^{-11} sec (for the sample of conc. 0.1 g/cc). The values determined by ESR measurements were used for $(\tau_c)_{\rm obs}$, and the methylene τ_c determined by NMR was also used as $(\tau_c)_{\rm R}$ for the estimation of $(\tau_c)_{\rm I}$ of the internal rotation of the spin label radical. This procedure, in which the same τ_c of the methylene carbon is used in estimating $(\tau_c)_{\rm I}$ in the ESR, is equivalent to assuming that the reorientation of the back-bone motion is identical even when the methoxy side-chain is replaced with the bulky labeling molecule. This is the assumption we wish to check. If the experimental results are analyzed and found to be consistent, then this will constitute experimental support of the assumption.

In order to make a more quantitative comparison between the correlation times estimated by ESR and NMR, two corrections should be taken into account. The first factor is the viscosity and the second factor is the bulkiness; the correlation time depends on these quantities as shown by the following equation,

$$\tau_c = \frac{4\pi a^3 \eta}{3kT} \qquad (6)$$

where η is the viscosity coefficient, and a is the radius of the spherical molecule. This equation is derived on the basis of the random motion of a spherical molecule suspended in an isotropic and continuous matrix. The spatial environment around the end of the side chain of PMMA, the methoxy and nitroxide group, violate this assumption. Both groups undergo rotations about fixed axes rather than three-dimensional random motion. However, this relation is still reasonable for a quantitative discussion, bearing this limitation in mind. Ideally both ESR and NMR experiments should be done using samples of the same concentration, but it was impossible to do so because of the lower sensitivity of NMR. Thus, the concentration of the NMR samples were inevitably higher and more viscous than the ESR samples. The observed values of η were inserted into the equation for comparison. Assuming a rotation about thc same axis, one may determine the radii of the spheres

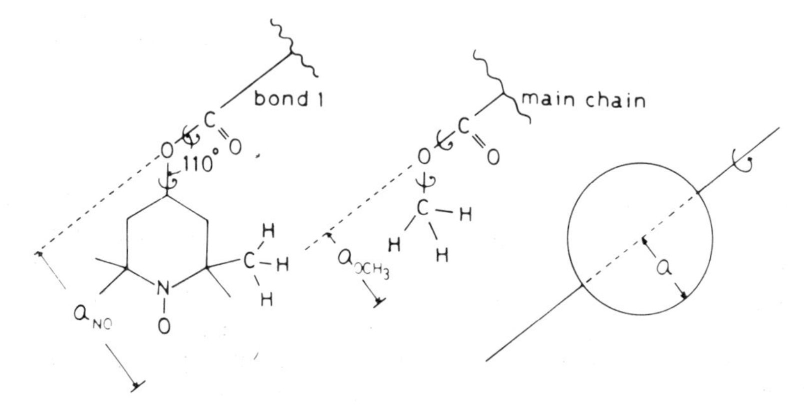

FIGURE 7 The effective rotational radii a_{NO}, a_{OCH_3}, and its spherical model. Reprinted with permission from K. Murakami and J. Sohma, *J. Phys. Chem.*, **82**, 2825 (1978). Copyright by the American Chemical Society.

approximated for these groups. In order to estimate the effective radius, an approximation was introduced, as shown in Fig. 7. In this way, the radius a_{NO} for the labeled molecule is taken as 4.98 Å and a_{OCH_3} for the unlabeled molecule as 1.69 Å. By inserting these values into Eq. (6) for τ_c, the ratio of the two τ_c's, which differ both in size and viscosity, was theoretically estimated at a temperature of 30°C as follows

$$R = \frac{\tau_c(a = 4.98\text{Å}, \eta = 0.92\text{cp})}{\tau_c(a = 1.69\text{Å}, \eta = 4.78\text{cp})} = 4.9 \tag{7}$$

In order to estimate the NMR τ_c from the observed ESR data, the ESR τ_c value should be multiplied by the ratio, R^{-1}, being a kind of conversion factor derived from corrections for size and viscosity effects. That is $(\tau_c)_{NMR} = R^{-1}(\tau_c)_{ESR} = (1/4.9)(1.4 \times 10^{-10}\text{ sec}) = 2.9 \times 10^{-11}$ sec. The τ_c experimentally determined by ^{13}C NMR is 3.8×10^{-11} sec; the agreement between these values is satisfactory, i.e., the experimental results obtained by both NMR and ESR are consistent. This consistency supports the assumption that little effect was induced by the introduction of the labeled radical on the polymer chain. It should be noted that the agreement is not obtained without correcting for label size. For estimating a correlation time for the unlabeled system from a spin label experiment, one has to correct for size differences. The above considerations suggest that perturbations due to spin labeling is limited only to the difference in size of the labeled and unlabeled molecules and that the environment around the labeled molecule is changed very little by the introduction of this foreign molecule.

REFERENCES

1. L.J. Berliner, ed., "Spin Labeling: Theory and Applications", Academic Press, New York, (1976).
2. A.T. Bullock, *et al.*; (a) *Europ. Polym. J.*, **7**, 445 (1971); (b) *J. Phys. Chem.*, **77**, 1635 (1973); (c) *J. Phys. Chem.*, **80**, 1792 (1976).
3. N. Kusumoto, S. Sano, N. Zaitsu, and Y. Motozato, *Polymer,* **17**, 448 (1976).
4. W.G. Miller, *et al.*, *J. Phys. Chem.*, **77**, 182 (1973).
5. P.L. Kumler and R.F. Boyer, *Macromolecules*, **9**, 903 (1976).
6. M. Shiotani and J. Sohma, *Polym. J.*, **9**, 283 (1977).
7. A.T. Bullock, G.G. Cameron, and V. Krajewski, *J. Phys. Chem.*, **80**, 1792 (1976).
8. D. Kivelson, *J. Chem. Phys.*, **33**, 1094 (1960).
9. A. Allerhand, D. Doddrell, and R. Komoroski, *J. Chem. Phys.*, **55**, 189 (1971).
10. Y. Inoue, A. Nishioka, and R. Chujo, *Makromol. Chemie.*, **168**, 163 (1973).

DISCUSSION

W. G. Miller (University of Minnesota, Minneapolis, Minnesota): You have used a sum of exponential decays to describe the molecular motion. Other workers studying the temperature dependence of ^{13}C T_1's, for example, Heatley in England, have found that motion giving rise to ^{13}C relaxations in polymers is better described by a combination of an exponential and a non-exponential decay (Hunt-Powles-Monnerie-Heatley model). Have you considered such a model? If not, why can you use a sum of exponential decays whereas other studies seem to indicate a non-exponential component?

J. Sohma: I know that a recovery of peaks is not simple and is expressed either by a sum of two or three exponential decays or by a non-exponential decay. In our case with these solutions of PMMA the recovery curve is simply represented by a single exponential curve, i.e., semilog plots of the peaks corresponding to different chemical shifts are quite linear with time, though the gradients of these lines are different from one chemical shift to another. We do not need to assume complex expressions to represent our results.

Spin Label Studies of Polymer Motion At or Near an Interface

Wilmer G. Miller, William T. Rudolf, Zorica Veksli*, Delbert L. Coon, Chia Chuan Wu, and Tai Ming Liang

Department of Chemistry, University of Minnesota, Minneapolis, MN 55455

The use of nitroxide spin labels to monitor polymer motion at or near an interface has been studied using several types of situations, including the penetration of low molecular weight diluents, the motion on and in surfactant stabilized latex emulsions, and the mobility of polymers adsorbed at a solid–liquid interface. Based upon temperture and solvent concentration studies on well-studied and well-defined systems, criteria are developed for choosing between anisotropic motion and a bimodal distribution of correlation times in highly concentrated polymer solutions. It is found that a problem with spectral uniqueness may exist, which can only be resolved through use of additional information. By the use of spin labels, scanning calorimetry, and other measurements, it is shown that nonsolvents may penetrate deeply into glassy, amorphous polymers though the diffusion may be slow. These studies indicate that diluent penetration does not occur by flow through channels or micropores in typical polymeric preparations. Spin label and spin probe studies give no indication of a core-shell morphology in plasticized latex emulsions; however, surfactant on glassy latex emulsions is tightly held to the latex surface with a very slow rate of exchange with non-latex associated surfactant. Studies with oxide surfaces show that poly(vinyl acetate) adsorbed on alumina and titanium dioxide has virtually no conformational freedom irrespective of contact with good or poor solvent, whereas some motional freedom exists when adsorbed on silica surfaces.

INTRODUCTION

The applications which one envisions for stable free radicals in synthetic polymer problems depends on one's background and experience. Our laboratory's involvement started around 1970 when we were interested in monitoring the motion of the rodlike, α-helical polypeptides by covalently attaching a nitroxide to one end of each helix, or in other studies to the monomer side chain.[1–6] In the limit of high molecular weight, motion of the rod no longer affects the nitroxide spectrum. A dilute solution of such "infinite" rods thus provides a well-defined model for a nitroxide at a solid–liquid interface. It also allows a comparison of theoretical spectral line shapes with experimental ones.

*Permanent Address: Institute Ruder Bošković, Zagreb, Yugoslavia.

In 1974 the scope of our work was broadened to include motion of a nitroxide tethered to a latex particle dispersed in a nonsolvent, and of a nitroxide labeled polymer adsorbed at a solid–liquid interface.[7-11] The latex-nonsolvent-interface idea, which seemed straightforward, proved difficult to interpret systematically though it was obvious that some nonsolvents penetrated the latex particle. This lead us to a variety of studies involving thermodynamically good solvents,[12,13] as well as nonsolvents. From this effort we have gained considerable experience in attempting to interpret nitroxide spectra in complicated situations, and in delineating the uses and limitations of nitroxides in synthetic polymer problems. As we have gained experience and broadened the nature of our studies some of the earlier work can be discussed more quantitatively and, in some instances, reinterpreted.

This communication is organized in four parts, each concerned with a particular aspect of motion at or near an interface. We will consider first the often difficult task of deciding on the type of molecular motion from analysis of the spectral line shape. This will be followed by studies on the penetration of low molecular weight species into amorphous polymers, on the mobility of surfactant and of plasticized polymer in surfactant stabilized latex emulsions, and on the mobility of polymers adsorbed at a solid–liquid interface.

MOLECULAR MOTION FROM SPECTRAL LINE SHAPE

Measurement of the nitroxide ESR spectrum is routine. However, deduction of the detailed molecular motions giving rise to the observed spectrum may be far from routine. In dilute solution in even moderately poor solvents a three line spectrum is always observed with labeled random coil polymers, which is generally interpreted in terms of isotropic motion,[14] though occasionally checked for motional anisotropy by comparison of the correlation time calculated from the linear and from the squared term of the line width dependence on the nitrogen spin state.[15,16] The calculated correlation times should be reasonably accurate, but one has virtually no information concerning the molecular motions contributing to the nitroxides' relaxation other than that there are a sufficient number of fast bond rotations to give 4π (or nearly 4π if anisotropy is detected) averaging. In studies on bulk polymers, a detailed understanding of the molecular motion may not be of interest.[17] If a correlation time is desired, it is generally calculated from the separation of the hyperfine extrema based on an approximate expression from Freed's group assuming isotropic motion.[18] Whether or not a nitroxide covalently bound to a polymer backbone in a solid matrix is likely to undergo isotropic motion is usually not discussed.

When considering the motion of a polymer attached nitroxide at or near a solid–liquid interface one must be concerned not only with motional anisotropy in a highly viscous medium, but also with the effect of the presence of unknown, variable amounts of low molecular weight diluents. The synthetic, helical polypeptide and the amorphous polymer-good solvent studies have been extremely helpful in interpreting nonsolvent penetration and adsorption data.

Some dilute solution spectra for end labelled, helical polybenzylglutamate as a function of molecular weight are shown in Figure 1. With short rods

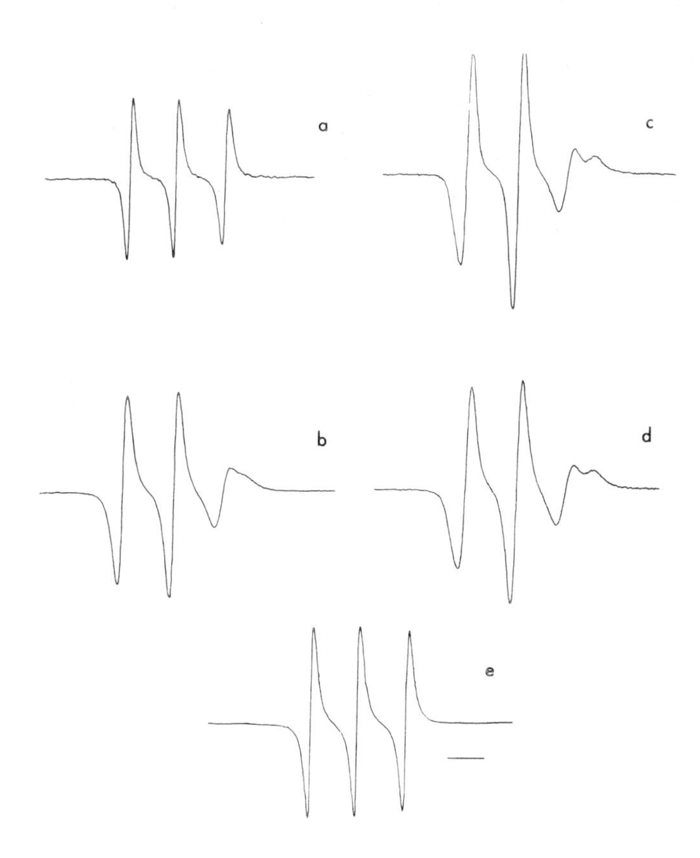

FIGURE 1 ESR spectra of one percent solutions of end labeled (a–d) or side chain labeled (e) helical poly(benzyl glutamate), PBLG, in dimethylformamide. The molecular weight was (a) 8000, (b) 22,000, (c) 122,000, (d) 180,000, and (e) 310,000. The temperature was 60°C (a–d), or room temperature (e). The mark in this and all other figures indicates 10G.

(helices) the rod motion significantly affects the nitroxide motion. The two highest molecular weights have almost identical spectra, indicating that the rod motion has become too slow to affect the spectrum. The calculations of Mason, Polnaszek, and Freed[19] have shown conclusively that the high molecular weight spectra correspond to a fast motion about the bond connecting the nitroxide ring to the helix (Figure 2a), i.e., it corresponds to a rigid limit spectrum for a fast anisotropic motion (the Hubbell-McConnell approximation[20]). It is quite clear that this differs from the rigid limit spectrum in the limit of no motion around any bond. Also shown in Figure 1 is the dilute solution spectrum for side chain labeled PBLG, which is independent of polymer molecular weight over a wide range. Although the spin label is the same, attaching the label to the "infinite rod" with three additional single bonds (Figure 2b) gives a simple three line spectra and effectively isotropic fast motion to the nitroxide. The spectra in Figure 1 give an excellent indication of the sensitivity to the length of the tether when the label is at the interface between a solid surface (in this case the "stationary" rod) and a low viscosity fluid. Additional computations using the Freed theory show that

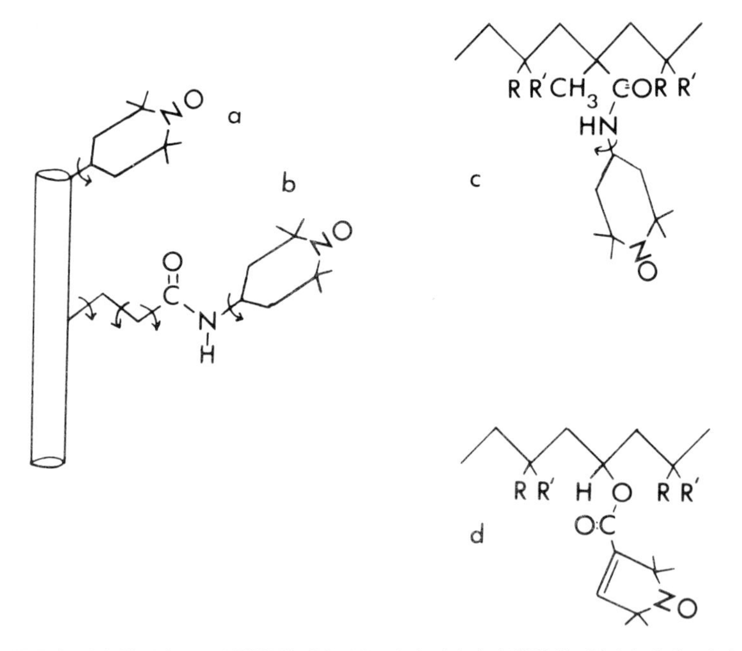

FIGURE 2 (a) End labeled PBLG, (b) side chain labeled PBLG, (c) labeled poly(methyl methacrylate) ($R = CH_3$, $R' = -COOCH_3$) or polystyrene ($R = H$, $R' = -C_6H_6$), and (d) labeled poly(vinyl acetate) ($R = H$, $R' = -OCOCH_3$). Except for (a), all polymers are labeled randomly.

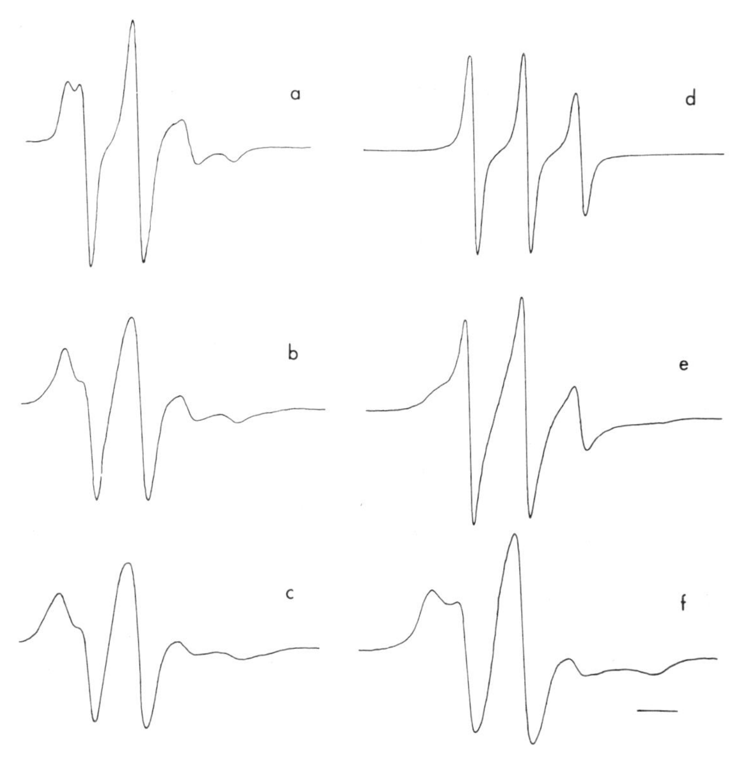

FIGURE 3 Effect of concentration on end (a–c) and side chain (d–f) labeled PBLG in dimethylformamide at room temperature. The polymer concentration was (a) 24%, (b) 60%, (c) 76%, (d) 74%, (e) 86%, or (f) 94%. The polymer molecular weight was 180,000 (a–c) or 310,000 (d–f).

spectra of the short rod, end labeled polymers can be simulated by allowing for the rod motion as well as the fast end bond rotation.[5,6]

Shown in Figure 3 are spectra for high molecular weight polymers as one goes to higher polymer concentration. With the end labeled polymer a low field doublet (sometimes unresolved) as well as the high field "doublet" is evident, and the outer hyperfine extrema become increasingly separated with increase in polymer concentration. This represents the effect of decreasing the motion about the terminal bond. The motion of the nitroxide on the longer tether, motionally narrowed, and effectively isotropic in dilute solution, becomes increasingly restricted, but in a quite different manner than with the tightly tethered one.

The temperature dependence of bulk, labeled PMMA[12] (Figure 2c), shown in Figure 5, is fairly representative of that observed with amorphous polymers. Below T_g a glass or powder spectrum is observed. The outer extrema separation decreases with increasing temperature, no clear low field doublet is evident, and a perceptible high field doubtlet exists at some temperatures. Using the Freed slow motion theory,[18,19] one can use the extrema separation to determine correlation times assuming either isotropic or highly anisotropic (single bond rotation, Figure 2c) motion. Assuming anisotropic motion yields correlation times consistent with PMMA motion deduced from dielectric, NMR, and dynamic mechanical data. Assuming isotropic motion, requiring on a molecular basis at least ester group if not main chain motion, is entirely inconsistent with other data, particularly ester group motion determined from dielectric data.[2,21] Nothing in the spectral line shape compels us to interpret the spectra with a particular model. The problem of uniqueness is serious. The existence of auxiliary data on this polymer is the crucial determinant.

The dilution of the bulk polymer with a thermodynamically good solvent, a few examples of which are shown in Figure 5, is valuable in understanding nonsolvent and interface motion. Its chief value lies in one's ability to vary the diluent composition in a continuous and known manner. As diluent is added the spectral line shape changes. If this change is a result of increased motion about the terminal single bond, the separation of the hyperfine extrema must decrease as in the dilution of the end labeled rods (Figures 1 and 3). But it does not. Instead, the spectrum appears to be a motionally narrowed three line spectrum superimposed on the original bulk polymer spectrum. In numerous

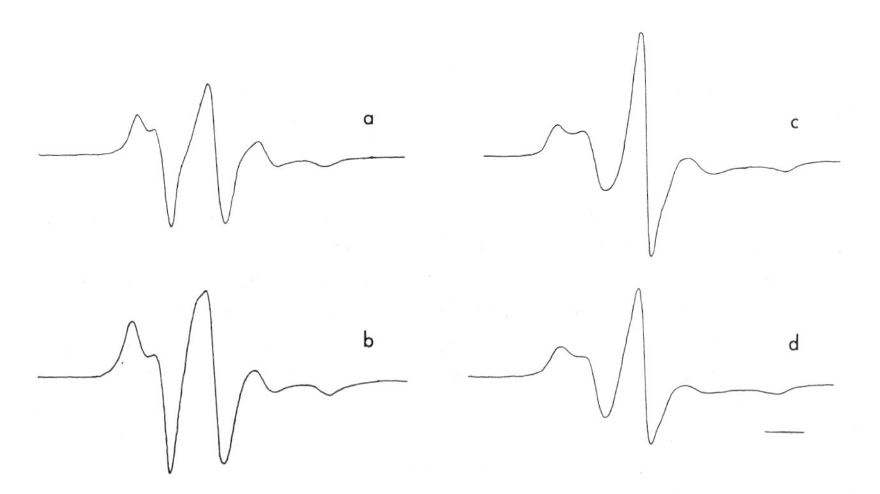

FIGURE 4 Examples of similarities in spectral line shape with motion that is highly anisotropic (a,b) or a superposition of a fast and a slow component (c,d). The anisotropic spectra are of end labeled PBLG (180,000) in dimethylformamide: (a) 9.4% solution at −22°C, (b) 24% solution at −48°C. The composite spectra are of (c) a 77% solution of labeled polystyrene in dimethylformamide at 25°C, and (d) a 86% solution of labeled poly(methyl methacrylate) in dimethylformamide at 25°C.

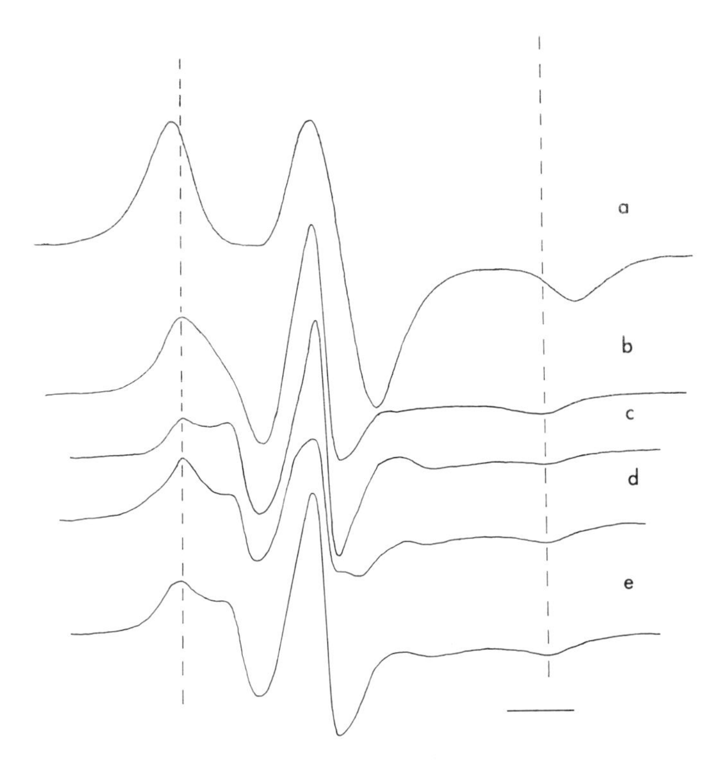

FIGURE 5 ESR spectra of labeled PMMA: (a) −140°C; (b) room temperature; (c) in the presence of 16.5 wt% dimethylformamide, (d) 32 wt% benzene, or (e) 32 wt% $CHCl_3$ at room temperature. Dashed lines are hyperfine extrema in dry polymer at room temperature.

instances these superposition spectra are qualitatively similar (Figure 4) to the spectra corresponding to highly anisotropic motion, showing both low and high field doublets, etc. However, the splitting of the doublets is far greater than that observed for known anisotropic motion, or calculated from the Freed theory,[6,19] and the outer extrema separation does not vary significantly with solvent composition. These criteria for deciding between highly anisotropic motion and a composite spectra are not yet expressible quantitatively. The deduction of a composite spectrum seems firm in this instance, though there may well be cases where a decision cannot be made from analysis of the spectral line shape. Under such circumstances the use of a pulse method[22] may be necessary.

The reason for the existence of a bimodal distribution of correlation times in certain concentration ranges can only be speculated on at present. But, having

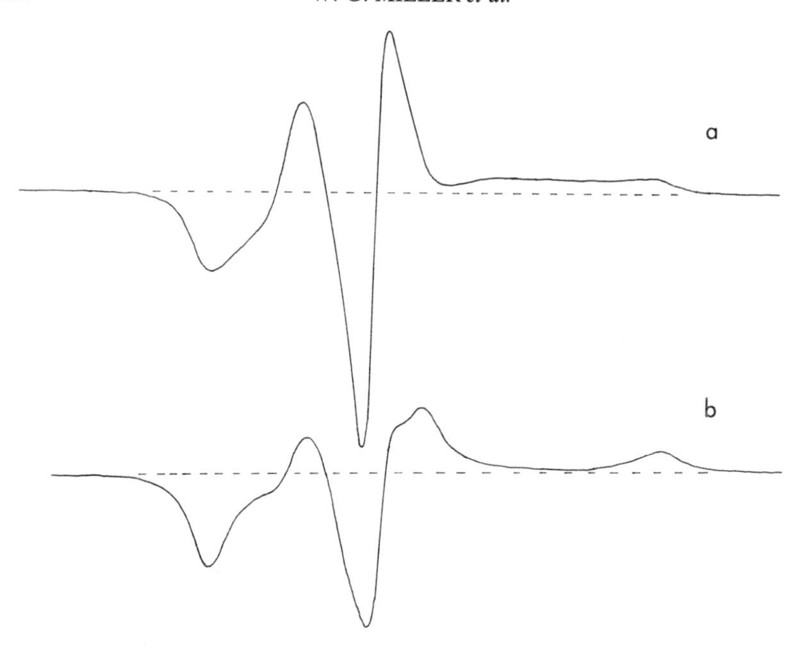

FIGURE 6 ESR spectra at room temperature of labeled polystyrene (a) before and (b) after annealing at 80°C.

concluded that the dilution spectra are composite, they may be decomposed into the fast and slow components by simulation.[12] A rather good fit to the experimental spectra can be obtained by assuming the slow component has the same spectrum as the dry polymer. One would certainly suspect that this is a considerable simplification in that there should be at least a distribution of correlation times. There is some evidence for this in that the simulation, particularly in the central region of the spectrum, does not give a truly quantitative fit. Better evidence that analysis of motionally slowed spectra in terms of a single correlation time is a simplification can be seen from Figure 6. Here a dry polymer was annealed 20° below T_g. Although the extrema separation is not changed, details in the line shape have changed, indicating a more uniform environment for the spin labels. However, not all preparations exhibit this behavior.

PENETRATION OF LOW MOLECULAR WEIGHT DILUENTS

Nonsolvent Penetration and Distribution

The penetration of low molecular weight diluents into polymeric films, coatings, laminates and composites has significant and frequently disastrous effects on the mechanical as well as chemical properties of these polymeric systems. Sometimes it is surface and interface interactions which are important, while in other instances bulk effects dominate. Our initial intent in using spin labels tethered to a latex surface to monitor motion at a polymer-nonsolvent interface ran into immediate difficulty when surface labeled and internally labeled latex spheres were compared.[7–9] In numerous instances nonsolvents for the polymer gave significant motion to the internally situated labels as well as to labels at the surface. Generally the ESR spectra, examples of which are shown in Figure 7, were composite spectra, using the criteria outlined in the previous section. The slow component to a good approximation was the same as the dry polymer, analogous to that observed with good solvents. Unlike the good solvent studies, where the composite spectra are observed primarily at solvent compositions which bring T_g close to the temperature of measurement, there was no obvious relationship between T_g

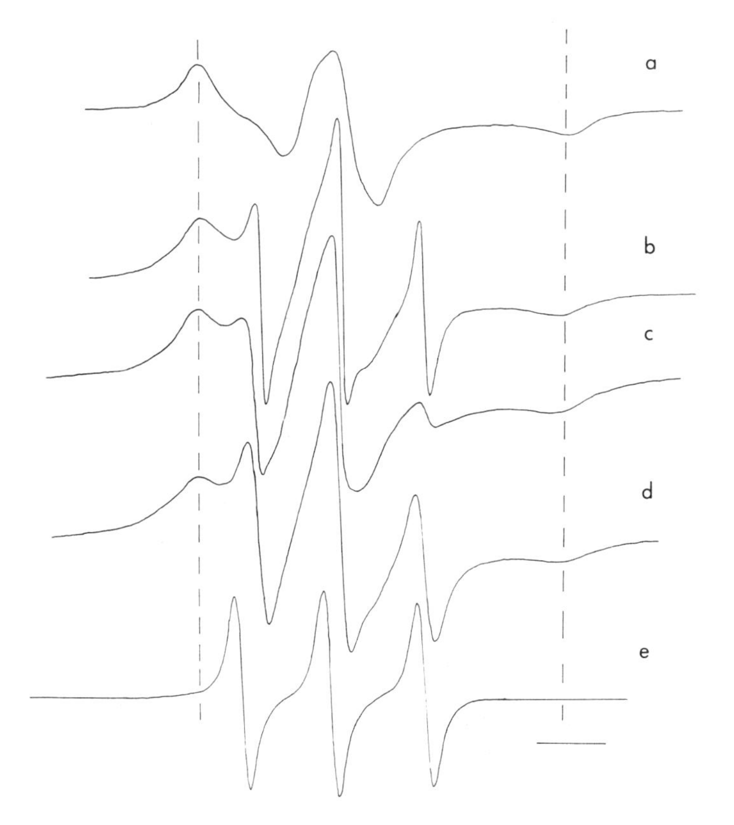

FIGURE 7 (a) ESR spectra of labeled polystyrene at room temperature when dry or in water, and (b) in the presence of pentadecane, (c) polyethylene oxide (20,000 mol. wt.), (d) decane, or (e) methanol. Dashed lines are hyperfine extrema of dry polymer.

and the amount of fast component. Water and methanol[23] have little effect on the T_g of polystyrene (PS) or PMMA, yet either labeled latex dispersed in water shows no fast component, but dispersed in methanol shows virtually no slow component.[9] Vapor sorption measurements, however, indicated a relationship between nonsolvent uptake and the amount of fast component present.

The question arises as to the origin of the fast and the slow component. It is conceivable that the nonsolvent penetrates only partway into the latex particle, and a strong gradient in diluent exists.[9,24] To answer this question we carried out several studies, choosing decane and polystyrene as the polymer-nonsolvent pair. Labeled polystyrene in decane had some fast component, and absorbed some decane in vapor sorption measurements.[9] The percent of fast component in samples equilibrated with decane at room temperature was dependent on the sample of labeled polymer. Samples equilibrated with decane above T_g and quenched at room temperature had a more constant amount of fast component.[25]

Bulk polystyrene of well defined particle size was equilibrated with liquid decane at room temperature. Differential scanning calorimetry measurements showed that decane lowered T_g the same amount (60°) with latex particles of 1090 Å, 2340 Å, and 5000 Å, and with 250 μ slabs.[26] Thus scanning calorimetry showed a single T_g which was not sample thickness dependent. This is a clear indication that the nonsolvent permeated the entire sample even though the sample is still glassy ($T < T_g$), and does not just form a solvent rich shell near the surface. However, the time necessary for T_g to be lowered from the dry value to the decane equilibrated value is very particle size dependent, ranging from the order of minutes with 1090 Å spheres to months for the 250 μ slab. This indicates a slow diffusion rate into the glassy polymer. During the diffusion of the decane the scanning calorimetry did not show two T_g's; instead, a broad range was observed. This suggests that during the diffusion of the nonsolvent into the glassy polymer there is not a sharp boundary between dry and diluent saturated polymer, as there is with a good solvent.[27] Electron microscopy also indicates that the latex surface is not plasticized by decane,[25] in agreement with the calorimetry results.

Further studies have been carried out to understand the chain dynamics in the presence of decane. It is well known that an amorphous polymer cooled below T_g has sharply restricted backbone motion[21] and becomes kinetically trapped in nonequilibrium states.[28] By annealing at a fixed temperature, the relaxation to the equilibrium state can be followed by volumetric[29] or calorimetric[30] measurements. This relaxation, which has been mathematically characterized,[29,30] is extremely temperature dependent. In fact, no samples of polystyrene or poly(methyl methacrylate) have likely been studied that are in equilibrium states at room temperature, not even emulsion polymerized and film formed samples.[25,26] The volume relaxation is due to small voids in chain packing at the molecular level when the backbone motion becomes very slow, as the chain motion cannot keep up with the thermal contraction. With this background in mind we carried out a study of enthalpy relaxation by scanning calorimetry of polystyrene in the presence of decane.[26] A decane equilibrated sample is brought above T_g, quenched to the annealing temperature, and the scanning calorimetry performed as a function of annealing time. The results show conclusively that the rate of relaxation of the polystyrene to its equilibrium state depends only on the degree of undercooling below its glass transition at the environmental conditions of interest, i.e., the relaxation rate of polystyrene equilibrated with decane at 25°C is the same as the relaxation rate of the dry polystyrene at about 83°C. The relaxation time at only 20° below T_g is greater than 10 years. Unlike the lowering of T_g, which depends on diluent diffusion into the sample, this relaxation is not particle size dependent. These results may explain the variable ratio of the content of fast to slow component of samples from different preparations equilibrated with decane at room temperature inasmuch as the dry samples are in varying states of relaxation (Figure 6 and Ref. 25). It would also explain why the fast to slow ratio is more uniform for samples taken above T_g and then quenched to room temperature. Inasmuch as we believe the fast component involves local mode backbone motion, the number of such motions may depend on the state of relaxation of the polymer. It seems clear from these studies that a nonsolvent can make available to the polymer high frequency, local mode backbone motions that are not equivalent to raising the temperature, and that the spin label technique is a sensitive indicator of this.

Solvent Channels and Mixed Solvents

The use of polymers as supports for transition metal catalysts, organic reagents, etc., is of increasing interest.[31] Penetration of solvents into such polymer supports is important in their use. In light of the results discussed in the preceeding section, it is evident that if enough solvent is imbibed to lower the polymer T_g to or below the working temperature, chain mobility becomes very rapid and the polymer reaches its equilibrium state quickly. It is thus clear that solvent channels cannot be invoked to explain spin label motion in systems clearly plasticized with a good solvent.[32] Even with nonsolvents, particularly those which leave the polymer a glass, where the greater incompatability might be expected to lead any extended faults or voids to act as diluent channels by capillary action, the size dependence of the time to lower the T_g of polystyrene in decane effectively rules this out. At best any such channels are not interconnected and hence channel or pore flow is not a mechanism for transporting the solvent throughout the polymer matrix, or for its habitation.

Penetration of diluents into spin labeled polymer beads in mixed solvents has been studied.[33] In such systems differential partitioning can be extreme. In order to show that we had surface as well as internally labeled latex, we dispersed the spheres in 1 M aqueous acetic acid, expecting the acid to destroy the surface nitroxides but not to affect the internal ones. When dispersed in water, the spectrum, as stated earlier, was the same as with the dry sample with both types of labeling. In 1 M acetic acid, the internally labeled polymer was initially unaffected, but eventually showed a significant three line component in the spectrum. This result, initially surprising, is easily explained. The estimated solubility parameter of acetic acid is not very different from PMMA. The PMMA latex simply extracted the acetic acid from the water and became plasticized. In a similar type of experiment, polystyrene was found to extract acetone from an acetone-water mixture even though acetone is a relatively poor solvent[34] for polystyrene.

SURFACTANT STABILIZED LATEX EMULSIONS

Concentration Gradient in a Polymer-Diluent Latex

In emulsion polymerization there is some evidence that as the polymerization proceeds the growing latex particle develops a concentration gradient in polymer and in monomer that has transferred to the growing latex but not yet reacted, giving rise to what is known as a core-shell morphology.[35] A spin label experiment was designed to investigate the concentration distribution in a latex particle, using a spin labeled polystyrene latex and benzene in place of monomer. On top of dilute aqueous suspensions of sodium dodecyl sulfate (SDS) stabilized latex, enough benzene was layered to produce 20–50 wt % solutions if all of the benzene transferred to the latex. The suspensions were gently stirred magnetically until the benzene droplet disappeared (about 2 days). Benzene is not taken up significantly in SDS micelles, the latex particles undergo visual change, and electron microscopy indicates that the uptake is fairly uniformly distributed.[36] The spin label spectrum was analogous to that previously observed with bulk polystyrene with the same benzene composition,[12] with no evidence of a solvent rich zone and a solvent poor zone. We conclude that in this range of diluent concentration the spin label experiments show no evidence of a core-shell morphology.

Surfactant Mobility on Latex Particles

In an emulsion polymerization one starts with an aqueous solution of micelles, typically SDS, and water insoluble monomer droplets. By diffusion of the monomer to the micelles, followed by polymerization, polymer latex particles are produced, typically tenths of microns in diameter, coated with the surfactant. A surfactant-like spin probe in a micelle is very mobile (Fig. 8a), indicating the interior of the micelle is quite fluid.[37] If water insoluble methyl stearate, spin labeled (Syva) at the third or fourteenth position (Fig. 8b), is equilibrated with surfactant stripped latex dispersed in water by ultrasonication, the spin probe slowly (hours) transfers to the latex surface and is found to be very highly immobilized. This is shown in Figure 8c, where the amount of spin probe has been kept low enough to avoid spin-spin interaction on the latex surface. The spin probe may be quickly stripped from the latex with a methanol-water wash, whereas a spin probe internally doped in a latex particle cannot be removed in this manner. The spin probe has consequently not become immobilized by diffusion into the latex. It is thus clear that the hydrocarbon portion of the spin probe is being held tightly to the latex surface. The primary driving force for this must be hydrophobic interactions, whereby

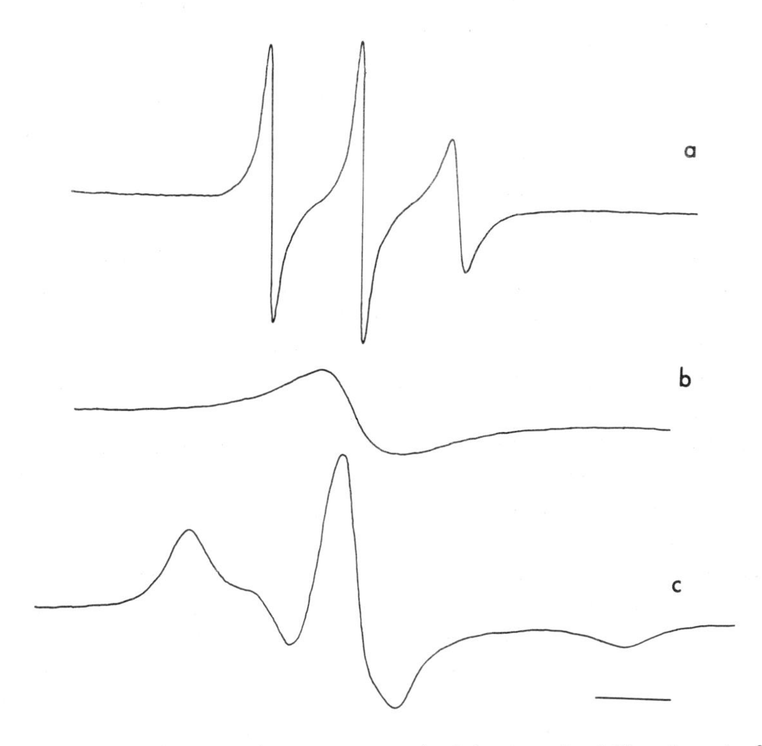

FIGURE 8 (a) ESR spectra (room temperature) of the water insoluble spin probe 2-(3-carboxypropyl)-4,4-dimethyl-2-tridecyl-3-oxazolidinyloxyl methyl ester in sodium dodecyl sulfate micelles, (b) as solid particles in water, and (c) on the surface of clean PMMA latex in water originally containing solid spin probe.

the observed state represents the minimum total contact between hydrocarbon and water.

If a SDS coated and stabilized latex is equilibrated with the same spin probe as above, but dispersed first in SDS micelles, transfer to the latex particle again takes place but the transfer half-time can take days. The transfer half-time was found to be different for PMMA than for polystyrene, indicating the nature of the latex surface is important. Similar results were obtained with carboxylate spin probes as well as with the esters. Transfer of the spin probe from the micelle is not likely to be rate limiting, considering the rates which have been determined for such processes.[38] Consequently the slow transfer must represent the slow turnover of surfactant from the surfactant coated latex. This may explain why it is such a long and difficult process to remove surfactant from latex emulsions by dialysis.

CONFORMATION AND MOBILITY OF POLYMERS ADSORBED AT A SOLID-LIQUID INTERFACE

Adsorption of polymers on solid substrates is of much theoretical interest as well as of practical importance.[39] Adsorption may take place from quite dilute solutions due to the many potential binding sites per polymer molecule. The conformation of the absorbed polymer molecule is considered to consist of trains of absorbed units interspersed with loops extending into the solvent, and a nonadsorbed tail at each end. Theoretical treatments of the fraction of the units adsorbed and the extension of the loops above the solid surface, depend on the polymer-solvent-surface interaction.[40] With a fixed polymer and adsorbent, the loop size is predicted to depend on the solvent.[40] Thickness of the adsorbed polymer layer has been estimated from isotherm plateaus, by ellipsometry and by flow through sintered glass. Thicknesses of the order of hundreds of angstroms are reported frequently.[39] Infrared spectroscopy has sometimes been used to determine the fraction of functional groups bound to the surface. Little is known concerning the detailed mobility or conformation of absorbed polymers.

The spin label technique seems potentially useful due to the low concentration of spin labels needed as well as the sensitivity of the spin labels to motion.[8,41] In order to keep the spin label density low to avoid spin-spin interactions, a fairly high surface area adsorbent must be utilized. Our initial efforts to use high surface area carbon black were frustrated due to the interfering ESR activity of carbon black.[8] We have turned to oxide surfaces due to the availability of fairly high surface area alumina, titanium dioxide and silica.

Activated alumina (Matheson, Coleman, and Bell) was heated at 200°C for

2 hours and stored in a desiccator. Titanium dioxide (Polysciences, .45 μ diameter) was used as received. Silica (Polysciences, 5–10 μ and 3–8 μ; Cabot, Cab–O–Sil M-5) was either used as received, or extracted with hot nitric acid and dried under vacuum.

Poly(vinyl acetate), PVAc, of molecular weights (M_w) 6.1×10^4, 1.9×10^5 and 6.0×10^5, was randomly labeled by ester exchange with 2,2,5,5-tetramethyl-3-pyrrolin-1-oxyl-3-carboxylic acid to give a spin labeled polymer (Fig. 2d) estimated to contain 1–10 labels per polymer molecule.[10] Any hydrolysis of the PVAc effects its adsorption.[42] Infrared spectroscopy (Perkin-Elmer 237) and differential scanning calorimetry (Perkin-Elmer DSC-2) indicated no detectable hydrolysis had occurred in sample preparation. The ESR spectrum of the spin labeled, dry polymer is shown in Fig. 9a.

Adsorption studies were made from four solvents, which, listed in order of increasingly good solvent power for PVAc, were: CCl_4, toluene, benzene, and $CHCl_3$. The solvents range from rather poor to very good, as CCl_4 is nearly a theta solvent whereas $CHCl_3$ is almost an athermal solvent. The spectrum of the polymer in dilute to moderate concentration indicated fast nitroxide motion (Fig. 9d). In dilute solution the correlation time was $5 \pm 1 \times 10^{-11}$ sec, except for CCl_4, which was a factor of four slower.

In a typical experiment one tenth gram of adsorbent was added to one milliliter of solvent containing ten milligrams of labeled polymer, sonicated for one minute, magnetically stirred for 48 hours (or 72 hours), centrifuged (sometimes), the supernatant withdrawn from above the adsorbent, diluted with fresh solvent, and the spectrum recorded of the surface adsorbed polymers (and of the supernatant). The polymer concentration corresponds to the plateau region of the isotherm. Polymer desorption is characteristically slow and the addition of fresh solvent to the polymer coated adsorbent produced no observable time effects (up to 30 minutes after addition) in the spectra.

When activated alumina was the adsorbent the ESR spectrum was similar for all molecular weights with each of the four solvents (Fig. 9b), i.e., the spectra were identical to the dry polymer spectra (Fig. 9a) except a tendency for the low and high field peaks to broaden in going from best to worst solvent. Inasmuch as the T_g for PVAc is $\sim 30°C$, only a small amount of solvent need be added to the polymer in the absence of adsorbent to produce measurable fast motion, analogous to the PMMA and PS results. It seemed possible that the spin label moiety NO itself might adsorb, though on statistical grounds all nitroxide adsorbing in this manner was remote. To check this possibility several small spin probes were dissolved in the solvents and equilibrated with the adsorbent. They were 2,2,5,5-tetramethyl-3-pyrrolin-1-oxyl-3-carboxylic acid, 2,2,6,6-tetramethyl-4-oxopiperidinooxyl, 4-amino-2,2,6,6-tetramethyl-

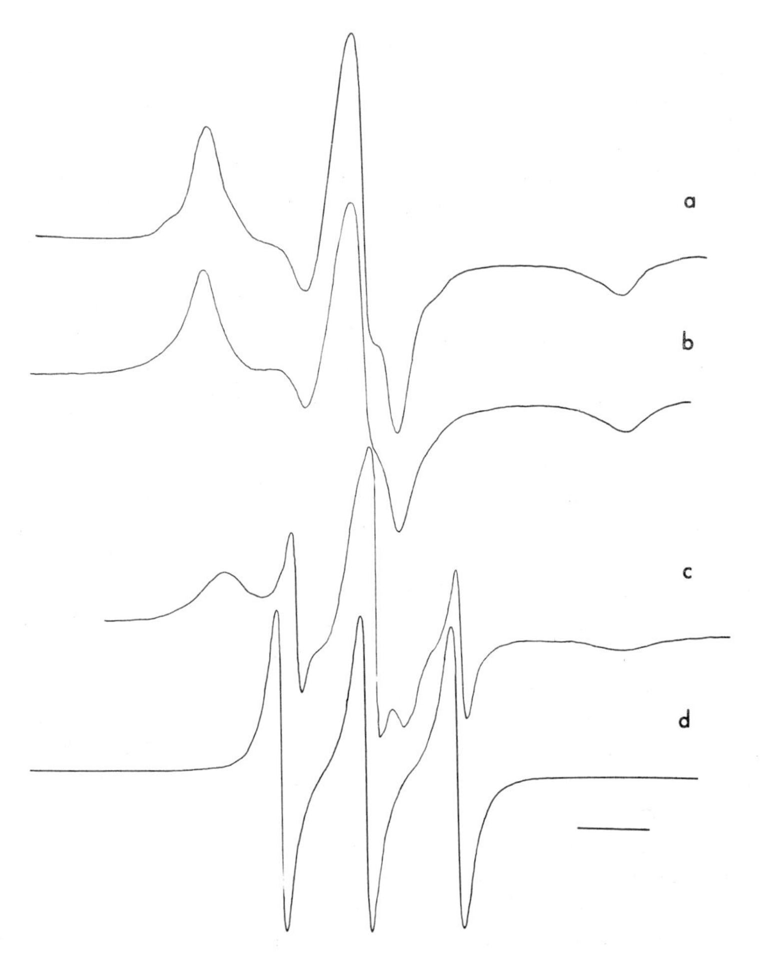

FIGURE 9 (a) ESR spectra (room temperature) of spin labeled poly(vinyl acetate) in a bulk sample, (b) when adsorbed from $CHCl_3$ on alumina, (c) when adsorbed from $CHCl_3$ on silica (3–8 μ), and (d) as a dilute solution in toluene. Polymer molecular weights were (a,b) 6.1×10^4, (c) 6.0×10^5, or (d) 1.9×10^5.

piperidinooxyl, and 4-acetamido-2,2,6,6-tetramethyl-piperidinooxyl. Indeed they were found to adsorb, but the adsorption depended on the spin probe. In order of decreasing adsorption tendency, they could be ranked as the acetamido, the carboxylic acid, the amino, and the oxo compounds. Since the adsorption was dependent on the functional group other than the NO moiety common to each, it was possible to conclude that the polymer segment adsorption was what forced the nitroxide labels to be immobilized. The fact that the spectrum of a polymeric monolayer on the adsorbent was virtually identical with the dry bulk polymer indicated that the alumina surface interacts very strongly with the polymer, and that a negligible fraction of the monomeric units are involved in mobile loops or tails. The polymer conformation must be quite flat, the thickness a few angstrons. Inasmuch as alumina is known to be a strong adsorber for polar species, these results may not be too surprising.

Initial studies[10] indicated that the two highest molecular weights, when adsorbed on TiO_2 behaved similar to the polymer on alumina, while the lowest molecular weight had some fast component, depending on the solvent. Further studies showed that if a paramagnetic impurity was removed from the adsorbent, all molecular weights had a small fast component in the spectra. This indicates that TiO_2 has a weaker interaction than alumina with the polymer.

When silica (5–10μ) was used as the adsorbent the initial results indicated a fast component. The percentage of fast component seemed to depend on molecular weight as well as solvent power. The results were initially interpreted[10] as indicating that the fast component came predominantly from tails. Studies with the spin probes again showed that the adsorption was not through the NO. However, by analysis of both supernatant and adsorbent, a significant fraction of the polymer could not be accounted for. This was eventually traced to a paramagnetic impurity which leached from the untreated glass and most likely reacted with some of the spin labels through an oxidation-reduction couple. Upon removal of the paramagnetic impurity by nitric acid treatment, all polymer could be accounted for. A fast component as well as a "bulk solid" component was again observed (Fig. 9c). However, the systematic study of the effect of chain length and solvent had to be repeated, and is still in progress. It is clear, however, that the binding sites for PVAc on silica are not as strong as on alumina or TiO_2.

ACKNOWLEDGMENTS

This work has been supported in part by grants from the Petroleum Research Fund (8749-AC5,6), administered by the American Chemical Society, from the Education Committee of the Rubber Division, American Chemical Society, and from the U.S. Public Health Service (GM-16922).

REFERENCES

1. E.L. Wee, Ph.D. Thesis, University of Minnesota, 1971.
2. E.L. Wee and W.G. Miller, *J. Phys. Chem.*, 77, 182 (1973).
3. C.C. Wu, Ph.D. Thesis, University of Minnesota, 1973.

4. W.G. Miller, C.C. Wu, E.L. Wee, G.L. Santee, J.H. Rai, and K.D. Goebel, *Pure Appl. Chem.*, **38**, 37 (1974).
5. Japan Polymer Society Symposium, Kyoto, Japan, Sept. 14–15, 1977; 175th Amer. Chem. Soc. National Meeting, Anaheim, Calif., March 12–17, 1978.
6. C.C. Wu, J. Zimmel, R. Mason, J. Dickson, and W.G. Miller, in preparation.
7. Z. Veksli and W.G. Miller, *Macromolecules*, **8**, 248 (1975).
8. W.G. Miller and Z. Veksli, *Rubber Chem. and Tech.*, **48**, 1978 (1975).
9. Z. Veksli, W.G. Miller, and E.L. Thomas, *J. Polym. Sci., Symposium* **54**, 299 (1976).
10. W.T. Rudolph, M.S. Thesis, University of Minnesota, 1976.
11. Z. Veksli and W.G. Miller, IUPAC Meeting, Rio de Janiero, July 22–26, 1974; W. Rudolph and W.G. Miller, IUPAC Meeting, Tokyo, Sept. 4–9, 1977.
12. Z. Veksli and W.G. Miller, *Macromolecules*, **10**, 686 (1977).
13. Z. Veksli and W.G. Miller, *Macromolecules*, **10**, 1245 (1977).
14. D. Kivelson, *J. Chem. Phys.*, **33**, 1094 (1960).
15. T.J. Stone, T. Buckman, P.L. Nardio, and H.M. McConnell, *Proc. Nat. Acad. Sci. USA*, **54**, 1010 (1965).
16. A.T. Bullock, G.G. Cameron, and V. Krajewski, *J. Phys. Chem.*, **80**, 1972 (1976).
17. P.L. Kumler and R.F. Boyer, *Macromolecules*, **9**, 903 (1976); P.L. Kumler, S.E. Keinath, and R.F. Boyer, *J. Macromol. Sci.-Phys.*, **B13**, 631 (1977).
18. S.A. Goldman, G.V. Bruno, and J.H. Freed, *J. Phys. Chem.*, **76**, 1858 (1972).
19. R.P. Mason, C.F. Polnaszek, and J.H. Freed, *J. Phys. Chem.*, **78**, 1324 (1974).
20. W.L. Hubbell and H.M. McConnell, *Proc. Nat. Acad. Sci. USA*, **63**, 16 (1969); **64**, 20 (1969); *J. Amer. Chem. Soc.*, **93**, 314 (1971).
21. D.W. McCall, *Natl. Bur. Stand. (US), Spec. Publ.* **No. 301**, 475 (1969); *Accounts Chem. Res.*, **4**, 223 (1971).
22. I.M. Brown, *J. Chem. Phys.*, **65**, 630 (1976).
23. R.P. Kambour, C.L. Gruner, and E.E. Romagosa, *J. Polym. Sci., Polym. Phys. Ed.*, **11**, 1879(1973); *Macromolecules*, **7**, 248 (1974).
24. E.G. Smith and I.D. Robb, *Polymer*, **15**, 713 (1974).
25. Z. Veksli, W.G. Miller, and E.L. Thomas, *Macromolecules*, in preparation.
26. D. Coon and W. G. Miller, *Macromolecules,* in press.
27. J. Crank and G.S. Park, "*Diffusion in Polymers*", (Academic Press, New York, 1968).
28. J.D. Ferry, "*Viscoelastic Properties of Polymers*", (John Wiley and Sons, New York, 1970).
29. A.J. Kovacs, *J. Polym. Sci.*, **30**, 131 (1958).
30. S.E.B. Petrie, *J. Poly. Sci. Part A-2*, **10**, 1255 (1972); A.S. Marshall and S.E.B. Petrie, *J. Appl. Phys.*, **46**, 4223 (1975).
31. C.C. Leznoff, *Chem. Soc. Rev.,* **3**, 65 (1974); C.G. Overberger and K.N. Sannes, *Angew. Chem.*, **13**, 99 (1974); C.V. Pitman and G.O. Evans, *Chem. Tech.,* 560 (1973).
32. S.L. Regen, *Macromolecules*, **8**, 689 (1975).
33. S.L. Regen, *J. Amer. Chem. Soc.*, **96**, 5275 (1974).
34. K.S. Siow, G. Delmar and D. Patterson, *Macromolecules*, **5**, 29 (1972).
35. M.R. Grancio and D.J. Williams, *J. Polym. Sci. Part A-1*, **8**, 2617 (1970); P. Keusch, R.A. Graff, and D.J. Williams, *Macromolecules*, **7**, 304 (1974).
36. Y. Talmon and W.G. Miller, *J. Colloid and Interface Sci.*, **67**, 284 (1978).
37. A.S. Waggoner, O.H. Griffith, and C.R. Christensen, *Proc. Nat. Acad. Sci. US*, **57**, 1198 (1967); G.P. Rabold, *J. Polym. Sci. Part A-1*, **7**, 1187 (1969).
38. R. Zana, *NATO Adv. Study Inst. Ser., Ser. C*, **18**, 133, 139 (1975).
39. Yu.S. Lipaton and L.M. Sergeeva, "Adsorption of Polymers", Wiley, New York, 1974; K.L. Mittal, ed., "Adsorption at Interfaces", Amer. Chem. Soc. Publ., Washington, D.C., 1975.
40. P. Mark and S. Windwer, *Macromolecules*, **7**, 690 (1974); C.A.J. Hoeve, *J. Polym. Sci. Part C*, **30**, 301 (1970); C.A.J. Hoeve, *J. Chem. Phys.*, **44**, 1505 (1966); A. Silberberg, *J. Phys. Chem.,* **66**, 1884 (1962); M. Lal in *"Adsorption at Interfaces"*, K.L. Mittal, ed., Amer. Chem. Soc., Washington, D.C., 1975, p. 161.

41. K.K. Fox, I.D. Robb, and R. Smith, *J. Chem. Soc. Faraday Trans. I.*, **70**, 1186 (1974); I.D. Robb and R. Smith, *Eur. Polym. J.*, **10**, 1005 (1974); A.T. Clark, I.D. Robb, and R. Smith, *J. Chem. Soc. Faraday Trans. I*, **72**, 1489 (1978); I.D. Robb and R. Smith, *Polymer*, **18**, 500 (1977).
42. J. Koral, R. Ullman, and F.R. Eirich, *J. Phys. Chem.*, **62**, 541 (1958).

DISCUSSION

G. G. Cameron (University of Aberdeen, Old Aberdeen, Scotland): In the case of your labeled PS the non-solvent that had the most striking effect on linewidth, i.e., in its ability to give motionally narrowed lines, was methanol. Is it likely that this effect can be attributed to preferential solvation of the MMA units (the labeled parts of the chain) by the polar alcohol?

W. G. Miller: It is possible that MMA units, actually amide units where the label is involved, are preferentially solvated. However, in order to obtain the motionally narrowed, three line spectra which we observe, backbone motion must also be involved. Furthermore, if preferential solvation is important, one would expect water to behave similarly, but no fast motion is seen in the presence of water.

G. G. Cameron: I find it surprising that in solutions containing 80% $CHCl_3$ there is still a slow motion component. Is this a reversible phenomenon, i.e., are you convinced that the polymer is fully dissolved?

W. G. Miller: We are convinced that the polymer is fully dissolved in the 80% $CHCl_3$ solution. The spectra observed when the sample was prepared by taking a dilute polymer solution and concentrating it to a 20% solution by solvent evaporation was the same as when the sample was prepared by adding solid polymer to the appropriate amount of solvent to give a 20% solution directly. At considerably higher polymer concentrations the method of sample preparation does affect the observed percentage of slow component, as we have reported (*Macromolecules*, **10**, 686, 1245 (1977)).

A. M. Bobst (University of Cincinnati, Cincinnati, Ohio): Are you using the Freed program which allows axially symmetric reorientation, i.e., the program where z′ is the axis of fast reorientation?

W. G. Miller: We are using the Freed slow motion program sometimes referred to as TILT, wherein the magnetic axes and the molecular rotation axes do not need to be coincident. The motion can be isotropic or anisotropic.

P. Törmälä (Tampere University of Technology, Tampere, Finland): According to the results of Shiotani and Sohma, their label radicals tumbled in solid PMMA considerably slower than your labels. Do you have any explanation for this?

W. G. Miller: Shiotani and Sohma (*Polymer J.*, **9**, 283 (1977)) studied PMMA labeled with a nitroxide ester, whereas, ours was bonded to the methacrylate with the corresponding amide. One might have expected the ester to show more rotational freedom than the amide; however, they report about an order of magnitude slower motion. Below 80°C we believe that the nitroxide motion is anisotropic and our correlation times are for a single bond rotation. The correlation times of Shiotani and Sohma were calculated assuming isotropic motion. Another factor is the measurement of A_{zz} in the low temperature limit, which is needed whether one interprets the motion as isotropic or anisotropic. We find A_{zz} to still be increasing even below −100°C, whereas Shiotani and Sohma report the low temperature limit is reached at +75°C. Bullock, Cameron, and Krajewski (*J. Phys. Chem.*, **80**, 1792 (1976)), who studied PMMA labeled with the same nitroxide as Shiotani and Sohma, find A_{zz} to be increasing even at −190°C. When these two factors are taken into consideration, the differences in correlation times between our work and that of Shiotani and Sohma do not seem so great.

R. F. Boyer (Midland Macromolecular Institute, Midland, Michigan): Commercial PMMA contains iso- and syndiotactic species which combine to form a complex. Probe molecules trapped in such complexes swollen in acetone may be less mobile than probe molecules in the swollen but non-complexed regions.

W. G. Miller: The formation of stereocomplexes in PMMA as an explanation of the bimodal distribution of relaxation times at certain polymer concentrations is certainly a reasonable suggestion, which we cannot rule out completely. However, we find (Z. Veksli and W.G. Miller, *Macromolecules*, **10**, 686 (1977)) the bimodal distribution persisting to a 3 to 1 (wt./wt.) ratio of solvent to polymer for PMMA in $CHCl_3$. Although PMMA stereocomplexes are known to exist in some solvents, they are not formed in $CHCl_3$ (H.Z. Liu and K.J. Liu, *Macromolecules*, **1**, 157 (1968); J. Spevacek and B. Schneider, *Makromol. Chem.*, **175**, 2939 (1974); **176**, 729 (1975); R. Buter, Y.Y. Tan, and G.P. Challa, *J. Polym. Sci., Polym. Chem. Ed.*, **11**, 2975 (1973); G.P. Challa, A. DeBoer, and Y.Y. Tan, *Int. J. Polym. Mater.*, **4**, 239 (1976); J. Spevacek and B. Schneider, *Polymer*, **19**, 63 (1978)). Inasmuch as we observed the bimodal distribution in three solvents in both PMMA and PS, it is by no means clear that it is related to stereocomplexes.

R. F. Boyer: Kanig has developed a theory for T_g which predicts that n-alkanes are very efficient plasticizers for lowering the T_g of polystyrene, even though they are basicly non-solvents (*Kolloid-Z. u. Z. Polymere*, **190**, 1 (1963)). Your results seem to be one of the best confirmations of his theory that I've yet seen.

W. G. Miller: We have used decane only as an example; we have looked at a variety of very poor solvents, observing similar results.

Molecular Dynamics of Nucleic Acids as Studied by Spin Labels

Albert M. Bobst

Department of Chemistry, University of Cincinnati, Cincinnati, Ohio 45221

The potential use of spin labeled nucleic acids, $1(N)_n$, to characterize by ESR some of the physico-chemical properties of nucleic acids including conformational transitions and interactions with large ligands will be discussed in this chapter.

1ST AND 2ND GENERATION SPIN LABELED NUCLEIC ACIDS

1st generation spin labeled nucleic acids, 1st $1(N)_n$, are obtained by chemically modifying a polynucleotide with various reagents containing a nitroxide moiety. Reagents used for this purpose have recently been reviewed[1]. In Fig. 1 a segment of a nucleic acid chain is shown, and the arrows next to the bases indicate the positions subject to electrophilic attack by spin labeling reagents. Because of the many reactive sites in nucleic acids, it has proved very difficult to site specifically label nucleic acid building blocks by a purely chemical approach unless the nucleic acid contains some rare bases with particularly reactive groups. Consequently, it is generally accepted that 1st $1(N)_n$ do not have the nitroxide moiety in a site specific position with respect to the building block or sequence in the nucleic acid chain. Second generation spin labeled nucleic acids, 2nd $1(N)_n$, on the other hand, contain nitroxide labels which exhibit site specificity with respect to the nucleic acid building block. They can be obtained enzymatically by incorporating a site specifically spin labeled nucleic acid building block into a polynucleotide as was recently shown for the incorporation of ppRUGT into poly(uridylic acid), $(U)_n$[2]. ppRUGT is the 5′-diphosphate of N–(1-oxyl-2,2,6,6-tetramethyl-4-piperidinyl)-0-(1-*β*-*D*-ribofuranosyl-uracil-5-yl)-glycolamide (RU-Glycol-amido-TEMPO = RUGT). In Fig. 2 an Ealing CPK atomic model of the 2nd $1(N)_n$ $(RUGT,U)_n$ is shown. $(RUGT,U)_n$ is a copolymer consisting of unmodified uridine and RUGT, a nitroxide containing uridine analog defined

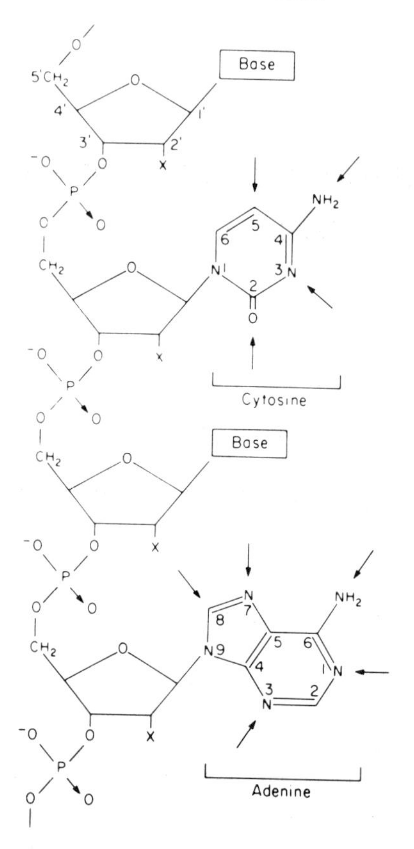

FIGURE 1 A random segment of a polynucleotide chain with a cytosine and an adenine base in the RNA molecule (X = OH) and DNA molecule (X = H). Arrows indicate the positions at which chemical modifications by site nonspecific labeling *via* electrophilic attack may occur in the bases. From Ref. 1, with permission.

above. The RUGT/uridine ratio is about 0.01 and the biophysical studies are usually done with $(\text{RUGT,U})_n$ of a weight-average molecular weight of 80,000–100,000.

ROTATIONAL CORRELATION TIMES OF THE NITROXIDE MOIETY IN NUCLEIC ACIDS

The theory for calculating rotational correlation times of the nitroxide moiety is well established[3]. In the case of spin labeled nucleic acids, the flexibility of the leg[4] between the nitroxide moiety and the nucleic acid chain allows the reporter group to tumble in the motional narrowing region. Experimental and

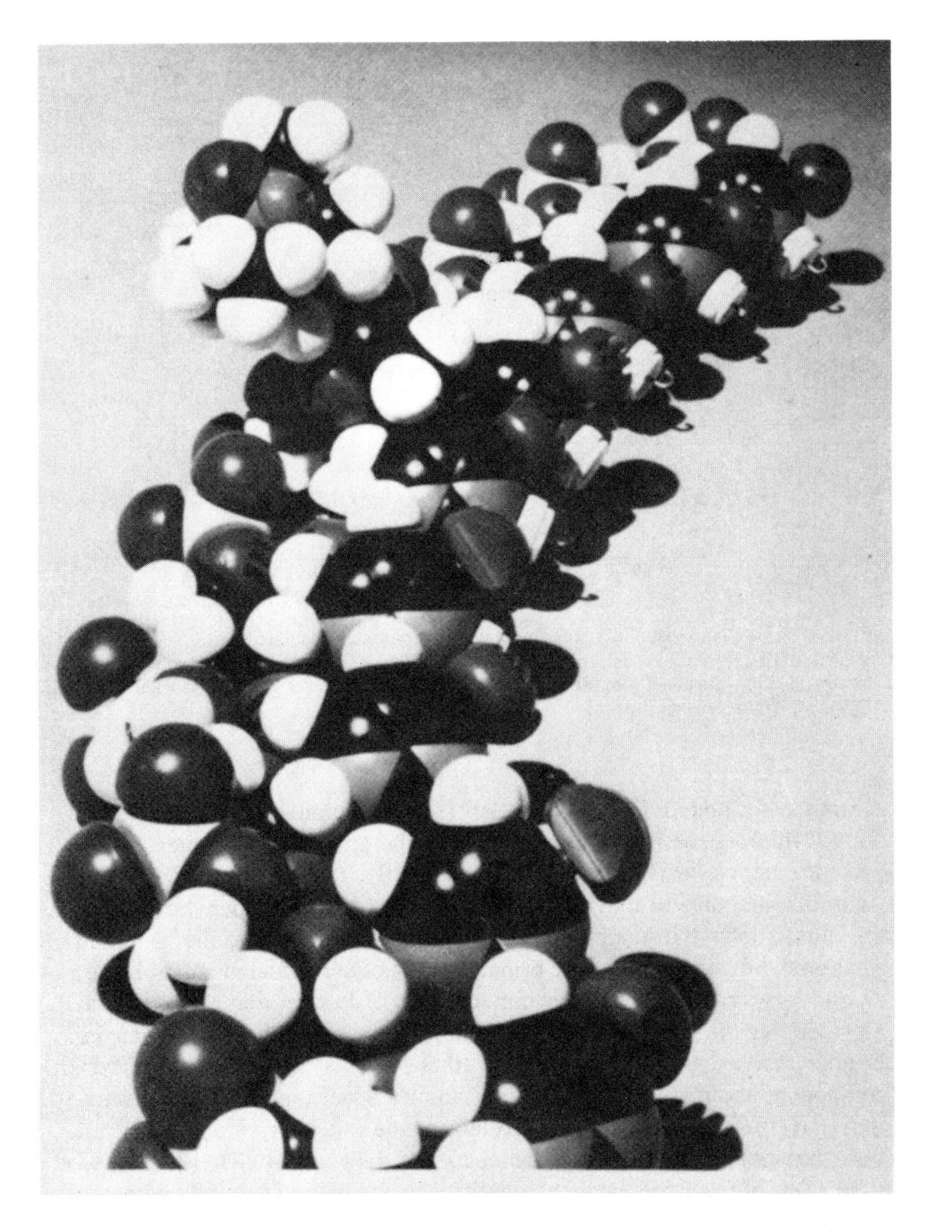

FIGURE 2 Molecular structure of $(RUGT,U)_n$: Ealing CPK atomic model showing a small stretch of the $(RUGT,U)_n$ chain (1 RUGT and 8 U residues); the nitroxide radical is visible in the upper left of the chain. From Ref. 2, reprinted by permission of the publisher, IPC Business Press Ltd. ©

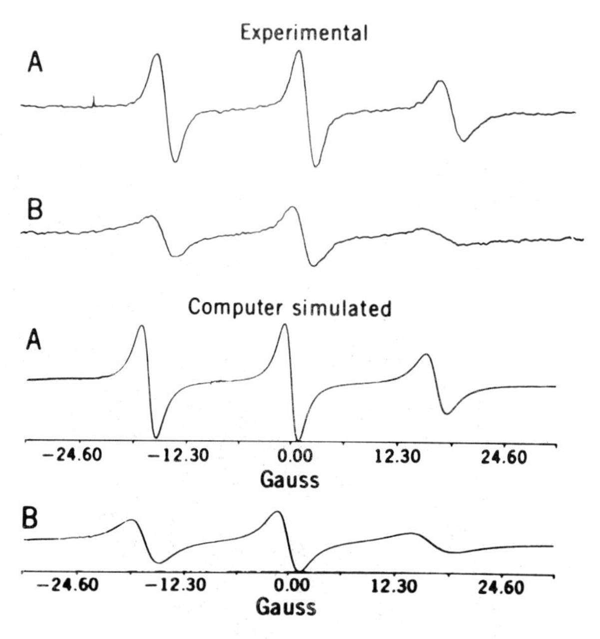

FIGURE 3 Comparison of experimental and computer-simulated ESR spectra of spin labeled $(dUfl)_n$ (spectra A) and spin labeled $(dUfl)_n \cdot (A)_n$ (spectra B) (at X-band with center of spectra at 3383 gauss). From Ref. 5, copyright 1978 by the American Association for the Advancement of Science.

computer simulated ESR spectra of 1st generation spin labeled $(dUfl)_n$, poly(2′-fluoro-2′-deoxyuridylic acid), and the spin labeled duplex $(dUfl)_n \cdot (A)_n$[5] are shown in Fig. 3. Spin labeled $(dUfl)_n \cdot (A)_n$ is obtained by mixing equimolar amounts of spin labeled $(dUfl)_n$ with $(A)_n$, poly(adenylic acid). For the simulations the tumbling times are based on an axially symmetric rotational diffusion model. The principal axes of the diffusion tensor R are x', y', and z' with the z' axis as symmetry axis of fast reorientation. It is also assumed that the x', y', and z' axes are either the same as the molecular axes x, y, and z or a cyclic permutation of them[6]. $R_{\parallel}$ is the rotational diffusion component about z', and $R_{\perp}$, about x' and y'. Spectrum A was simulated by setting $R_{\parallel} = R_{\perp}$, and selecting a correlation time value of $\tau_R = 3.5 \times 10^{-10}$ sec. For the more rigid double stranded spin labeled $(dUfl)_n \cdot (A)_n$ a good simulation of the ESR data was possible by assuming an axially symmetric Brownian rotational diffusion with $z' = y$. Spectrum B was simulated with the values 6.3×10^{-10} and 22.1×10^{-10} sec for $\tau_{R_{\parallel}}$ and $\tau_{R_{\perp}}$, respectively. From Fig. 3 it is apparent that it is possible to monitor by ESR a structural transition from double stranded to single stranded nucleic acids through nitroxide

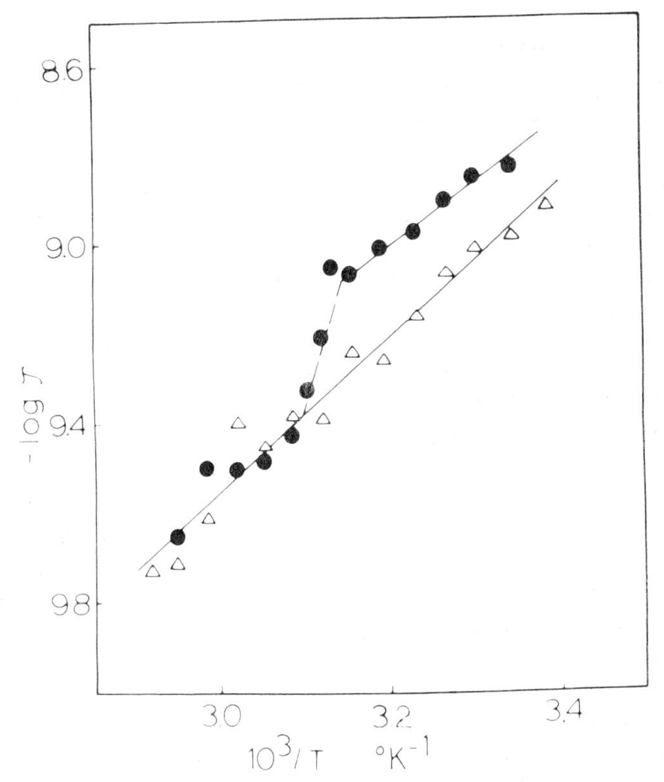

FIGURE 4 Dependence of $-\log \tau$ on inverse absolute temperature for spin labeled $(A)_n \cdot (U)_n$ (●-●), and for spin labeled $(A)_n$ (△-△) at a total polymer concentration of 8.9×10^{-4} M and 3.58×10^{-4} M, respectively. From Ref. 8, reprinted by permission of publisher.

radicals covalently attached to the nucleic acid chain. Such a transition is shown in Fig. 4.

CONFORMATIONAL TRANSITIONS OF $1(N)_n$

By 1967 the potential use of ESR to follow order–disorder transitions with spin labeled nucleic acids had been established[7]. Subsequently, many studies were done using spin labels to monitor conformational properties of nucleic acids. For a historical overview see the chapter by Bobst[1]. The temperature dependent transition of 1st generation spin labeled $(A)_n$ and spin labeled $(A)_n \cdot (U)_n$, a duplex consisting of spin labeled $(A)_n$ and $(U)_n$, is shown in the form of an Arrhenius plot in Fig. 4[8]. A linear relationship is observed for the melting of the single stranded $(A)_n$ with an activation energy of 8.3 kcal/mol. Fourier-transform NMR studies on single stranded $(A)_n$ gave a similar activation energy for the internal motion of the sugar–phosphate backbone and base sugar moiety[9]. The spin melting of the duplex reveals the existence of a sharp transition close to 50°C flanked on both sides by linear segments. The temperature at the midpoint of the break has been referred to as the spin denaturation temperature (T_m^{sp})[8]. From the two segments one can calculate the activation energies 8.3 and 6.3 kcal/mol, which correspond to the single and double stranded nucleic acid, respectively. After the transition, it is apparent that the activation energy, as well as the value of the calculated tumbling times, corresponds to those of the single stranded spin labeled $(A)_n$ supporting the hypothesis that T_m^{sp} reflects the temperature-dependent transition from double to single strands. In addition, the value of T_m^{sp} is very close to that of T_m^{OD}, the latter being the midpoint transition observed in temperature dependent optical density studies. As stated earlier[1], however, the absolute value of the activation energies must be treated with caution. Namely, for most Arrhenius plots correlation times are calculated with the simple formalism which assumes isotropic reorientation. However, it has been observed that the nucleic acid systems display some degree of anisotropic reorientation, particularly before the transition. In addition, more than one type of motion can be present before, as well as after the transition, in complex hetero-polymeric nucleic acids.

The occurrence of a step wise change in mobility upon increasing the temperature of spin labeled duplexes of homopolymers has been noticed elsewhere[10]. On the other hand, no discrete measurable change in nitroxide mobility has usually been observed for transitions in naturally occurring nucleic acids of varying base composition over a narrow temperature range. In such instances the transition is characterized by a single breakpoint referred to as T_{crit}[1]. Such T_{crit} have been reported for some tRNA's[11] and for the 2 to 1 transition of calf thymus DNA[12,13]. It has been observed that a T_{crit} usually occurs several degrees below T_m^{OD}, and it is believed that in these systems various sets of different spin label motions are present. Such a situation could arise if the nitroxide label destabilizes hydrogen bonding between complementary strands of varying base composition to different degrees.

Of particular interest is the spin melting of some naturally occurring DNA's which seem to reveal the existence of a transition at about 20°C by ESR[12,13]. The 20°C transition also coincides with an ESR transition observed through surface probing of DNA with the noncovalently bound nitroxide radical TEMPOL[14]. So far, this low temperature transition in DNA has only been observed by ESR, and while its origin is not yet known, it has been suggested that structurally bound water might play an important role in this transition. However, the possibility that the transition is a result of the chemical DNA modification can presently not be excluded[13]. This issue will probably be clarified with 2nd generation spin labeled nucleic acids.

Recently, it was also observed that strand rearrangements of the type shown below can be monitored by ESR[13].

$$2\ 1\,(A)_n \cdot (U)_n \rightleftharpoons 1(A)_n \cdot 2(U)_n + 1(A)_n$$

In this case $-\log \tau$ values are no longer calculated since two sets of unrelated motions will definitely be present after the rearrangement reaction. From a plot of the empirical ratio h_0/h_{+1} versus the inverse absolute temperature several linear segments are obtained. Two of the segments have a steep slope in a temperature range which corresponds rather closely to the $2 \rightarrow 3$ rearrangement and the $3 \rightarrow 1$ transition observed by UV melting studies.

INTERACTION OF $1(N)_n$ WITH LARGE LIGANDS

Binding of a large nucleic acid binding ligand such as poly-L-lysine to a spin labeled nucleic acid causes a drastic change in the ESR lineshape. This is shown in Fig. 5 where 1st $1(A)_n$ contaminated with a small amount of free

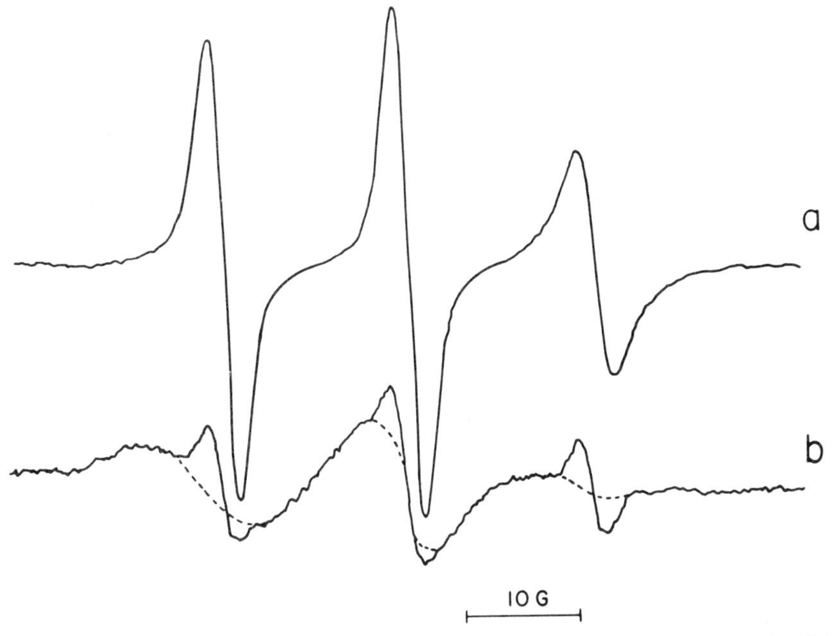

FIGURE 5 ESR control for the presence of unbound nitroxide radicals in a spin labeled polynucleotide. a) Spin labeled $(A)_n$ contaminated with a small amount of free nitroxide radicals; b) after complex formation with poly-L-lysine of a free spin contaminated spin labeled $(A)_n$ (solid line) and of an extensively purified spin labeled $(A)_n$ (broken line). From Ref. 1, with permission.

nitroxide radical is complexed with poly-L-lysine. The freely mobile spin label is readily distinguished from the hindered one, and this effect is very valuable for detecting free spin contaminations in $1(\mathrm{N})_n$[1] as well as spin labeled duplexes[13]. Conformational changes of DNA were reported during the formation of a complex between RNA polymerase from *E. coli* and DNA isolated from bacteriophage T2[15,16]. Finally, $1(\mathrm{N})_n$ were used to determine the relative affinity of proteins for nucleic acids by ESR[17].

In conclusion, valuable information about the molecular dynamics of biopolymers such as nucleic acids can be gained from the spin labeling approach. In particular, spin labeled nucleic acid homopolymers are found to be extremely useful for gaining insight into the dynamics of such macromolecules.

ACKNOWLEDGMENT

This work was supported in part by CA 15717, awarded by the National Cancer Institute, DHEW.

REFERENCES

1. A.M. Bobst, "Spin Labeling II", (Academic Press, New York, 1979, L.J. Berliner, ed.), Chap. 7.
2. A.M. Bobst and P.F. Torrence, *Polymer*, **19**, 115 (1978).
3. See chapters by P.L. Nordio (p. 5) and J.H. Freed (p. 53) in "Spin Labeling: Theory and Applications", (Academic Press, New York, 1976, L.J. Berliner, ed.).
4. D. Wallach, *J. Chem. Phys.*, **47**, 5258 (1967).
5. A.M. Bobst, T.K. Sinha, and Y.C.E. Pan, *Science*, **188**, 153 (1975).
6. S.A. Goldman, G.V. Bruno, C.F. Polnaszek, and J.H. Freed, *J. Chem. Phys.*, **56**, 716 (1972).
7. I.C.P. Smith and T. Yamane, *Proc. Nat. Acad. Sci. U.S.*, **58**, 884 (1967).
8. Y.C.E. Pan and A.M. Bobst, *Biopolymers*, **12**, 367 (1973).
9. K. Akasaka, *Biopolymers*, **13**, 2273 (1974).
10. W.J. Caspary, J.J. Greene, L.M. Stempel, and P.O.P. Ts'o, *Nucleic Acid Res.*, **3**, 847 (1976).
11. M. Caron and H. Dugas, *Nucleic Acid Res.*, **3**, 35 (1976).
12. E.M. Mil', S.K. Zavriev, G.L. Grigoryan, and K.E. Kruglyakova, *Dokl. Akad. Nauk SSSR*, **209**, 217 (1973).
13. A.M. Bobst, A. Hakam, P.W. Langemeier, and S. Kouidou, *Arch. Biochem. Biophys.*, submitted.
14. B.I. Sukhorukov, B.I. and L.A. Kozlova, *Biofizika*, **15**, 539 (1970).
15. S.G. Kamzolova, A.I. Kolontarov, L.I. Elfimova, and B.I. Sukhorukov, *Dokl. Akad. Nauk SSSR*, **208**, 245 (1973).
16. B.I. Sukhorukov, S.G. Kamzolova, A.I. Kolontarov, and A.I. Petrov, *Biofizika*, **18**, 377 (1973).
17. A.M. Bobst and Y.C.E. Pan, *Biochem. Biophys. Res. Commun.*, **67**, 562 (1975).

DISCUSSION

R. F. Boyer (Midland Macromolecular Institute, Midland, Michigan): Does h_0/h_{+1}, which indicates transitions when plotted against 1/T, suffer from the problems of anisotropic motion of the label or probe?

A. M. Bobst: The h_0/h_{+1} ratio is a purely empirical parameter which was found to give steep slopes in the case of some nucleic acid homopolymers, the midpoints of which coincided well with T_m's obtained from UV meltings. The problem of anisotropic motion in such polymers has to be solved by tedious computer simulation of each spectrum taken at a given temperature.

R. F. Boyer: You indicated that the activation energy was greater for the single strand than for the double strand. It should be pointed out that the activation energy of a typical polymer in the glassy state is less than it is above the glass transition.

A. M. Bobst: This raises the question of what is meant by order and disorder. In our systems we know which state is more ordered and we observe a lower activation energy for the more ordered state.

R. F. Boyer: Is the single strand a random coil?

A. M. Bobst: It depends on the particular system, poly (U) is a completely random coil, while poly (A) has quite a bit of helical structure present.

G. G. Cameron (University of Aberdeen, Old Aberdeen, Scotland): Spin labeled polyethylene gives a much lower activation energy than a spin probed polyethylene over the same temperature range, by roughly a factor of two.

SPIN PROBE STUDIES

Spin Probe Studies in Polymer Solids

A. L. Kovarskii, A. M. Wasserman, and A. L. Buchachenko

The Institute of Chemical Physics of the Academy of Sciences of the USSR, Moscow, 117334, USSR

Spin probe techniques are widely used in physical chemistry studies of polymers (compatibility, crystallization, mechanical deformation, structural micro-inhomogeneity, structuring, thermo-oxidative degradation, etc.).[1,2] One of the important applications of the spin probe method is the investigation of the frequencies of the rotational and translational motion of low molecular weight particles, and in particular, stable radicals, in solid polymers. This aspect is of special importance for the study of the kinetics of solid-phase reactions.[3]

The problems regarded as being of primary importance are the following: the mechanisms of rotational and translational motion of a spin probe; the relationship between probe motion and the molecular motions in the polymer; the scale of the molecular motion of polymer segments which control the motion of a spin probe.

1. ROTATIONAL MOTION

We have demonstrated that the mechanisms of rotational motion of a spin probe are essentially different in the two states of polymers, below and above T_g.[4]

Fig. 1 shows the temperature dependence of the correlation time for probe rotation in a number of polymers, in Arrhenius equation coordinates. Similar behavior was observed earlier in many polymers. A feature typical for these curves is the inflection point in the neighborhood of the glass transition temperature. The rotational activation energies in these two ranges are quite different: 5 to 20 kcal/mol above T_g and 1 to 2 kcal/mol below T_g (Table 1). These peculiarities are quite general and are independent of the choice of probe.

Fig. 2 (a and b) compares spin probe rotational correlation times in two polymer systems with the relaxation times measured by NMR and dielectric

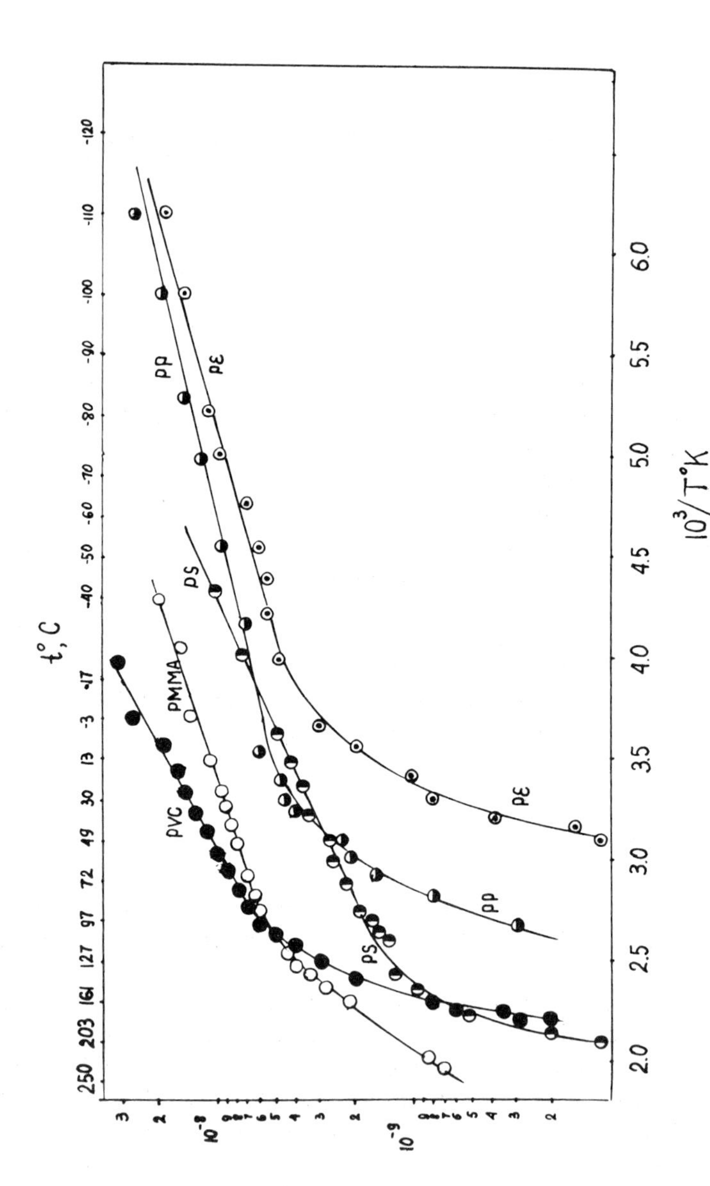

FIGURE 1 Arrhenius plot of τ_c for probe I (TEMPO) in a series of polymers.

TABLE 1

Characteristic temperatures and activation energies for polymers, according to spin probe data.

				above T_g		below T_g	
Polymer	Radical	T_g, °C	T_n, °C	E, kcal/mol	τ_0, s	E, kcal/mol	τ_0, s
PMMA	A	70–105	70	5	5.0×10^{-12}	1.0	2.0×10^{-9}
	B	—	80	5	2.0×10^{-11}	1.0	4.0×10^{-9}
PVC	A	74	78	10	6.9×10^{-15}	2.1	4.2×10^{-10}
PS	A	100	104	20	1.1×10^{-19}	1.9	1.6×10^{-10}
PP	A	0	10	11	5.0×10^{-17}	0.9	1.9×10^{-9}
	B	—	10	11	3.2×10^{-16}	0.9	2.9×10^{-9}
PE	A	−20	−43	10	1.6×10^{-17}	1.5	3.2×10^{-10}
NR	A	−70	−48	6.5	5.4×10^{-15}	2.7	3.0×10^{-11}

Radicals:

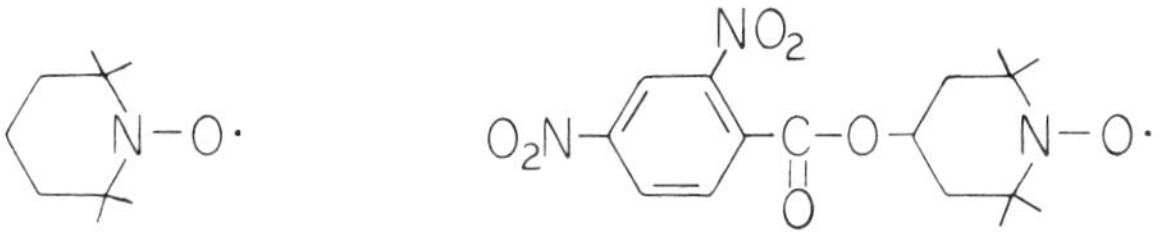

A

B

and mechanical relaxation techniques. In the high temperature range the probe motion frequency approaches the characteristic frequencies of the α, β, and γ processes.

The frequencies of the α and β processes at the glass transition temperature are two to five orders of magnitude lower than the probe relaxation frequencies. In the low temperature range, the frequencies of the γ process and of probe rotation differ by up to five orders of magnitude. The activation energy of probe rotation is much smaller than the activation energy for the main relaxation processes. Similar frequency relationships are observed for all polymers.

The mobility of a probe and the motions of chain segments are closely related in the high temperature range.[1,2,5] It should be added that there is a satisfactory agreement of frequencies in the range of the intersection of the recorded (or extrapolated) temperature dependence curves for the α, β, and γ processes, i.e., at temperatures where short chain segments play the part of independent kinetic units.

In the range below T_g, it is significant that the probe rotational frequencies are high (10^7–10^8 s^{-1}) and the activation energies extremely low (1–2 kcal/mol). As a rule, the probes are not "sensitive" to dynamic trans-

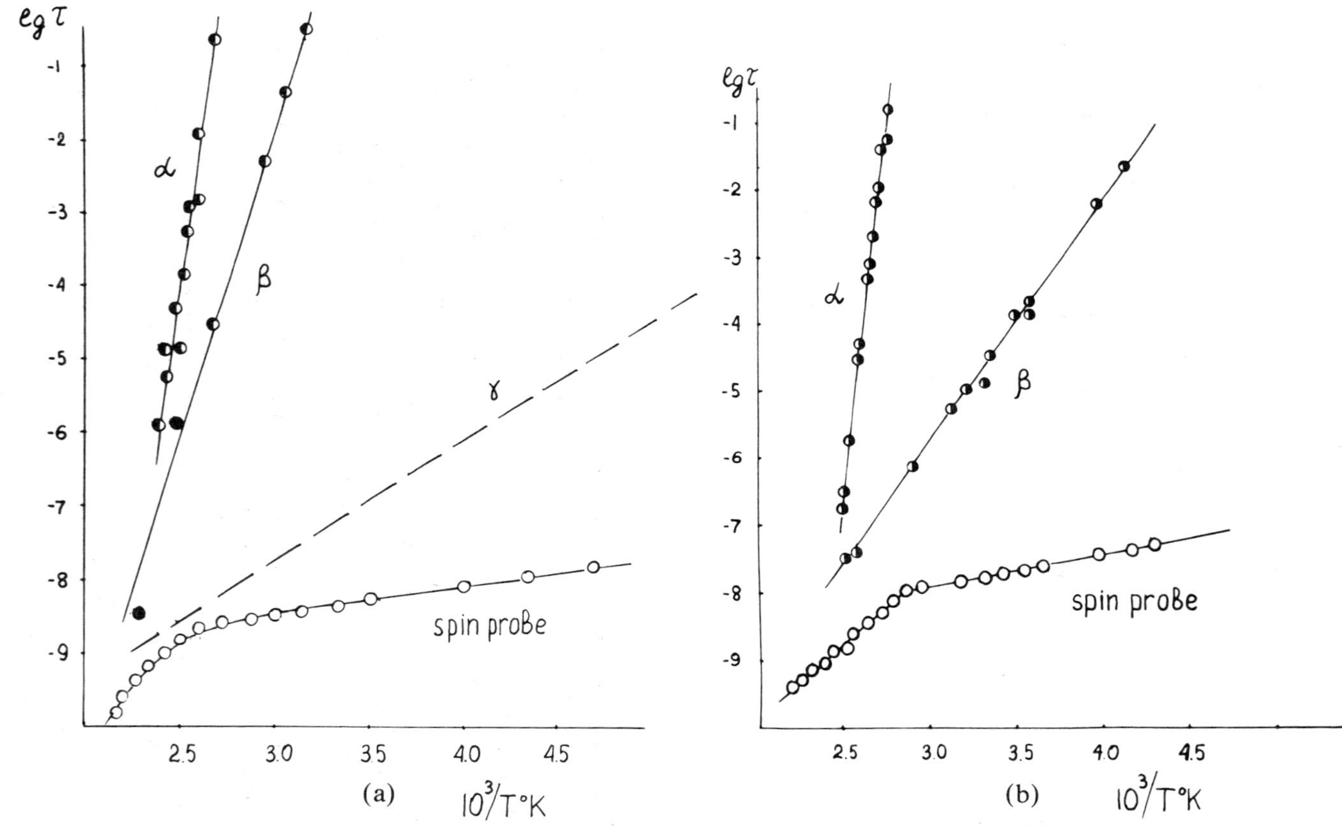

FIGURE 2 Frequency diagram of data obtained for (a) polystyrene, (PS), and (b) poly(vinyl chloride), (PVC), by the methods of dielectric relaxation (◑), mechanical relaxation (◐), NMR (●), and spin probe (○).

formations of a polymer in this temperature range. It appears that an explanation must be sought in terms of modifications of the motional mechanism. According to well-known data on permeability to gases, and sorption of vapors, glass-like polymers are porous sorbents, owing to the looseness of their molecular packing.[6] The required freedom of motion of a radical is ensured by the existence of "holes" in a polymer with dimensions comparable to those of a probe. Below T_g, probe rotation is definitely taking place within these structural defects. In order to model the process, we investigated spin probe rotation in synthetic zeolites with a channel diameter of 8 Å. Rotation of probes in these zeolites is unhindered ($\sim 10^9\ s^{-1}$ at 25°C), with an activation energy of 2.5 kcal/mol, i.e., coinciding with that for polymers below T_g.

The effect of the looseness of packing on the rotational mobility of a spin probe was investigated using porous polymers produced by crosslinking dissolved polystyrene with monochlorodimethylether. The total pore volume in these polymers and the glass transition point are known to increase with increased degrees of crosslinking.[7] As shown in Table 2, the rotational frequency of a probe in these polymers rises concurrently with the increase in the degree of crosslinking and the total pore volume.

Spin probes in polymers at temperatures below T_g are localized in "hole"-type structural defects and consequently can rotate independently of the molecular motions of the macromolecules. In these conditions, the molecular mobility of a spin probe depends mainly on the size of the "hole" in which it is located. Loose packing is typical for many glass-like polymers. Being normally introduced into a polymer at high temperatures or from a solution below the glass transition point, the probe itself is a defect. A very small excess volume is required for reorientation of such a particle. By studying the effect of hydrostatic pressure (up to 2.5 kbar) on the rotational mobility of a probe in

TABLE 2

Correlation time of probe rotation in PS-based porous polymers with different degrees of crosslinking.

Degree of crosslinking, %	W_0^*, cm^3/g	T_g, °C	$\tau \times 10^9$ s $T = 25$°C	E, kcal/mol, low temperature range	$\tau_0 \times 10^9$ s
0	—	80	7.2	2.9	5.8
4	—	140	6.2	2.8	5.4
25	0.21	160	2.9	2.8	4.2
66	0.44	160	2.1	2.7	2.8
100	0.51	280	2.4	2.8	3.8

* W_0 denotes total pore volume derived from density measurements.

polyethylene, we find the rotational activation volume of a spin probe, Δv, to be $\sim$20 cm^3/mol, the intrinsic radical volume being 160 cm^3/mol.[8] This signifies that the excess volume required to provide the freedom for probe reorientation is only about 8 to 10% of the probe volume.

Consequently, rotation of a spin probe in the low temperature range (below T_g) usually follows the "hole" mechanism and is characterized by a low activation energy. In reality, the motion of a spin probe in these conditions is determined by the static free volume of the polymer, and proceeds independently of the molecular dynamics in the polymer. At temperatures above T_g, the spin probe motion is perturbed by molecular motions of the polymer chain and its segments, resulting in increased frequency and activation energy of probe rotation. In this situation, probe rotation is determined not only by the static free volume, but also by dynamic fluctuations due to molecular motions of macromolecules and their elements. As temperature rises, the relative contribution of the first mechanism diminishes while that of the second increases, so that at high temperatures the spin probe motion proceeds mostly via the second mechanism and is controlled by the small-scale molecular motion of macromolecular segments having dimensions comparable with those of the spin probe.

The change in the relative contribution of these two motion mechanisms as a result of changing temperature produces a new value of the activation energy; this in turn results in anomalously high apparent values of experimentally measured activation energy. The well-known compensation effect, i.e., the linear relationship between the pre-exponential factor and the activation energy, is a corollary of the temperature dependence of the activation energy. The compensation effect is often experimentally observed in polymers by spin probe and spin label techniques.

A few words are in order here concerning the amplitudes of spin probe rotation. The recently developed methods of analysis of EPR spectra make it possible to determine whether a radical in the system rotates jump-wise (by large angles of rotation) or continuously (by small angles of rotation).[9–11]

The quantity $\Delta H_{-1}/\Delta H_{+1}$, plotted in Fig. 3 as a function of ΔH_{+1}, enables one to find the amplitude of the spin probe rotation. Table 3 lists correlation times of probe rotation in a number of polymers at the glass transition temperature.

We notice that in polymers with a higher frequency of probe rotation, probe motion can be described satisfactorily by the jump model, and in polymers with a lower frequency of probe rotation, by the continuous diffusion model (Fig. 3a). We find, therefore, that rotational frequency is related to amplitude. For the same reason, the probe motion is described better in terms of the continuous diffusion model when probe size increases (Fig. 3b).

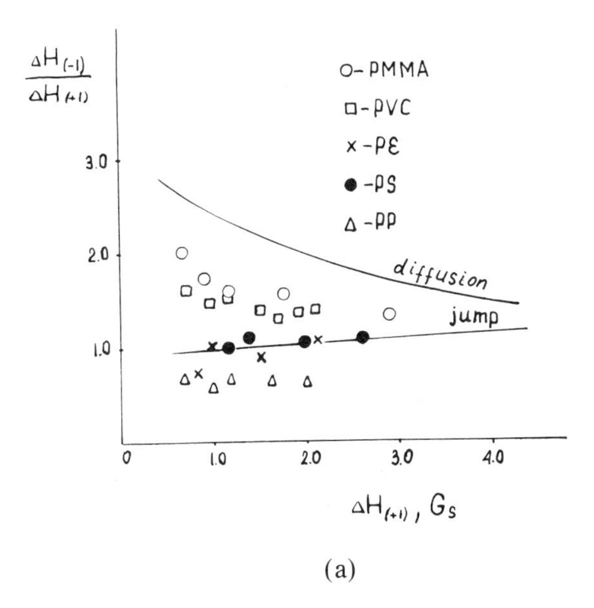

(a)

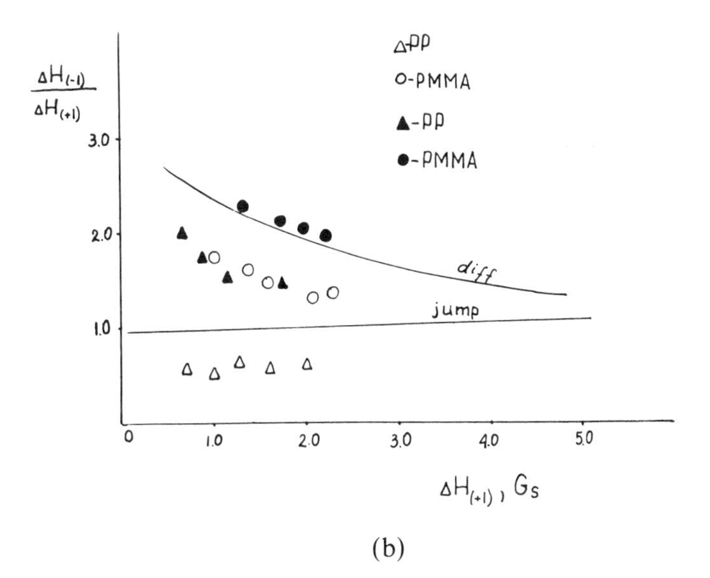

(b)

FIGURE 3 (a) $\Delta H_{-1}/\Delta H_{+1}$ as a function of ΔH_{+1} for probe I (TEMPO) in various polymers. (b) $\Delta H_{-1}/\Delta H_{+1}$ as a function of ΔH_{+1} for probe I (TEMPO) (open symbols) and the probe BzONO (filled symbols) in polypropylene and poly(methyl methacrylate).

TABLE 3

Correlation time of probe rotation at the glass transition point for a number of polymers, and the rotation models (according to Kuznetsov).

Polymer	$\tau \times 10^{-9}$ s	Model characteristics
PP	6.0	jump
PE	5.0	jump
PS	3.2	jump
PVC	10.0	diffusion
PMMA	9.0	diffusion

2. TRANSLATIONAL MOBILITY OF SPIN PROBES

The translational mobility of probes was investigated in various rubbers in the 20 to 90°C temperature range.[12] The activation energy and pre-exponential factors for rotation and translation are listed in Table 4.

Both processes, viz., rotation and translation, are described by the Arrhenius equation, and the relationship between the coefficients of rotational and translational diffusion can be presented in the form

$$D_{tr} = \alpha \cdot D^{\beta}_{rot}$$

where $\beta = E_{tr}/E_{rot}$, and $\alpha = D^0_{tr}/(D^0_{rot})^{\beta}$.

In a series of related polymers (rubbers, for instance) the values of α and β are quite close.

In most liquids the rotational and translational diffusion are known to be closely related, and governed by a common mechanism. Each reorientation of

TABLE 4

Activation energy and pre-exponential factors for rotation and translation of the probe TEMPO in polymers above T_g.

Polymer	E_{rot}, kcal/mol	$\tau_0 \times 10^{16}$ s	E_{tr}, kcal/mol	D_0, cm²/s
(1) NR	6.5	54	12.8	8.0×10^{1}
(2) SRN-40	9.0	5.6	18.4	1.6×10^{4}
(3) SRS-50	8.7	6.5	17.7	8.0×10^{3}
(4) PIB	10.7	0.25	17.8	1.1×10^{3}
(5) BR	8.0	23	12.3	2.3×10^{-1}

(1) Natural rubber, (2) synthetic nitrile rubber, (3) synthetic styrene rubber, (4) polyisobutylene, (5) butyl rubber.

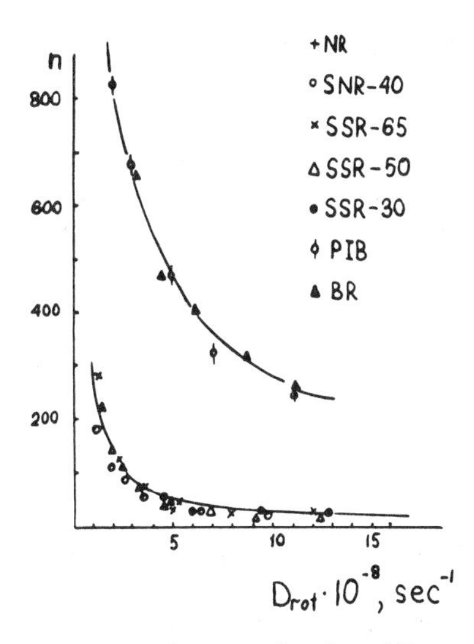

FIGURE 4 n as a function of D_{rot}.

a particle in a liquid corresponds to its translation by a distance approximately equal to its diameter. The rotational and translational activation energies in liquids are approximately equal. In polymers this rule is violated. Fig. 4 plots the quantity $n = \tau_{tr}/\tau_{rot}$ (where $\tau_{tr} = x^2/6D$) as a function of D_{rot}; n shows how many times a radical changes its orientation over the translation distance of 5 Å which is approximately equal to its diameter. When D_{rot} changes from 1×10^8 to 1×10^9 s^{-1}, n changes from 300 to 10 in polybutadienes, and from 800 to 250 in PIB and BR. The following conclusions can be drawn. First, rotation in polymers is a much faster process than translation of a particle. Second, as temperature increases, the ratio of the frequencies of the two types of motion diminishes, so that in the high temperature range the behavior of a low molecular weight particle becomes identical to that of a particle in a liquid. Finally, the ratio of frequencies of two types of motion depends on polymer structure.

As temperature decreases, n becomes a steeper function of D_{rot}, i.e., rotation and translation are independent, with no interrelation between them at $T < T_g$. Rotational frequencies remain sufficiently high below T_g (10^7–10^8 s^{-1}), but translation frequencies become extremely low. These data confirm the conclusion that rotation of particles in polymer glasses proceeds owing to the static free volume, while restructuring of the nearest-neighbor surroundings becomes a necessary condition for translation diffusion, i.e., dynamic fluctuations of free volume are required. The two forms of motion become correlated only in the high temperature range, when molecular motions of macromolecules and their elements induce dynamic fluctuations ensuring rotational and translational mobility of spin probes.

The anisotropy of rotation, i.e., differences in frequencies of rotation about the principal molecular axes, is an important characteristic of molecular rotation of a spin probe. We have suggested[13] characterizing the rotational anisotropy by the parameter ε, which is related to the linewidth of the EPR triplet spectrum of a nitroxyl radical by the expression

$$\varepsilon = \frac{T_2^{-1}(+1) - T_2^{-1}(0)}{T_2^{-1}(-1) - T_2^{-1}(0)} = \frac{[I(0)/I(+1)]^{1/2} - 1}{[I(0)/I(-1)]^{1/2} - 1}$$

where $T_2(m)$ and $I(m)$ denote the linewidths and line intensities, for the corresponding magnetic quantum number of nitrogen ($m = \pm 1, 0$), respectively.

The parameter ε allows a theoretical derivation as well, provided the components of the g- and A-tensors of the radicals and their orientation with respect to the principal axes of the tensor of rotational diffusion are known. The rotational anisotropy, i.e., $N = \tau_\perp/\tau_\parallel$, can be determined by comparing the theoretically calculated and experimentally measured values of ε.

We investigated the rotational anisotropy of the nitroxyl radicals I–V (see below), having nearly axially symmetric tensors of rotational diffusion and in which the principal axes of the hyperfine coupling tensor and the g-tensor coincide with those of the rotational diffusion tensor.

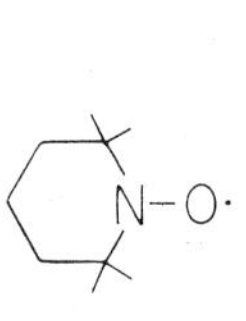

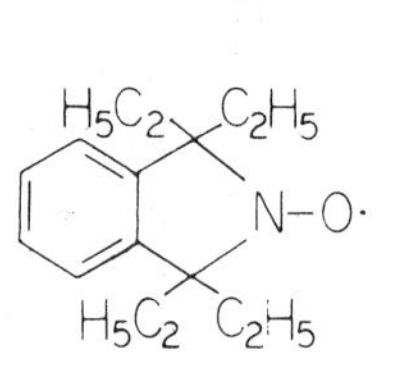

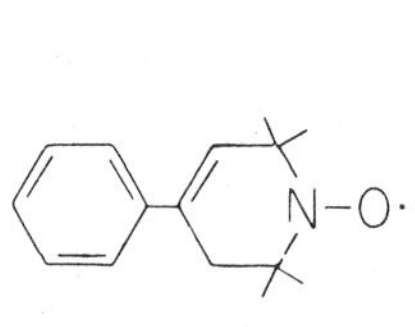

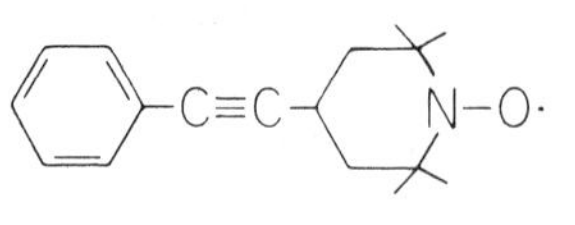

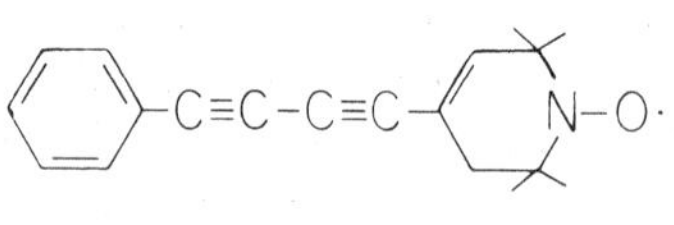

By comparing the theoretical and experimental values of ε in natural rubber and ethyl benzene, we derived the ratio $\tau_\perp/\tau_\parallel$, where $\tau_\perp$ and $\tau_\parallel$ are the

TABLE 5

The values of $N = \tau_{\perp}/\tau_{\parallel}$.

Radical	Natural rubber	Ethyl benzene
I	2–9	1–2
II	1.5	1
V	6.0	5.0

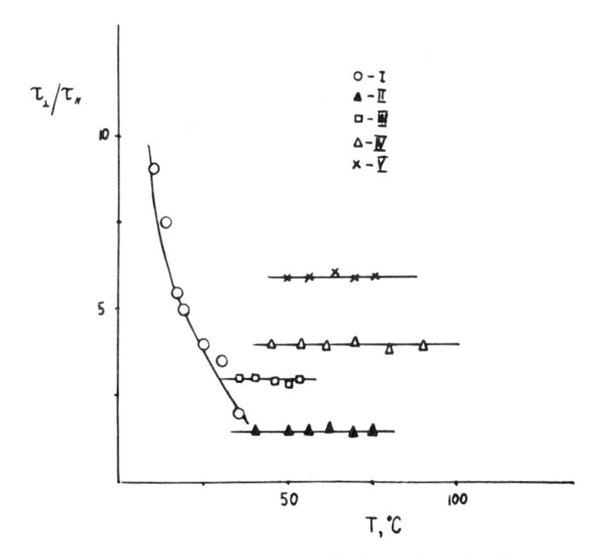

FIGURE 5 Temperature dependence of $\tau_{\perp}/\tau_{\parallel}$ for spin probes in natural rubber.

correlation times relative to the axes of symmetry of the radicals (Table 5).

As should have been expected, the rotational anisotropy is greater the more prolate the rotational ellipsoid of a radical. Fig. 5 plots $\tau_{\perp}/\tau_{\parallel}$ as a function of temperature for a number of radicals. It is clear that the rotational anisotropy for large radicals and for ellipsoid-shaped ones is independent of temperature, and determined only by the radical shape.

For the small radical, the rotational anisotropy is a function of temperature, and is therefore determined by the anisotropy of the intermolecular interaction potential. Activation energies of rotation for this radical around the principal axes differ considerably: $E_{\parallel} = 9.8$ kcal/mol, $E_{\perp} = 20$ kcal/mol, $\tau^{\circ}_{\parallel} = 1 \times 10^{-17}$ s, $\tau^{\circ}_{\perp} = 8 \times 10^{-19}$ s.

REFERENCES

1. A.L. Buchachenko, A.L. Kovarskii, and A.M. Wasserman in "Advances in Polymer Science", (Z.A. Rogovin, ed.), p. 33, Halsted Press, Wiley, New York, (1974).
2. P. Törmälä, and J.J. Lindberg in "Structural Studies of Macromolecules by Spectroscopic Methods", (K.J. Ivin, ed.), p. 255, Wiley, New York, (1976).
3. N.M. Emanuel, V.A. Roginskii, and A.L. Buchachenko, "Problems of Radical Reaction Kinetics", Uspekhy Chimiy (USSR), (in press).
4. A.L. Kovarskii, J. Plaček, and F. Szöcs, *Polymer*, **19**, 1137 (1978).
5. A.M. Wasserman, A.L. Buchachenko, A.L. Kovarskii, and M.B. Neiman, *Vysokomolek. Soedin., USSR,* **10a,** 1930 (1968).
6. S.A. Reitlinger, "Permeability of Polymer Materials", Chimiya, Moskow, (1974).
7. M.V. Tsilipotkina, A.A. Tager, V.A. Davankov, and M.P. Tsurupa, *Vysokomolek. Soedin., USSR*, **18b**, 874 (1976).
8. A.M. Wasserman, A.A. Dadaly, A.L. Kovarskii, A.J. Koshuhar, and V.I. Irzhak, *Doklady, ANSSSR*, **237**, 130 (1977).
9. J. Freed, G. Bruno, and C. Polnaszek, *J. Phys. Chem.*, **75**, 3385 (1971).
10. A. Kuznetsov and B. Ebert, *Chem. Phys. Lett.*, **25**, 342 (1974).
11. N.N. Korst and L.I. Antziferova, *Uspekhy Phys. Nauk, (USSR)*, **126**, 67 (1978).
12. A.M. Wasserman, I.I. Barazhkova, L.L. Yasina, and V.S. Pudov, *Vysokomolek. Soedin., USSR*, **19**, 2083 (1977).
13. A.M. Wasserman, A.N. Kuznetsov, A.L. Kovarskii, and A.L. Buchachenko, *Zhurnal Strukt. Khimiy*, **12**, 609 (1971).

ESR Spin-Probe Studies of Polymer Transitions

PHILIP L. KUMLER

Chemistry Department, State University of New York, College at Fredonia, Fredonia, New York 14063

The use of paramagnetic nitroxides, both as spin-probes and spin-labels,[1] to study molecular motion was first applied to biological polymers.[2] The application of both spin-probes and spin-labels to synthetic macromolecules has been developed much more recently but is gaining increasing attention.[3] The pioneering work in the area of ESR studies of synthetic polymers was performed by Stryukov and Rozantsev[4] in the Soviet Union and by Rabold[5] in the United States. A recent review article by Buchachenko and co-workers[6] presents a good overview of the theory and scope of the technique and summarizes, as well, most of the Russian work prior to 1974. The potential of the ESR technique for studying molecular relaxations in high polymers was suggested by Rabold[5] and extended significantly by Boyer.[7] We began, in 1974, a series of studies to investigate the scope and utility of the ESR spin-probe technique as a tool for the study of molecular relaxations in a variety of polymeric materials. This paper will summarize the highlights of our work over the last four years (1974–1978),[8–19] and will be presented in the following sequence.

(a) Introduction and General Principles
(b) Correlation of T_{50G} with T_g
(c) Effect of Molecular Weight and Molecular Weight Distribution
(d) Free Volume Effects; Oligomeric *vs* Polymeric Glasses
(e) Block Copolymers
(f) Multiple Environments Observed by a Single Probe
(g) Probe Size Studies
(h) Activation Volumes
(i) Crystalline Polymers
(j) Multiple Probe Experiments

INTRODUCTION AND GENERAL PRINCIPLES

All of our studies have utilized nitroxides of the general formula $R_2N—O$ as the paramagnetic probe; structures and molecular weights of the various nitroxides we have employed are presented in Figure 1, which also includes any trivial names that we have used for these molecules.

All of the nitroxides shown in Figure 1 exhibit a three-line ESR spectrum whose detailed line-shape, etc. is a function of the tumbling frequency of the probe in a particular matrix. Various theoretical approaches have been used to describe the total line-shape of the nitroxide spectrum.[2b] Most of our experimental work has used an empirical line-shape parameter, T_{50G}[20] that can be directly determined from the ESR spectra of a nitroxide-doped polymer by recording the spectra at a series of temperatures over a wide temperature range. T_{50G} is the temperature at which the peak-to-peak separation of the outer lines of the three-line spectrum is equal to 50 gauss. In a typical experiment[21] a polymer is doped with a low concentration of an appropriate nitroxide, and the ESR spectra are determined at a series of temperatures. At low temperatures the nitroxide is highly immobilized and a spectrum similar to that shown in Figure 2a results; as the temperature is increased, the nitroxide experiences increased tumbling frequency resulting in an ESR spectrum similar to Figure 2b. At sufficiently high temperatures, the experimental spectrum will resemble closely that shown in Figure 2c, which is characteristic of a rapidly-tumbling

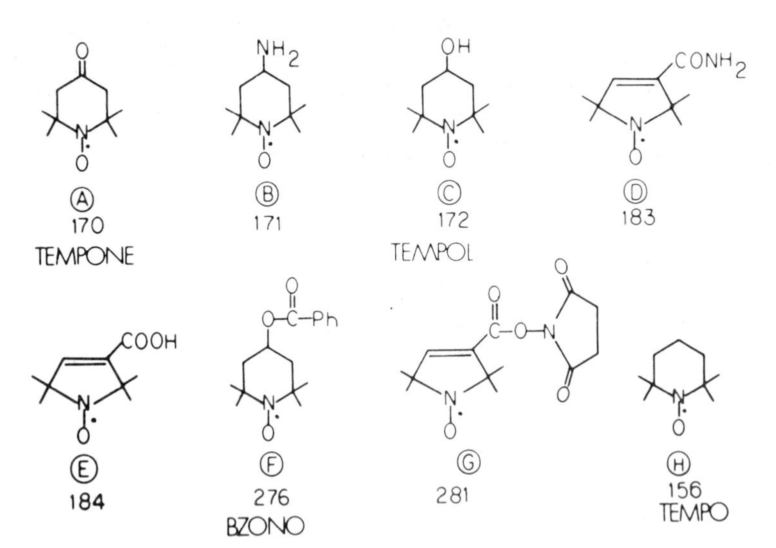

FIGURE 1 Nitroxide probes used in spin-probe studies. Circled numbers indicate letter code used in text. Molecular weights (in amu) given for each structure; trivial names used indicated for appropriate structures.

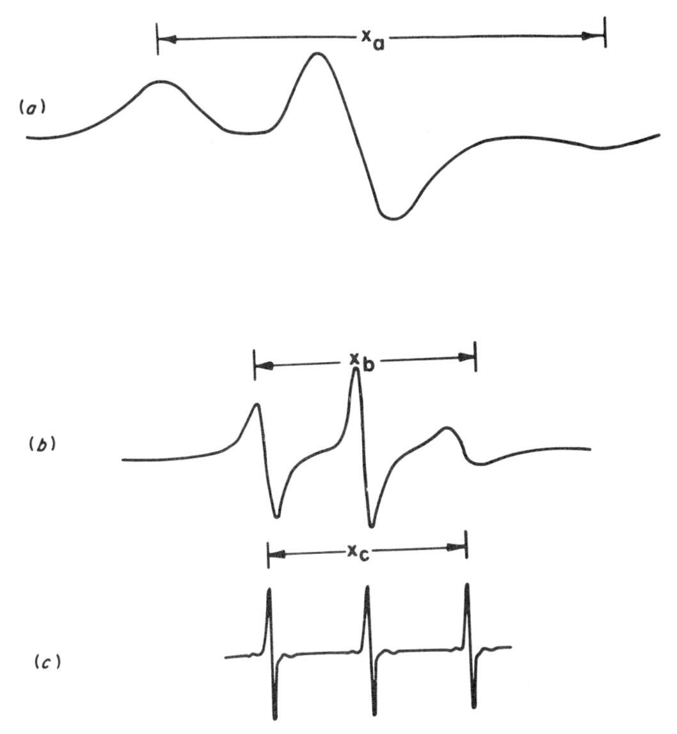

FIGURE 2 Representative spectra for nitroxides exhibiting varying rates of rotational reorientation: (a) highly immobilized, slow tumbling frequency, (b) intermediate mobility, intermedite tumbling frequency, (c) rapidly tumbling nitroxide, fast tumbling frequency. X_a, X_b, and X_c indicate the measured extrema separation for each spectrum.

nitroxide. The extrema separation in gauss (X_a, X_b, and X_c in Figure 2) is directly determined from the experimental spectra and is plotted as a function of temperature. A schematic representation of the types of extrema separation versus temperature plots that result is shown in Figure 3. In most cases the upper limit of extrema separation observed is 60–65 gauss and corresponds to a highly immobilized nitroxide; the lower limit observed is typically 35–40 gauss and corresponds to a rapidly-tumbling nitroxide, and approaches the value observed for a nitroxide in a fluid solution of low viscosity.

CORRELATION OF T_{50G} WITH T_g

Rabold[5] was the first to suggest that there may be a correlation between T_{50G} and either the glass temperature (T_g) or the melt temperature (T_m) in

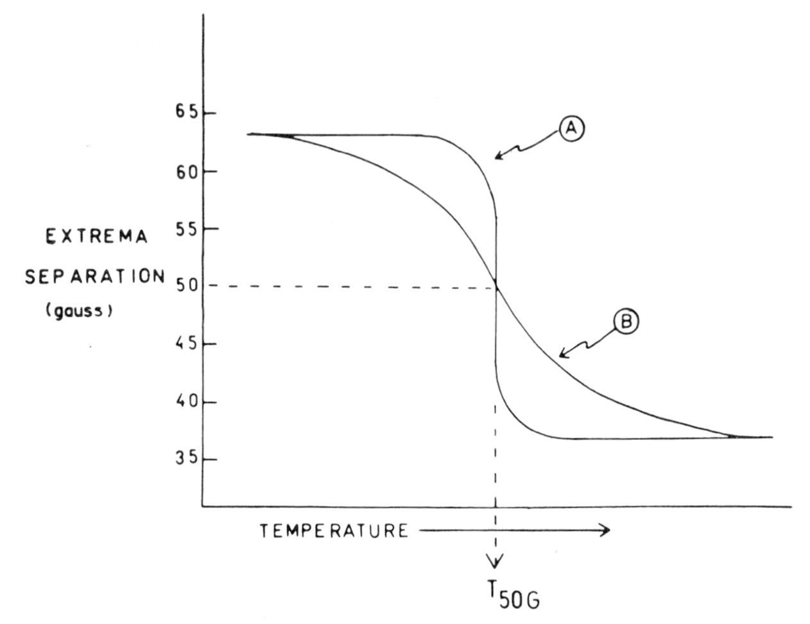

FIGURE 3 Schematic plot of extrema separation *vs* temperature. The two curves represent extremes of the type of experimental plots actually observed; A represents a "sharp" plot and B represents a "broad" plot. Reprinted with permission from ref. 14. Copyright by the American Chemical Society.

synthetic polymers. This hypothesis was extended by Boyer[7] who suggested a direct linear correlation of T_{50G} with T_g.

We have examined, during the course of our studies, a wide range of polymers and random copolymers with glass temperatures ranging from −150°C to 145°C, using BzONO as probe, and have shown that there is, indeed, a correlation between T_{50G} and T_g. For establishing this correlation we have utilized polymers for which there is little or no controversy about the glass temperature and which have been examined by more than one technique. This series of studies[8,14] has led to the construction of a T_{50G}–T_g correlation plot shown in Figure 4 which allows determination of the glass temperature for most polymers. It was initially suggested[5,7] that such a correlation would be linear. This linear dependence is, in fact, true for polymers having glass temperatures in the range of 0–150°C. For glass temperatures below 0°C, there is considerable curvature in the T_{50G}–T_g plot.

The curvature at low T_g and the approximate linearity at high T_g, while initially surprising to us, are a direct result of two characteristic features of the glass transition. The observed transition temperature, as measured by a

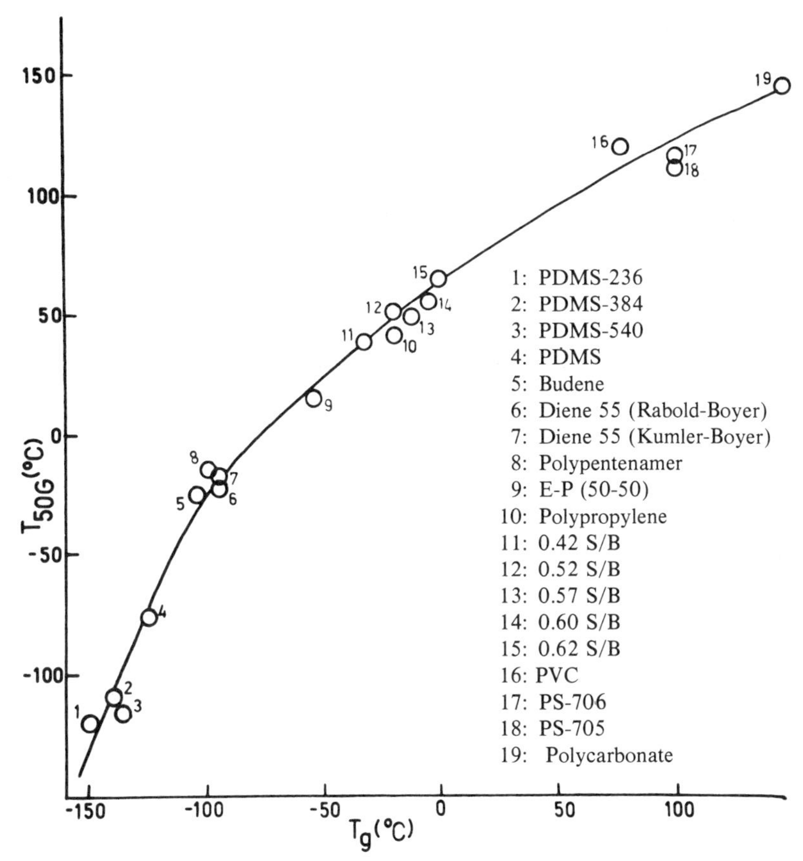

FIGURE 4 Correlation plot for T_{50G} *vs* T_g. T_{50G} values determined from spin-probe studies; T_g values are accepted literature values. The smooth curve is a computer-fit best curve for the observed data. Reprinted with permission from ref. 14. Copyright by the American Chemical Society.

particular method, is a function of the frequency of the test method; this frequency dependence has been treated theoretically by Williams, Landel, and Ferry.[22] The WLF equation predicts a dramatic dependence of the apparent activation enthalpy, ΔH_a, on frequency, especially at low temperatures. At temperatures significantly above T_g, ΔH_a tends to be constant. The dependence of ΔH_a on T_g has been discussed by Boyer[23] and many examples of relaxation maps have been presented by McCrum, Read, and Williams.[24]

From the above considerations, it is obvious that the transition temperature observed in ESR spin-probe experiments should, in fact, be a function of frequency, apparent activation enthalpy, and glass temperature. Assuming simple Arrhenius behavior over a limited frequency range, as described by equation (1):

$$f_{max} = A \exp(-\Delta H_a/RT) \tag{1}$$

we have derived a relationship (equation 2), that describes the observed transition temperature, T_{50G}, in terms of both

$$T_{50G} = T_g/(1-0.03T_g/\Delta H_a) \tag{2}$$

the glass temperature and the apparent activation enthalpy for the glass transition. This theoretically derived equation predicts quite well the observed T_{50G}–T_g correlation. We have also shown that this equation will allow the estimation of apparent activation enthalpies, once T_{50G} and T_g values are known.

Using similar considerations, Törmälä and co-workers[25] later proposed that ΔT, defined as $T_{50G}-T_g$, can be related to a reference correlation temperature (T_c), by the following expression:

$$\Delta T = T_g/\exp(T_g/T_c) \tag{3}$$

and that T_c is equal to 173K. Incorporation of this equation into the Arrhenius treatment leads to equations (4) and (5), for T_{50G} and ΔH_a, respectively.

$$T_{50G} = T_g[1+(1/\exp(T_g/T_c)] \tag{4}$$

$$\Delta H_a = R \ln\left(\frac{f_{T50G}}{f_{T_g}}\right) T_g\ [1 + \exp(T_g/T_c)] \tag{5}$$

Either of these approaches allows the correlation of ESR derived data with that gathered by dynamic mechanical methods and confirms the validity of the ESR spin-probe method as a reliable method for the study of molecular relaxations.

An important limitation of this technique, when applied to the glass transition, is that it assumes that the size of the nitroxide probe is larger than the polymer segment experiencing increased motion at T_g. This limitation presents some interesting experimental and theoretical implications of the relative size of the probe compared to the size of the polymer segment and data directed toward these implications have been derived in a number of studies.

EFFECT OF MOLECULAR WEIGHT AND MOLECULAR WEIGHT DISTRIBUTION

It was not intuitively obvious to us whether the glass transition temperature determined by ESR spin-probe studies and the T_{50G}–T_g correlation, would be responsive to variations in molecular weight of the host polymer. Many polymers exhibit a very dramatic dependency of T_g on molecular weight and the versatility of the ESR method would be much greater if a similar dependency was observed in the T_{50G} values. As a model polymer we selected polystyrene to examine this question because of: (a) the availability of well-characterized narrow molecular weight distribution samples covering a wide range of molecular weights, (b) considerable variation in the T_g of polystyrene with varying molecular weight,[26] (c) the availability of considerable data on the effects of both molecular weight and molecular weight distribution on the T_g of polystyrene,[26] and (d) the fact that many theoretical considerations of the glass transition phenomenon have been developed from studies of polystyrene.[26]

A series of styrene oligomers and polymers—varying in molecular weight from 104 (styrene monomer) to 171,000—was investigated by the ESR spin-probe method using BzONO as the paramagnetic probe.[10,16] The T_g of each of the relatively monodisperse samples was determined from the observed T_{50G} value using the correlation curve described above. The T_g values determined in this way did vary smoothly as a function of molecular weight and the observed T_g values were in excellent agreement with similar data determined by more conventional techniques such as dilatometry,[27,28] DTA,[29] and DSC.[30] A summary of our data is shown schematically in Figure 5, which includes similar data derived by other experimental techniques. It is obvious that our data is in excellent agreement with that derived by other methods and that the T_g determined by ESR is sensitive to molecular weight.

It was not, however, possible from this data to determine whether the ESR technique was responding to variations in $\bar{M}_n$ or $\bar{M}_w$, because all of the samples examined had a very low polydispersity index ($\bar{M}_w/\bar{M}_n$ less than 1.1). Essentially identical curves would result from plotting either $\bar{M}_n$ or $\bar{M}_w$ versus observed T_g values. It is well-known that the T_g of polystyrene varies as a function of $\bar{M}_n$ rather than $\bar{M}_w$[26] and it was of obvious interest to see if the T_g values determined by ESR spin-probe studies exhibited similar dependency.

We resolved this question by examining the T_g behavior (by the spin-probe method) of two different series of binary blends of polystyrene and a thermally prepared polydisperse polystyrene sample. Figure 6 presents a summary of our data. The T_g values determined by ESR (from the T_{50G}–T_g correlation) are plotted against both $\bar{M}_w$ and $\bar{M}_n$. It is obvious that the ESR determined T_g is a function of the number-average molecular weight rather than the weight-

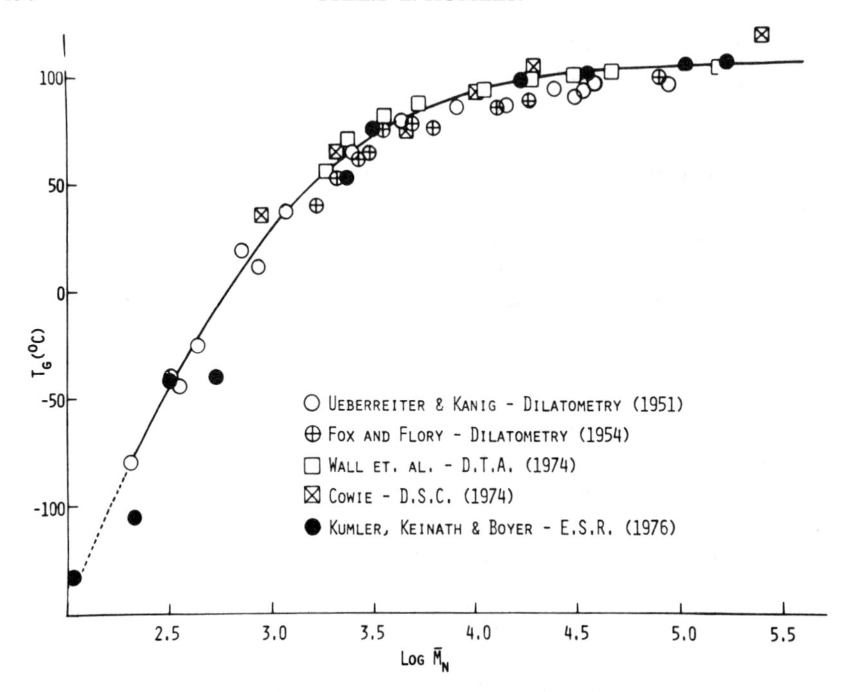

FIGURE 5 Glass temperature of polystyrene as a function of $\bar{M}_n$ as determined by various methods. See text for appropriate references to both ESR spin-probe data and data from other methods indicated. Reproduced with permission from ref. 10.

average molecular weight. It is also heartening that the individual T_g values for the blends are in excellent agreement with the T_g for a comparable molecular weight monodisperse sample (solid line in Figure 6). We have further confirmed the validity of the above results by independently determining the T_g of the blends by DSC. As shown in Figure 7, there is again excellent agreement between the T_g values determined by ESR and those determined by DSC. These results are especially gratifying because of the difficulty of accurately determining T_g values on samples possessing broad molecular weight distributions by some of the more classical techniques.

While the $\bar{M}_n$ of low molecular weight polymers can be determined by a variety of methods such as vapor phase osmometry, boiling point elevation, freezing point depression, gel-permeation chromatography, etc., there are frequently problems with each of these methods. Our results suggest that the ESR technique may provide a new tool for $\bar{M}_n$ determination, even though it is a non-absolute method since it requires calibration.

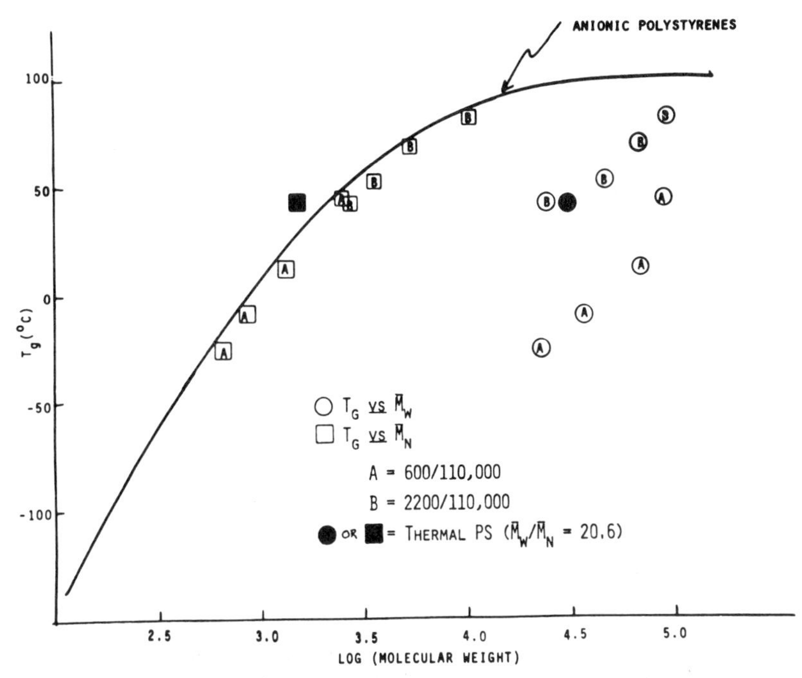

FIGURE 6 Glass temperature of polystyrene binary blends as a function of molecular weight. The smooth curve represents data for "monodisperse" polystyrene samples. Circles represent T_g plotted *vs* $\bar{M}_w$; squares represent T_g plotted *vs* $\bar{M}_n$. A and B represent data from two different series of blends using the monodisperse samples with the molecular weights indicated on the figure. The solid circle and square represent data points for a thermally prepared polydisperse polystyrene ($\bar{M}_w/\bar{M}_n = 20.6$). Reproduced with permission from ref. 10.

An important corollary arises from our molecular weight studies. The value of ΔH_a must decrease with molecular weight as T_g decreases, since we use Figure 4, which is represented by equation (2) to calculate T_g. This suggests, then, that ΔH_a depends only on T_g whether T_g is affected by chemical structure or by molecular weight.

FREE VOLUME EFFECTS: OLIGOMERIC VS. POLYMERIC GLASSES

During our studies of the molecular weight and molecular weight distribution dependence of T_g as determined by ESR, we detected some results that may offer some valuable insight into the nature of the glassy state in low molecular

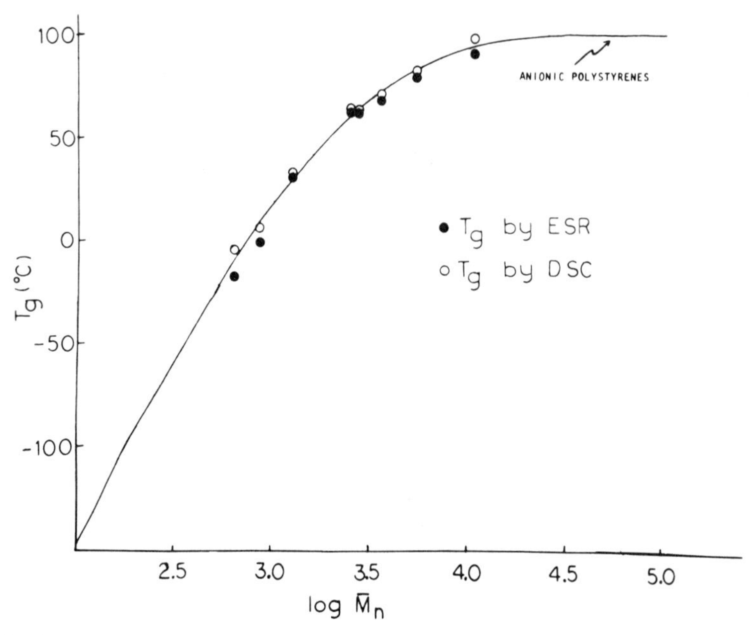

FIGURE 7 Comparison of T_g data for polystyrene blends as determined by ESR spin-probe studies (solid circles) and by DSC (closed circles). The smooth curve represents data for "monodisperse" polystyrenes, as shown in Fig. 6. Reprinted from ref. 16 by courtesy of Marcel Dekker, Inc.

weight, oligomeric, and polymeric glasses.[10,12,16] A careful examination of the extrema separation *vs* temperature plots for anionic polystyrenes of varying molecular weight revealed two correlated trends. As the molecular weight of the sample decreased the sigmoidal-shaped curve became significantly "sharper" and the upper limit observed for the extrema separation increased. Both phenomena are summarized in Figure 8 which shows extrema separation versus temperature plots for five low molecular weight members of the series of anionic polystyrenes.

We have examined this phenomenon in more detail and Figure 9 indicates the conventions that we have used in our study. G_{max} is the maximum value observed for the extrema separation and ΔT is the temperature interval between extrema separations of 60 gauss and 40 gauss. Using these conventions, the polystyrene data shown in Figure 8 can be summarized by saying that as the molecular weight increased, ΔT increased while G_{max} decreased. The original study has been expanded to include the entire molecular weight range of polystyrene samples studied during the molecular weight dependency studies. Our results of this extended study are summarized

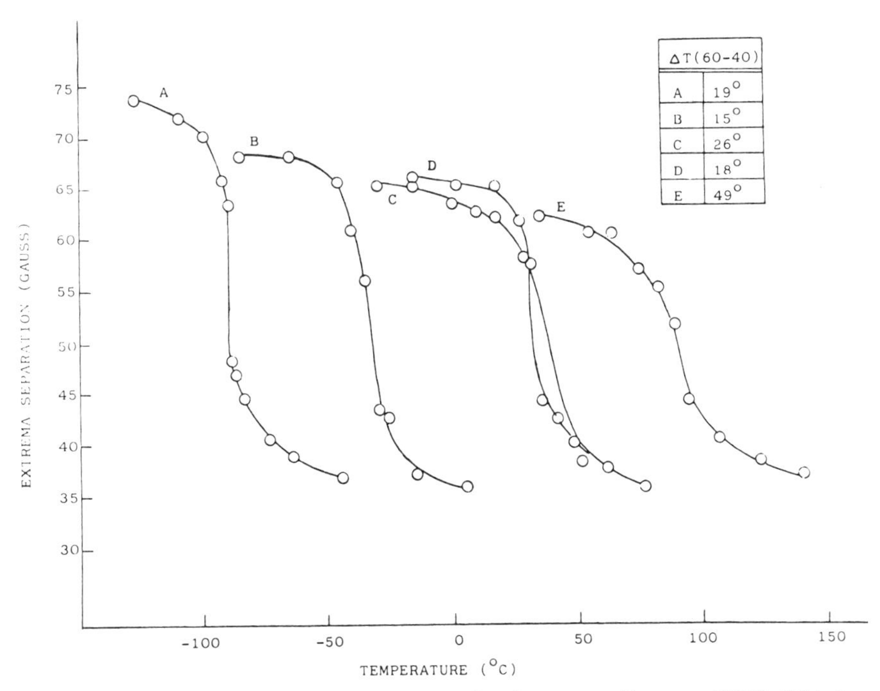

FIGURE 8 Extrema separation *vs* temperature plots for styrene oligomers; $\Delta T(60\text{–}40)$ is the temperature range over which the extrema separation narrows from 60 to 40 gauss; (a) monomer, (b) dimer, (c) trimer, (d) PS-600, (e) PS-2100. Reprinted from ref. 16 by courtesy of Marcel Dekker, Inc.

in Figure 10, and can be satisfactorily ascribed to the effects of free volume and its distribution. The smaller ΔT at low molecular weight indicates that vitrification occurs over a narrower temperature interval, while the larger G_{max} at low molecular weight indicates a more rigid glass.

We feel that both of these phenomena are due, at least in part, to the increase in free volume at T_g, as the molecular weight increases. In addition to the increasing free volume with molecular weight it is expected that the distribution of free volumes will become broader as chain length increases.[31,32] These phenomena could well account for the increase in ΔT and the decrease in G_{max} as the molecular weight of the glass increases. In addition to suggesting the utility of this technique for studying the glassy state and the vitrification process, we have indicated some precautions concerning the interpretation and application of this technique.[12]

In addition to examining oligomeric and polymeric glasses we have applied the ESR spin-probe technique to the study of a small selection of conventional low-molecular weight glass-forming liquids.[12] The observed results are quite

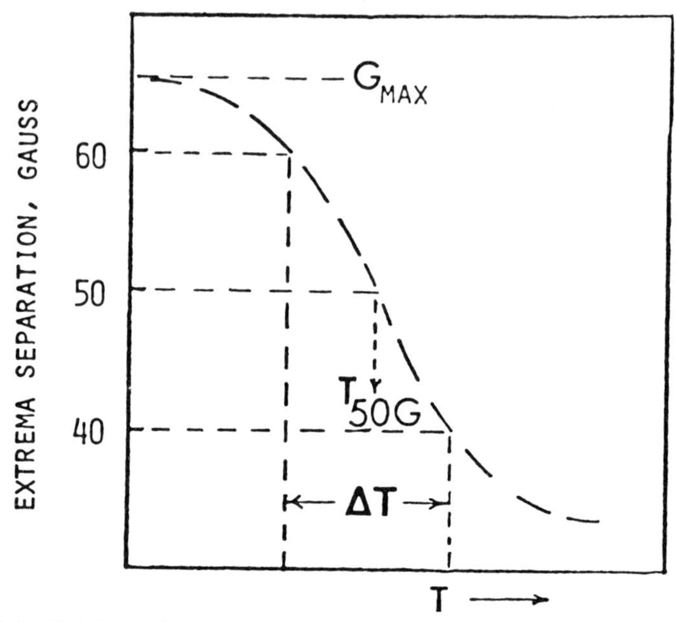

FIGURE 9 Definition of terms and conventions used to discuss key features of extrema separation *vs.* temperature plots for oligomeric and polymeric glasses. Reproduced with permission from ref. 12.

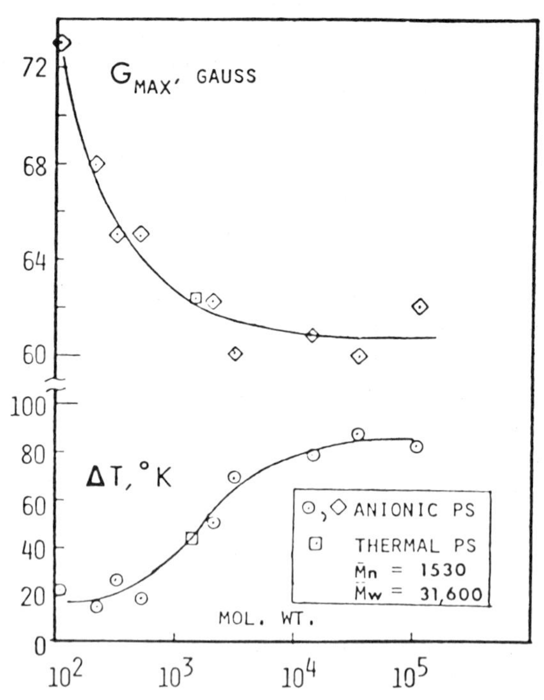

FIGURE 10 Variation of G_{max} and ΔT (defined in Fig. 9) with the logarithm of molecular weight for the indicated polystyrene samples. Reproduced with permission from ref. 12.

similar to that of the styrene oligomers, i.e., large G_{max} and small ΔT. It seems obvious that the spin-probe technique offers a valuable method for the study of the glassy state and the vitrification process.

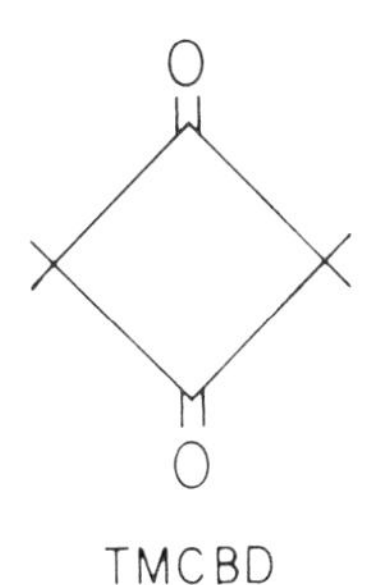

TMCBD

The G_{max} value usually expected for a highly immobilized nitroxide is 64 gauss, based upon the classical experiment of Griffith, Cornell, and McConnell[33] in which the di-t-butyl nitroxide radical was introduced as a substitutional impurity into crystalline 2,2,4,4-tetramethylcyclobutane-1,3-dione (TMCBD). The fact that we find G_{max} values greater than 64 gauss implies that oligomeric glasses may be "more rigid" than the host crystal used by McConnell and co-workers. Because of the symmetry present in TMCBD one might expect the possibility of solid state rotation well below the crystalline melting point. We have not located any literature data to verify such a rotational transition. Our results, combined with those of McConnell indicate that di-t-butyl nitroxide in TMCBD may experience an environment allowing considerable mobility. We have found other cases where nitroxides exhibit G_{max} values greater than 64 gauss.[5,34,35]

BLOCK COPOLYMERS

In view of our success in applying the ESR spin-probe technique to studies of the glass transition in homopolymers and random copolymers, it was of obvious interest to apply this technique to a study of multiphasic block copolymers. We have examined a wide series of diblock, triblock, and radial block copolymers.[9,11,17] The block copolymers examined (with one exception) contain polystyrene as the "hard" component with a variety of polymers—polybutadiene, polyisoprene, hydrogenated polybutadiene, hydrogenated polyisoprene, and poly(dimethylsiloxane)—as the "soft" component.

A schematic representation of the type of behavior to be expected from biphasic block copolymers in an ESR spin-probe experiment is shown in Figure 11. Such a plot should exhibit three distinct regions: (a) Below T_1 the

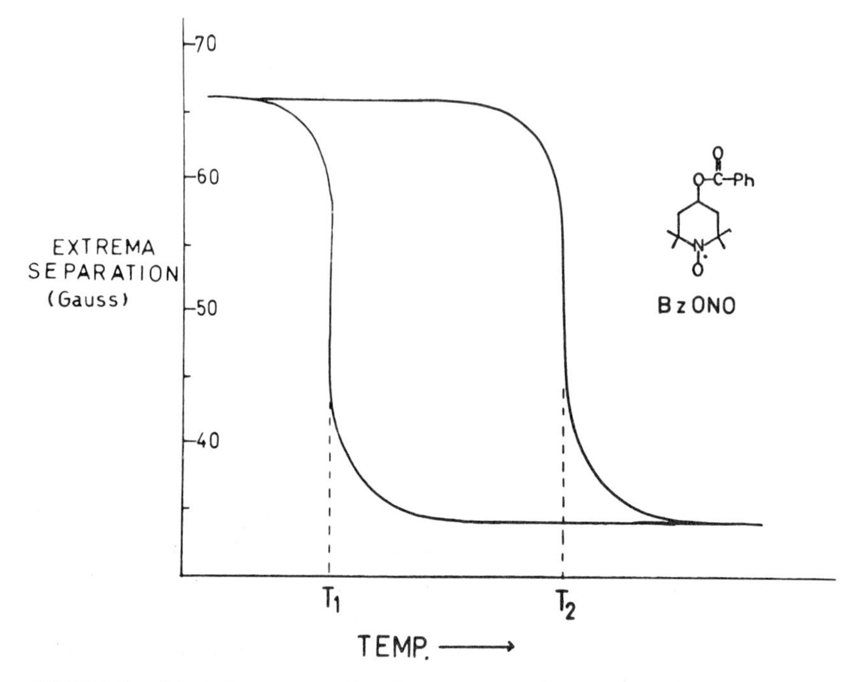

FIGURE 11 Schematic representation of extrema separation *vs* temperature plot for a biphasic block copolymer. Below T_1, the nitroxide is highly immobilized; above T_2, the nitroxide is rapidly tumbling in both phases. Between T_1 and T_2, the observed spectrum would be a superposition of a "rigid limit" spectrum and that of a rapidly tumbling nitroxide. Reproduced with permission from ref. 17.

nitroxide is highly immobilized in both phases and a single three-line spectrum with a separation of 60–65 gauss should be observed; (b) above T_2 the nitroxide is rapidly tumbling in both phases and a single three-line spectrum characteristic of a rapidly tumbling nitroxide (separation of 30–35 gauss) should be observed; (c) at temperatures between T_1 and T_2 the observed spectrum should be a superposition of two three-line spectra, one characteristic of a highly immobilized nitroxide (60–65 G separation) and the other typical of a nitroxide experiencing increased tumbling frequency (separation less than 60–65 G). Results such as these have, in fact, been observed for most of the block copolymers that we have examined. Typical examples of experimental spectra observed in the three temperature regions described are shown in Figures 12 and 13 for a styrene–siloxane diblock and a styrene–butadiene–styrene triblock, respectively. Extrema separation versus temperature plots for the same two polymers are shown in Figures 14 and 15.

In virtually all of the cases we have examined, the T_g determined for the

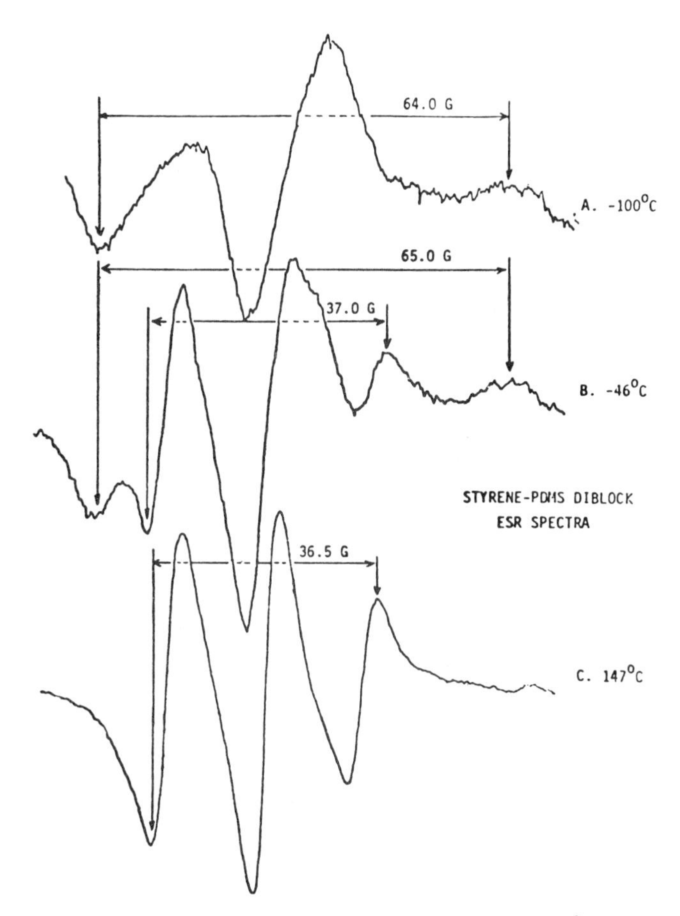

FIGURE 12 Representative experimental spectra at three temperatures for a styrene–siloxane diblock copolymer. Extrema separation are indicated on the figure. Reproduced with permission from ref. 9.

"soft" phase (from the T_{50G}–T_g correlation) is in very good agreement with the T_g expected for a comparable molecular weight homopolymer. Significant departures from the more or less ideal results seen in Figures 12–15 were found when the percent polystyrene in the block and/or the molecular weight of the styrene block were low.

T_g's determined for the "hard" phase by this technique almost universally deviated from what would be expected for the comparable molecular weight homopolymer. The T_g's determined by ESR were consistently lower than expected. This rather anomalous behavior led to a careful and extensive

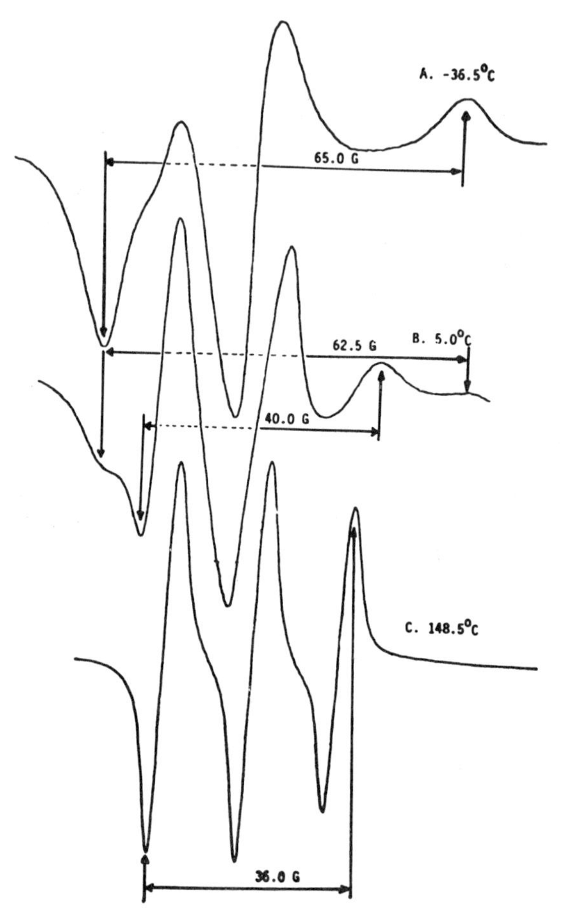

FIGURE 13 Representative experimental spectra at three temperatures for a styrene–butadiene–styrene triblock copolymer. Extrema separation are indicated on the figure. Reproduced with permission from ref. 9.

consideration of factors that might be responsible for the anomaly. We presented and discussed, both in a qualitative and quantitative fashion, the expected contribution of a number of factors including molecular weight of the hard phase, differences in solubility parameter for the two phases, percent of hard phase present, and the effect of crystallinity.

All of these considerations, including a comparison of T_g's determined by ESR with T_g's from dynamic mechanical methods for similar block copolymers, led us to conclude that in the cases we examined the ESR technique was responding to events in the interphase rather than events in the

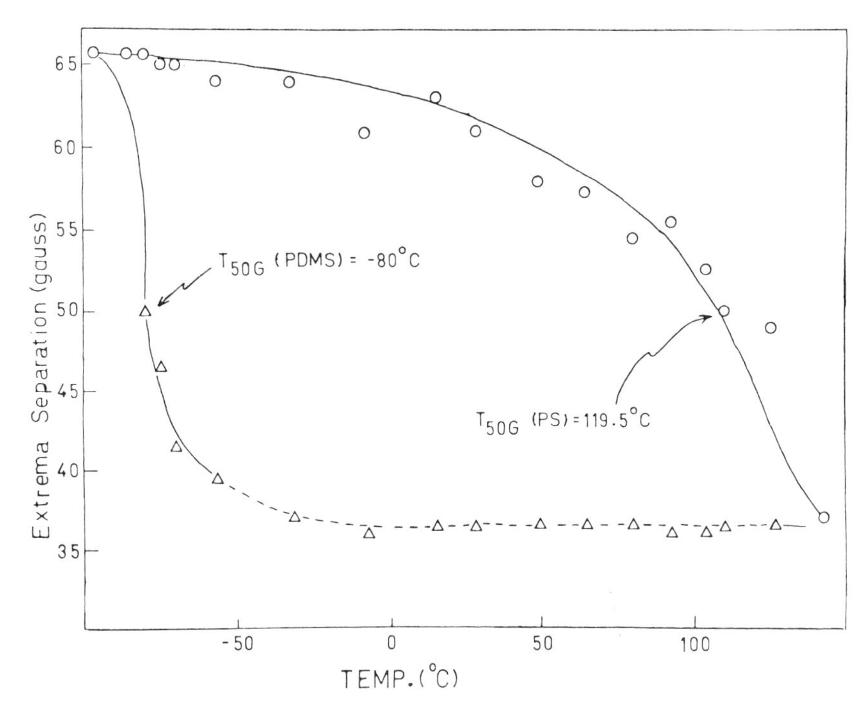

FIGURE 14 Extrema separation *vs* temperature plot for the styrene–siloxane diblock indicated in Fig. 12. The experimentally determined T_{50G} values for the two phases are indicated. Reproduced with permission from ref. 9.

pure hard phase. The only exceptions to this generalization were observed in a styrene-siloxane diblock and a styrene-isoprene radial block. In the case of the styrene–siloxane diblock a combination of factors (reasonably high molecular weights for each block, high weight percent styrene, and a large difference in solubility parameters for the two phases) resulted in a block copolymer possessing very little, or no, interfacial material and the two T_g's determined by the ESR technique were totally consistent with comparable molecular weight homopolymers. The styrene–isoprene radial block was exceptional in that it was the only one of the 26 block copolymers examined that did not exhibit an extrema separation plot similar to that shown in Figure 11. We were unable to detect any evidence for nitroxide molecules existing in more than one discrete phase over the entire temperature range studied and the extrema separation plot was typical of that observed for homopolymers and random copolymers. The single T_g determined was essentially identical with that of a polyisoprene homopolymer of comparable microstructure. We believe that in this case

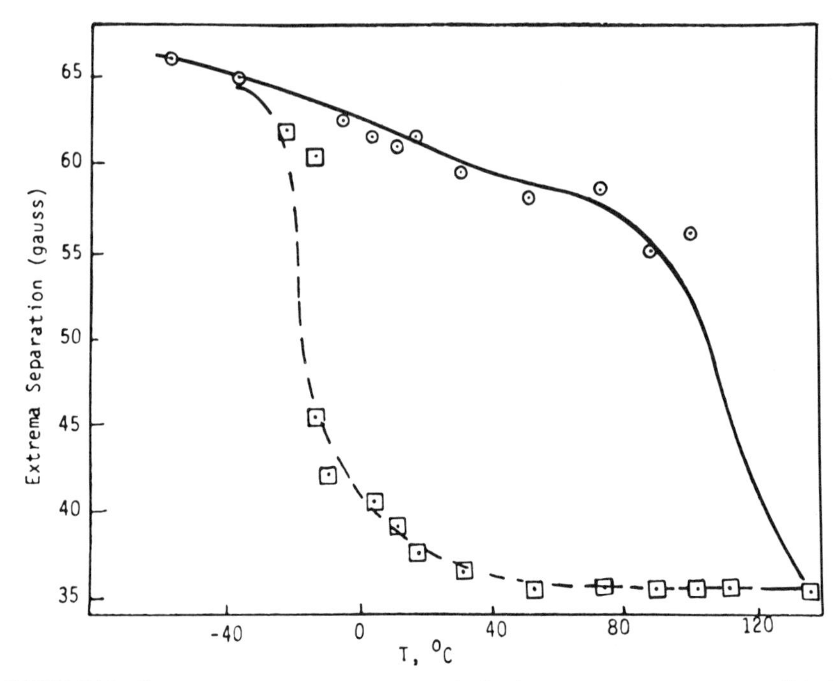

FIGURE 15 Extrema separation *vs* temperature plot for the styrene-butadiene-styrene triblock copolymer indicated in Fig. 13. Reproduced with permission from ref. 9.

(because of a low percent styrene content and low molecular weight of the polystyrene) the hard phase does not exist as a separate entity and the single T_g determined represents that of a pure polyisoprene phase or a mechanical blend of PS and PI (with a low percent styrene content) that would be experimentally indistinguishable from a pure polyisoprene phase. All of the other cases examined were intermediate between these two extremes, and we feel the results are consistent with observations of probe in pure soft phase and in interfacial material. Such an explanation is totally consistent with both dynamic mechanical data for block copolymers and theoretical treatments of the nature and extent of interfacial material expected to be present. We have suggested therefore, that the ESR technique, coupled with independent measures of the amount of interfacial materials, offers a unique tool for the study of interfacial material in multiphasic systems.

MULTIPLE ENVIRONMENTS OBSERVED BY A SINGLE PROBE

It would not be surprising for a single probe to partition itself between more than one environment in a polymer matrix possessing more than one distinct phase or environment. Such a distribution of the probe could, in theory, be detected in an ESR spin-probe experiment under suitable experimental conditions. The results reported above for multiphasic block copolymers represent the epitome of this type of behavior. We have occasionally observed similar behavior in both our reported and as yet unreported experiments. A particularly striking example of this type of behavior in a homopolymer was provided by a study of polypropylene.

We have examined, by the spin-probe method using BzONO, both atactic and isotactic polypropylene.[12] Atactic polypropylene exhibited ESR spectra and extrema separation *vs* temperature plots quite similar to other amorphous homopolymers; see Figure 16. Isotactic polypropylene, however, exhibited both spectra and extrema separation *vs* temperature plots consistent with the presence of probe molecules in two distinct environments. The motion of the probe in the second environment could be observed and "followed" up to a temperature of approximately 200°C, at which point the probe began to decompose. The extrema separation plot for this sample is shown in Figure 17. Although the exact nature of this second environment has not been elucidated at this time, a number of suggestions for the nature of this environment have

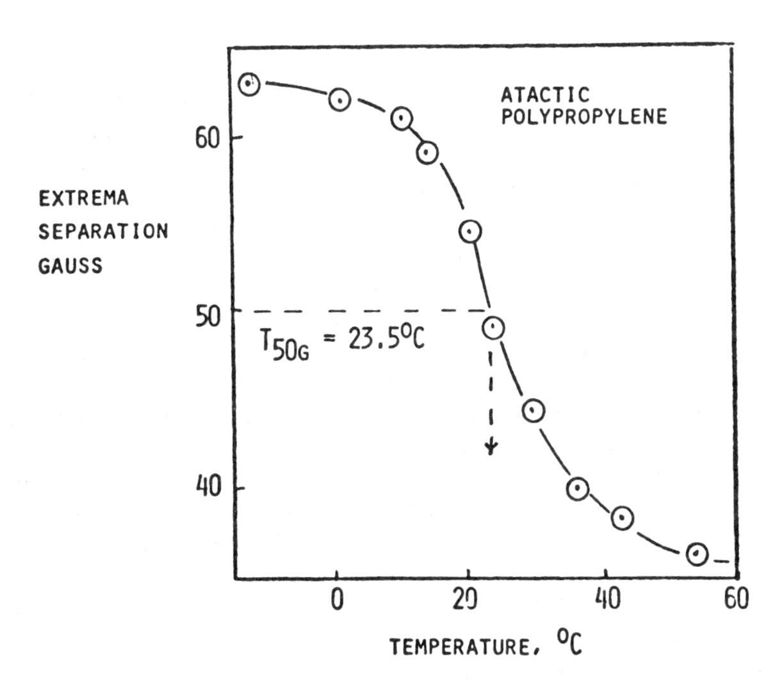

FIGURE 16 Extrema separation *vs* temperature plot for atactic polypropylene. The experimentally determined T_{50G} value is indicated. Reproduced with permission from ref. 12.

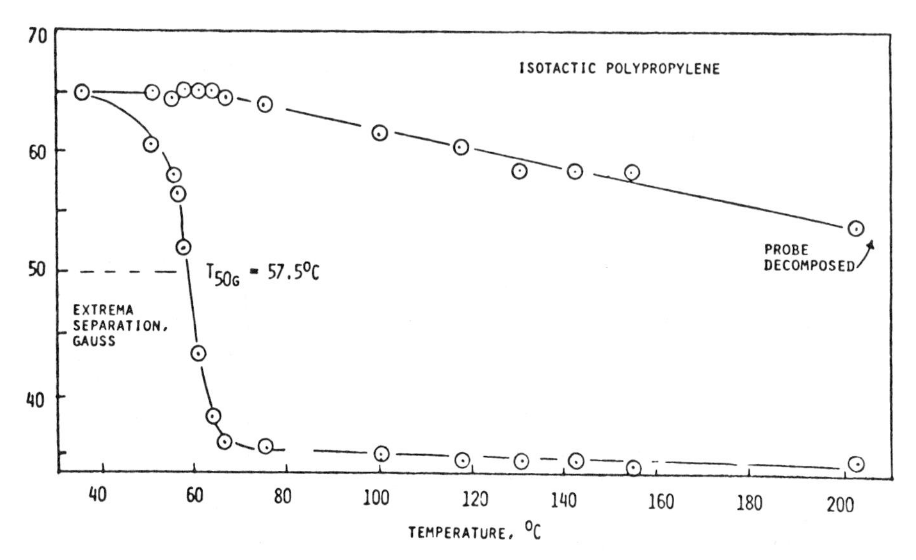

FIGURE 17 Extrema separation *vs* temperature plot for isotactic polypropylene. The T_{50G} value that could be determined from the overlapping spectra is indicated. Reproduced with permission from ref. 12.

been made.[12] We are convinced, however, that it is not an artifact for four distinct reasons: (a) It can be detected in isotactic PP but not in atactic PP. (b) Similar "anomalous" results seen in our laboratories have typically been associated with semi-crystalline polymers. (c) A series of probe size experiments, using probes A through F shown in Figure 1, showed evidence for dual environment, in all cases, for isotactic polypropylene. (d) Neither isotactic polybutene-1 nor isotactic poly-4-methylpentene-1 showed any indication of dual environments; both of these polymers are chemically very closely related to polypropylene but are expected to be totally amorphous. We currently believe that all such "dual environment" results are due to the presence of two distinct amorphous environments, one exhibiting a considerably higher transition temperature than the other.

Commercial isotactic polypropylene is believed to be a block copolymer of long isotactic and short atactic segments. The latter are amorphous whereas the former will contain amorphous material in the form of loose loops under restraint by crystallites to which they are connected. Hence, there are two environments. Moreover, the loose loops are not destroyed until the polymer melts.[36] This explanation for the dual environment observed in the ESR experiments with isotatic polypropylene is due to Boyer.[37]

PROBE SIZE STUDIES

The ESR spin-probe method depends upon observing the temperature-dependent ESR spectra of a paramagnetic "guest" in a diamagnetic "host". The observed spectra are expected to be dependent upon, and to reflect, the magnitude and type of motion experienced by the polymeric matrix upon traversing a particular relaxation. A variety of studies from both our laboratories[8–10,12–15] and others[38] has suggested that the ESR technique is sensitive to the size, shape, and functionality of the particular probe utilized.

In an attempt to gain more insight into the relation between probe size and observed ESR spin-probe results we have examined over the last few years a series of polymers using probes of varying size and functionality (see Figure 1). The polymers examined in our laboratories were: poly(2,6-dimethyl-1,4-phenylene oxide), PDMPO; polystyrene, PS; poly(vinylidene fluoride), PVF_2; isotactic polypropylene, i-PP; polyisobutylene, PIB; polydimethylsiloxane, PDMS; and dian polycarbonate, poly(2,2-bis(4′-hydroxyphenyl)propane carbonate), PC.

Consideration of our probe-size studies and those of others[38] have led us to conclude that in any particular spin-probe experiment, the paramagnetic nitroxide may be "seeing" the glass transition, may be "seeing" a sub-glass transition, or may simply be responding to increased free volume in the glassy state due to thermal expansion. In general, the particular relaxation observed or detected in a spin-probe experiment is dependent upon the relative size of the probe molecule compared to the size of the polymer segment experiencing increased motion upon traversing a particular relaxation.

If the volume of the paramagnetic probe is "large" relative to the activation volume at T_g of the polymer, then the probe will "see" the glass transition. If the volume of the probe is comparable to, or smaller than, the volume segment of the polymer experiencing increased motion at the glass temperature, the ESR experiment will be responsive to relaxations with activation volumes smaller than that of the glass transition (typically sub-glass transitions); alternatively, in such a case, the probe may simply experience increased tumbling frequency as a result of thermal expansion in the glassy state if there are no prominent secondary relaxations.

In addition to our work, a particularly striking example is provided by some early work of Gross.[39] She examined the response of several probes of varying size in dian polycarbonate. We suggested[13] that the two largest probes were indeed responding to the glass transition, but that the smallest probe that she utilized was responding to the prominent sub-glass relaxation that has been shown[40] to involve motion of the carbonate group coupled with some motion of the adjoining benzene rings.

We believe that an answer to the question of which particular transition (glass transition, sub-glass transition, etc.) a particular probe is "seeing", or in fact if it is simply responding to thermal expansion of the matrix can only be resolved by examining relaxation maps when they are available. Figure 18 shows a schematic relaxation map for a polymer possessing two secondary relaxations (β and γ) on which we have inserted values of $1000/T_{50G}$ for five probes (P_1–P_5) of varying size. We suggest that in such a case the two largest probes (P_1 and P_2) will respond to T_g, P_4 will respond to the β relaxation, and probes P_3 and P_5 will be responding to thermal expansion. A specific example of the application of these ideas is provided by work on poly(2,6-dimethylphenylene oxide), PDMPO, from both our laboratories and those of Törmälä,[41] and is summarized in Figure 19.

The relaxation map shows the three main transitions in PDMPO and was constructed using literature data for low frequency dynamic mechanical spectra and activation energies. Both our probe size studies (using probes A–F of Figure 1) and the single data point from Törmälä's work are shown. Törmälä's experiment was probably observing the β process, our probe F is

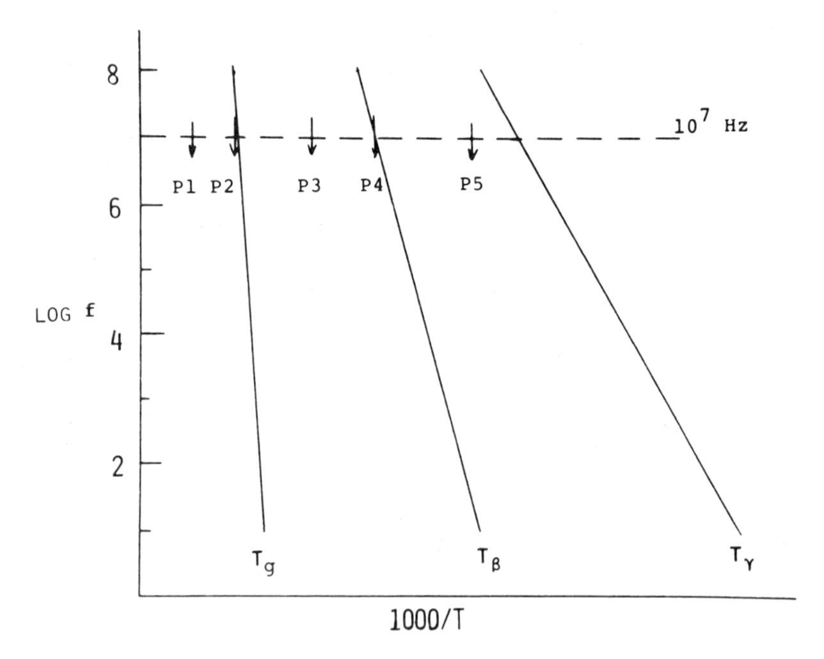

FIGURE 18 A schematic relaxation map for a polymer possessing two secondary relaxations in addition to the glass relaxation. The arrows indicate hypothetical values of $1000/T_{50G}$ for five probes (P_1 through P_5); P_1 is the largest probe and P_5 is the smallest. Reproduced with permission from ref. 13.

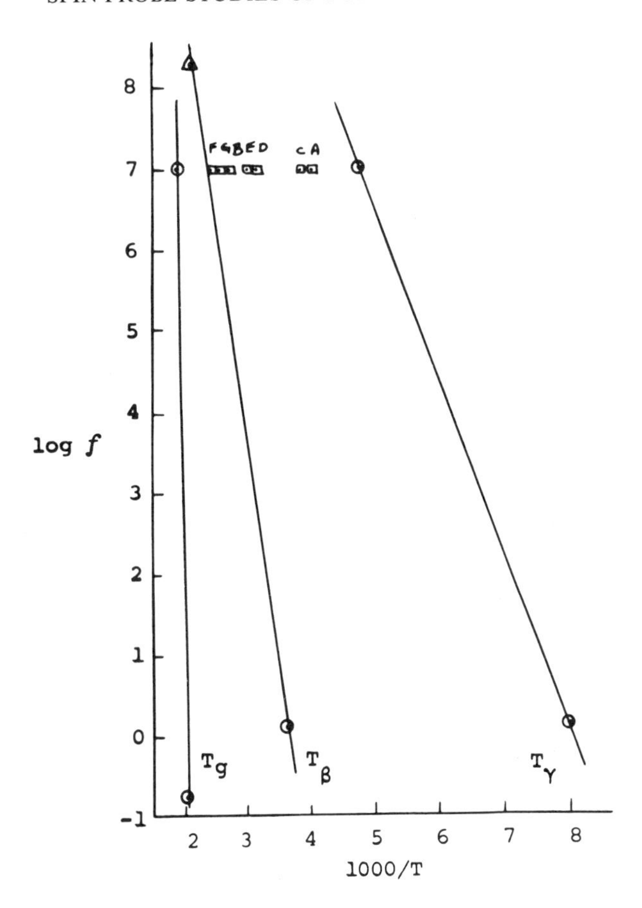

FIGURE 19 Relaxation map for PDMPO (see text). Data points from spin-probe studies of both Törmälä and Kumler-Boyer are indicated. Reproduced with permission from ref. 13.

probably "seeing" the same β relaxation, and our other size probes are probably responding to thermal expansion. The proximity of the T_g and T_β lines at high frequencies should also be noted, and is likely to be a general situation. At frequencies of the ESR method a probe might not be able to resolve the T_g from the T_β, and may in fact respond to a merged T_g–T_β transition. From considerations such as these, we have in fact suggested that the Gross study of dian polycarbonate[39] may be just such a case; we suspect that with her largest probes she was observing a merged T_g–T_β relaxation.

It was, in retrospect, rather fortuitious that we chose a relatively large probe (F in Figure 1) for attempting to correlate T_{50G} with T_g.[8,14] This probe is apparently large enough to be responding to T_g rather than sub-glass transitions, for all homopolymers and copolymers used to establish the T_{50G}–T_g correlation. It should also be noted, however, that this correlation curve will only be valid for estimation of T_g provided that two conditions are satisfied: (a) the correlation is valid only for BzONO; (b) reliable values for T_g will only result when the size of this probe is "large" relative to the volume of the appropriate polymer segment, i.e., the probe is truly responding to T_g rather than a relaxation possessing a smaller activation volume.

These same limitations, however, suggest that it may be possible to selectively study a particular molecular relaxation (beta, gamma, delta, etc.) by a judicious choice of the size of the paramagnetic probe.

ACTIVATION VOLUMES

Our experimental results, and those of others, pointed out very dramatically the need for a reliable independent estimate of the molecular volumes involved at both glass and sub-glass transitions. Kusumoto[42] had previously suggested an equation relating T_{50G} to T_g which included a fitting parameter "f", defined as the ratio of probe volume (V_p) to the volume of the polymer segment experiencing motion (V_s). The parameter "f" was, however, empirically determined from T_{50G} *vs* T_g data for a series of polymers. We preferred to develop an estimate of the activation volume from data totally independent of the ESR experiment.

Utilizing an approach initially developed by Eby,[43] we have calculated activation volumes ($\Delta V^{\ddagger}$) at T_g for ten amorphous polymers from isothermal compressibility data and apparent activation enthalpies. A direct linear relationship is observed between the glass temperature T_g and the log of the activation volume at T_g. Activation volumes have similarly been calculated for sub-glass transitions in a series of nine polymers representing both amorphous and crystalline polymers. Activation volumes at these sub-glass transitions, while consistently smaller than activation volumes at T_g, do not vary in any systematic manner with either T_g or the transition temperature associated with the sub-glass relaxation. A summary of the activation volume data, at T_g, is shown in Figure 20. The semi-logarithmic nature of the activation volume—T_g relationship is quite obvious. It should also be noted that the right-hand ordinate of this plot indicates the diameter of a spherical object possessing the volume shown on the left-hand ordinate. The availability of data such as this will allow a more quantitative examination of the effect of relative probe size for studying a particular transition in a particular polymer. An examination of the relatively few cases for which the necessary data (activation volumes and reasonable estimates of the probe size) are available confirms the validity of

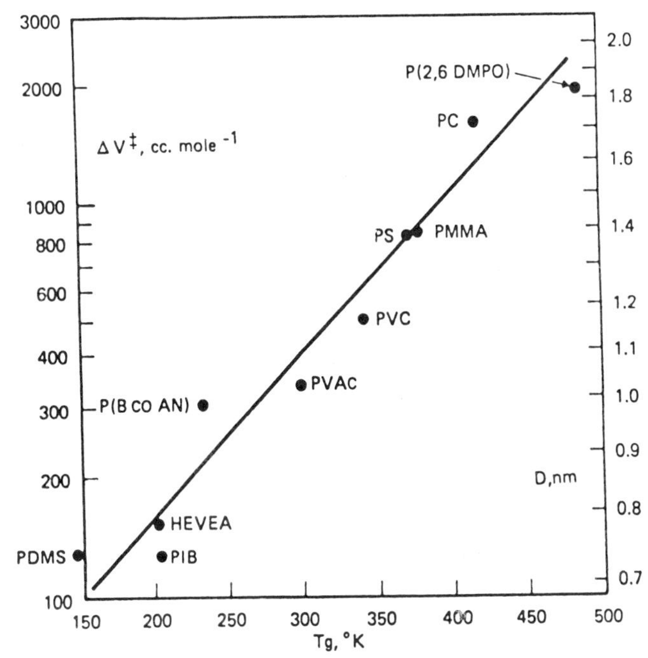

FIGURE 20 Activation volumes at T_g as a function of T_g. The right hand ordinate indicates diameter (in nm) of a spherical molecule having the volume shown opposite on the left hand ordinate. The left hand ordinate is activation volume (in $cm^3\ mol^{-1}$) calculated as indicated in the text. Reprinted with permission from ref. 15. Copyright by the American Chemical Society.

this approach. The tumbling behavior of nitroxide probes of different sizes, as observed by ESR spectrometry in a limited number of polymers, is qualitatively and even semiquantitatively related to activation volumes.

CRYSTALLINE POLYMERS

An extremely controversial area is the determination of glass transition temperatures in partly or highly crystalline polymers. Such controversy is perhaps epitomized by the case of polyethylene. The glass temperature of amorphous polyethylene has been assigned (and staunchly and vociferously defended) a variety of temperatures covering a range of approximately 100°. A wide variety of techniques have been applied to this polymer, by a variety of investigators, in attempts to unambiguously assign the T_g value.[44] It is perhaps not surprising that we have utilized ESR spin-probe techniques to investigate this highly controversial area. Expanding upon earlier work of Boyer[7] we have carried out four separate series of experiments.

We originally examined, by the spin-probe technique using BzONO, a series of four homopolymers of ethylene.[14] Two NBS standard samples (one a linear polyethylene and the other a branched polyethylene) gave T_g values (determined from the T_{50G}–T_g correlation curve) of −73 and −72°C, respectively. Two high molecular weight commercial samples exhibited, however, T_g's of −53 and −40°C. Although the precise reason for the variation observed in the four samples could not be established, a number of possible reasons for the discrepancies (variation in method of synthesis, variation in degrees of crystallinity, morphological variations, and differing thermal histories) were suggested. Following the lead of other investigators using other techniques, we have applied the ESR method to two different series of random copolymers of ethylene.

Three ethylene-vinyl acetate copolymers and four ethylene–propylene copolymers have been studied in our laboratories.[14] The ethylene content of these samples varied from 55 weight percent to 88 mole percent. The results of these two copolymer series supplement very nicely earlier data, derived by more classical techniques, gathered by Illers[45] and by Kontos and Slichter.[46] Our ESR data for both series agree very well with the curves for either copolymer series extrapolated to 100% ethylene content by Illers using the Gordon-Taylor copolymer equation. A graphical superposition of our ESR data onto the earlier data of both Illers and Kontos-Slichter is shown in Figure 21. We believe that these results demonstrate conclusively the suitability of the ESR technique for studying ethylene copolymers containing a high percent ethylene content and support the assignment of the transition occuring at approximately −80°C to the glass temperature of amorphous polyethylene.[47] Most recently we have examined a well characterized series of ethylene–hexene copolymers by the spin-probe method[48] and these results are again consistent with the glass transition of amorphous polyethylene occurring at −80°C.

Cowie and McEwen[49] have recently settled the controversy concerning the value of T_g for polyethylene by investigating T_m, T_g, and $T<T_g$ for a series of partially hydrogenated cis-polybutadienes with T_g varying from 160 K for cis-PBD to an extrapolated value of 198 K for amorphous polyethylene. This confirms and strengthens our work on the use of spin-probes in highly crystalline polymers such as polyethylene.

MULTIPLE PROBE EXPERIMENTS

Based upon all of our experimental work reported above, we recently proposed[13] a new approach for the study of polymer transitions by ESR. The

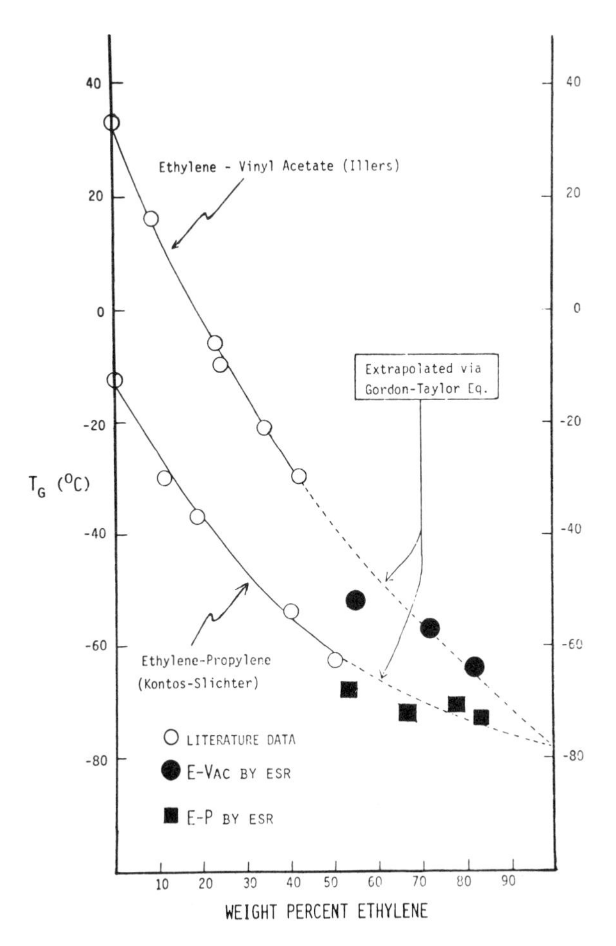

FIGURE 21 T_g data for ethylene copolymers by ESR; comparison with literature data. (See text). Reprinted with permission from ref. 14. Copyright by the American Chemical Society.

fundamental idea was to incorporate more than one probe in a single polymer and to examine the ESR spectra over a wide temperature range. In the ideal case, different probes would respond to different transitions.

This proposal was a logical outgrowth of several independent events: (a) the probe size studies by Gross on dian polycarbonate, (b) our familiarity with relaxation maps, (c) our study of activation volumes, (d) our single probe work in multiphasic block copolymers which developed our facility at detecting and interpreting overlapping spectra arising from the motion of the probe in two distinct environments, and (e) recognition that the temperature difference

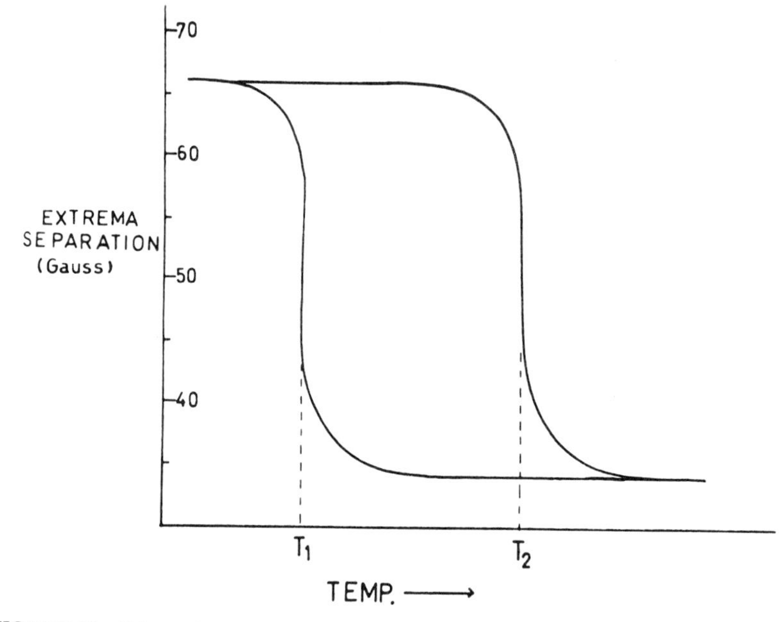

FIGURE 22 Schematic representation of an extrema separation *vs* temperature plot for a spin-probe experiment using two probes of varying size in a single polymer. Reproduced with permission from ref. 17.

observed for T_{50G} between small and large probes in our probe size studies of polystyrene was as large as the temperature difference between T_{50G}'s for the two phases of many block copolymers we examined.

We feel there would be a number of advantages of such a multiple probe method to study polymers exhibiting multiple relaxations. Only a single doping process is required. Only one set of temperature dependent spectra need to be recorded and interpreted although a wider range of temperature (relative to any single probe study) will have to be utilized. We also feel that potentially more useful information concerning molecular relaxations in polymers can be acquired by such experiments, even though it may be difficult, or impossible to predict *a priori* which probe may respond to which relaxation.

We have made a preliminary examination of the feasibility of such multiple probe studies of polymers possessing multiple relaxations by a series of dual-probe experiments on three different polymers. A schematic representation of the type of behavior that would be expected, based upon our ideas and thinking, is shown in Figure 22; the obvious similarity to a single-probe experiment in biphasic block copolymers (Figure 11) should be noted.

A summary of these dual-probe experiments is presented in Table 1, and an

TABLE 1

Dual-probe experiments (T_{50G} °C)

Polymer	TEMPO*	TEMPONE	BzONO	TEMPO & BzONO		TEMPONE & BzONO	
Polystyrene	−59 (−13)	3	120	−46	114	24	93
Polycarbonate	−40 (27)	—	145	2	129	24	106
Poly(dimethyl phenyleneoxide)	−92 (14)	−23	130	—	—	—	—

*For reasons which are currently unclear to us, spectra of TEMPO-doped samples seem to show a superposition of a weak, but distinct, three-line nitroxide spectrum suggesting the probe is present in two environments. Values in parentheses are T_{50G} values for the weak superposed spectra.

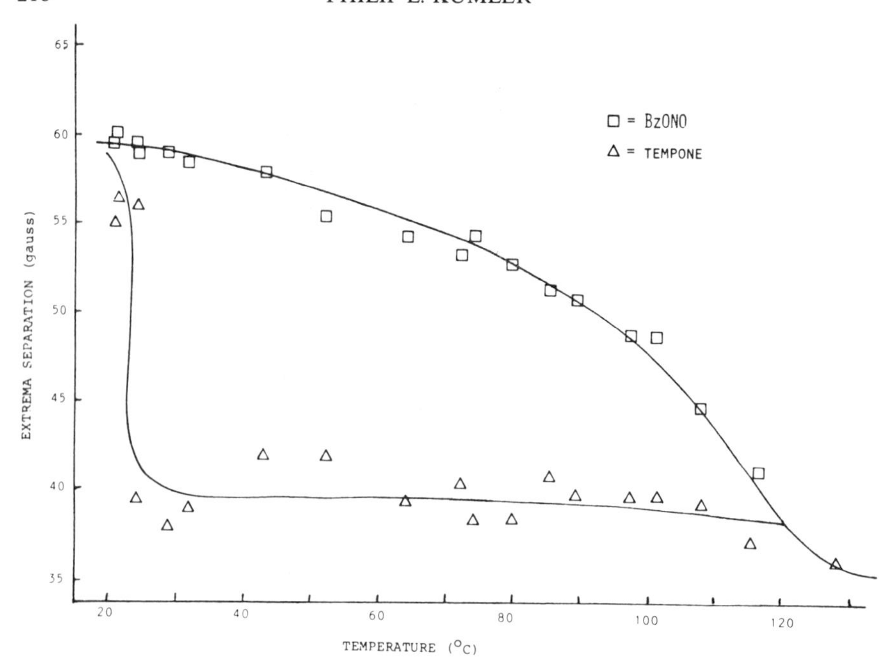

FIGURE 23 Experimental extrema separation *vs* temperature plot for a dual probe experiment (using BzONO and TEMPONE) in a high molecular weight polystyrene sample.

extrema separation versus temperature plot for one such dual-probe experiment (BzONO + TEMPONE in high molecular weight polystyrene) is shown in Figure 23. From the data presented, it is obvious that such dual-probe experiments are, in fact, feasible. It is possible to simultaneously follow the motion of each probe independently. Although detailed interpretations of the exact significance of such dual probe experiments must await further experimental investigation, some trends are obvious. In the dual-probe experiments reported, the T_{50G} for the BzONO probe is depressed (relative to the single probe experiment with BzONO) by the presence of the other probe (either TEMPO or TEMPONE), and TEMPONE depresses this T_{50G} more than TEMPO. In contrast, the T_{50G} for either TEMPO or TEMPONE in the dual-probe is greater than that observed for the corresponding single-probe experiments. It is obvious that the presence of a second probe has a distinct effect on the observed T_{50G} of both probes. Further experiments designed to clarify the nature of this interaction and to further elucidate molecular interpretation of dual-probe experiments are currently in progress in our laboratories.

ACKNOWLEDGEMENTS

None of this work would have been possible without the very valuable collaboration of Raymond F. Boyer of Midland Macromolecular Institute. Steven E. Keinath, also of Midland Macromolecular Institute, has served as a collaborator and coauthor for much of the work. The technical assistance of the following Saginaw Valley State College students is also acknowledged: George Bousfield, Mike Gebler, Frank Walles, and Brent Polak. Much of the later work has been facilitated by the superb technical assistance of Lisa Richards Denny. This work was made possible by generous donations (ESR spectrometer, operating supplies, financial support) by the Dow Chemical Company. Continuing access to the ESR spectrometer, donated to Saginaw Valley State College while I was a faculty member there, is due to the cooperation of Professor George Eastland. The continuing hospitality extended to me at Midland Macromolecular Institute is also gratefully acknowledged.

NOTES AND REFERENCES

1. For a discussion of the distinction between spin-probe and spin-label experiments, see ref. 14.
2. (a) I.C.P. Smith, "Biological Applications of Electron Spin Resonance", H.M. Swartz, J.R. Bolton, and D.C. Borg, Ed., Wiley-Interscience, New York, N.Y., 1972.
   (b) "Spin Labeling: Theory and Applications", L.J. Berliner, Ed., Academic Press, New York, 1976.
3. See ref. 19 for a recent summary of all spin-probe and spin-label experiments on synthetic macromolecules.
4. V.B. Stryukov and E.G. Rozantsev, *Vysokomol. Soedin., Ser. A*, **10**, 626 (1968).
5. G.P. Rabold, *J. Polym. Sci., Part A-1*, **7**, 203 (1969).
6. A.L. Buchachenko, A.L. Kovarskii, and A.M. Wasserman, "Advances in Polymer Science", Z.A. Rogovin, Ed., Wiley, New York, N.Y., 1974, pp 37 ff.
7. R.F. Boyer, *Macromolecules*, **6**, 288 (1973).
8. P.L. Kumler and R.F. Boyer, *Polymer Preprints*, **16(1)**, 572 (1975).
9. S.E. Keinath, P.L. Kumler, and R.F. Boyer, *Polymer Preprints*, **16(2)**, 120 (1975).
10. P.L. Kumler, S.E. Keinath, and R.F. Boyer, *Polymer Preprints*, **17(2)**, 28 (1976).
11. P.L. Kumler, S.E. Keinath, and R.F. Boyer, *Polymer Preprints*, **18(1)**, 313 (1977).
12. S.E. Keinath, P.L. Kumler, and R.F. Boyer, *Polymer Preprints*, **18(2)**, 456 (1977).
13. P.L. Kumler and R.F. Boyer, *Polymer Preprints*, **19(1)**, 612 (1978).
14. P.L. Kumler and R.F. Boyer, *Macromolecules*, **9**, 903 (1976).
15. R.F. Boyer and P.L. Kumler, *Macromolecules*, **10**, 461 (1977).
16. P.L. Kumler, S.E. Keinath, and R.F. Boyer, *J. Macromol. Sci.–Phys.*, **B13**, 631 (1977).
17. P.L. Kumler, S.E. Keinath, and R.F. Boyer, *Polym. Eng. & Sci.*, **17**, 613 (1977).
18. P.L. Kumler and R.F. Boyer in "Midland Macromolecular Monographs", #4, D.J. Meier, Ed., Gordon & Breach Science Publishers Ltd., New York, 1978.
19. P.L. Kumler, "Paramagnetic Probe Techniques" in "Methods of Experimental Physics: Polymer Physics", R.A. Fava, Ed., Academic Press, New York, In Press.
20. Rabold originally called this parameter $T_{\Delta W=50}$, see ref. 5.
21. See ref. 14 for experimental details of nitroxide doping, concentration utilized, ESR operating parameters, etc.

22. M.L. Williams, R.F. Landel, and J.D. Ferry, *J. Am. Chem. Soc.*, **77**, 3701 (1955).
23. R.F. Boyer, *Rubber Chem. Technol.*, **36**, 1303 (1963).
24. N.G. McCrum, B.E. Read, and G. Williams, "Anelastic and Dielectric Effects in Polymeric Solids", Wiley, New York, 1967.
25. D. Braun, P. Törmälä, and G. Weber, *Polymer*, **19**, 598 (1978).
26. R.F. Boyer in "Encyclopedia of Polymer Science and Technology", H.F. Mark, N.G. Gaylord, and N. Bikales, Eds., Wiley, New York, 1970; Vol. 13, pp 128–447.
27. K. Ueberreiter and G. Kanig, *Z. Naturforsch.*, **6A**, 551 (1951); *J. Colloid Sci.*, **7**, 569 (1952).
28. T.G. Fox and P.J. Flory, *J. Polym. Sci.*, **14**, 315 (1954); *J. Appl. Phys.*, **21**, 581 (1950).
29. L.A. Wall, Roestamsjah, and M.H. Aldridge, *J. Res. Nat. Bur. Std.*, **78A**, 447 (1974).
30. J.M.G. Cowie, *Eur. Polym. J.*, **11**, 297 (1975).
31. T.G. Fox and S. Loshaek, *J. Polym. Sci.*, **15**, 371 (1955).
32. A. Bondi, "Physical Properties of Molecular Crystals, Liquids, and Glasses", Wiley, New York, 1968; Table 13.4, p 385.
33. O.H. Griffith, D.W. Cornell, and H.M. McConnell, *J. Chem. Phys.*, **43**, 2909 (1965).
34. T. Nagamura and A.E. Woodward, *Biopolymers*, **16**, 907 (1977).
35. M.C. Lang, C. Noël, and A.P. Legrand, *J. Polym. Sci. (Phys.)*, **15**, 1329 (1977).
36. J.R. Knox (Amoco; Napierville, Ill.), Plenary Lecture, Spring Meeting, American Physical Society, March 21–24, 1977, San Diego, Calif.; no printed abstract.
37. R.F. Boyer, private communication.
38. (a) Ref. 5; (b) Ref. 6; (c) Ref. 39; (d) Ref. 41; (e) I. Bulla, P. Törmälä, and J.J. Lindberg, *Finn. Chem. Lett.*, 129 (1974).
39. S.C. Gross, *J. Polym. Sci. A-1*, **9**, 3327 (1971).
40. S. Matsuoka and Y. Ishida, *J. Polym. Sci.*, **C-14**, 247 (1966).
41. A. Savolainen and P. Törmälä, *J. Poly. Sci. (Phys.)*, **12**, 1251 (1974).
42. N. Kusumoto, S. Sano, N. Zaitsu, and Y. Motozato, *Polymer*, **17**, 448 (1976).
43. R.K. Eby, *J. Chem. Phys.*, **37**, 2785 (1962).
44. Extensive references to these data can be found in refs. 7 and 14.
45. K.-H. Illers, *Kolloid Z. Z. Polym.*, **190**, 16 (1963).
46. E.G. Kontos and W.P. Slichter, *J. Polym. Sci.*, **61**, 61 (1962).
47. For extensive discussions of the controversial glass temperature of polyethylene, see: (a) R.F. Boyer, *J. Polym. Sci.*, *Symposium*, **#50**, 189 (1975); and, (b) R.F. Boyer and R.G. Snyder, *J. Polym. Sci. (Polym. Lett.)*, **15**, 315 (1977).
48. R.F. Boyer, L.R. Denny, P.L. Kumler, B. Michel, R. Spitz, A. Douillard, and A. Guyot; Manuscript in Preparation.
49. J.M.G. Cowie and I.J. McEwen, *Macromolecules*, **10**, 1124 (1977).

DISCUSSION

N. Kusumoto (Kumamoto University, Kumamoto, Japan): You showed an equation for deriving T_g from ESR data. In this equation ΔH_a is included. Is ΔH_a from data other than by ESR? What about the probe size effect in this equation?

P. L. Kumler: ΔH_a is the apparent activation enthalpy at T_g. Our derived equation does not define where the ΔH_a comes from. We have shown, however, that we can get very good agreement between T_{50G} calculated from this equation and the ESR observed T_{50G} using literature data for ΔH_a. It should be noted as well, that this derived equation should allow estimation of

ΔH_a once the T_{50G} and T_g values have been determined. T_{50G} values calculated from this derived equation will only be valid if the probe size is "large" relative to the activation volume at T_g of the polymer. Such a condition was satisfied for probe BzONO and the polymers used for construction of the T_{50G}–T_g correlation curve.

D. J. Meier (Midland Macromolecular Institute, Midland, Michigan): In your equation relating T_{50G} and T_g via ΔH_a, can one use the general (universal) form of the WLF equation to eliminate ΔH_a?

P. L. Kumler: We have elected to use simple Arrhenius behavior, rather than the WLF treatment, for two reasons. (a) It is simpler and usually valid for T_g in the frequency range from 1 to 10^7 Hz. (b) Secondary relaxations frequently follow Arrhenius-type behavior over an even broader frequency range than does the glass transition.

D. J. Meier: What are the consequences of the spin probe having a size comparable to the thickness of the block copolymer interfacial region, as far as measurements of interfacial properties are concerned, i.e., does the probe give some sort of average of the interfacial properties?

P. L. Kumler: We believe that the ESR spin-probe technique will respond to interfacial material, if it is present in significant amounts, and the estimated T_g will reflect the composition of this interlayer material. As the thickness of the interfacial layer approaches the dimensions of the probe molecule, the probe will begin to "see" interfacial material rather than pure phase. The composition of the interfacial material probably approximates a blend (of varying composition) of hard and soft components; it will possess a range of T_g values and the T_g determined by the ESR technique will correspond to the T_g of the "average" composition.

W. G. Miller (University of Minnesota, Minneapolis, Minnesota): Your change in A_{zz} in going from high to low molecular weight polystyrene was an increase of 5–6 gauss. How do you explain this large effect? If it is due to an increase in free volume, I guess one would presume that an increase in free volume would correspond to a decrease in polar environment. From the data shown by Berliner, one would expect that the change in A_{zz} should go in the opposite direction. How do you explain this? Could it possibly be that not all of your data are rigid limit values?

P. L. Kumler: Although we cannot, at this time, explain the magnitude of the increase in A_{zz} or G_{max} at low temperatures, we believe that it means that there is less free volume in a low molecular weight oligomer than there is in a similar high molecular weight polymer. The data suggest the probe molecule is "more immobilized" in the low molecular weight glasses than it is in higher molecular weight systems. This effect is probably not a "polar" effect because we have seen it in low molecular weight glasses of considerably varying polarity. We do believe that our data are true rigid limit values because in most cases the spectra were recorded all the way down to liquid nitrogen temperature.

L. J. Berliner (Ohio State University, Columbus, Ohio): These large splittings (e.g., 72 gauss) have been reported in cases where there is either protonation of the N–O group or interaction of the N–O group with another Lewis acid.

P. Törmälä (Tampere University of Technology, Tampere, Finland): The differences between peak-to-peak separations of probes in styrene and polystyrene may be partially explained by the different characteristics of the δ-relaxation in these two systems.

G. G. Cameron (University of Aberdeen, Old Aberdeen, Scotland): Are you confident that the differences in T_{50G} values obtained with single and dual probe experiments cannot be ascribed to experimental errors which originate from the superposition of fast and slow spectra?

P. L. Kumler: I don't think that the magnitude of the "shift" observed can be due to experimental error from interpreting overlapping spectra. Also, the fact that TEMPO and TEMPONE give consistent but different shifts argues against this interpretation. Our experiences with single probe experiments in block copolymers (with little or no interfacial material) can also be cited as evidence against such a suggestion. In these cases the overlapping spectra are very similar to the multiple probe experiments, and the T_g values estimated for the two pure phases are in excellent agreement with literature data for comparable molecular weight homopolymers.

ESR Studies of Bulk Polymers Using Paramagnetic Probes Differing in Size and Shape

NAOSHI KUSUMOTO

Department of Industrial Chemistry, Faculty of Engineering, Kumamoto University, 860 Kumamoto, Japan

The physical meaning of the narrowing of the extrema separation of the ESR spectra in the spin-probe method for the study of solid polymers using the nitroxide radical is discussed in detail in relation to the motional mode of the probe. The theory relating the narrowing temperature to the glass transition temperature of host polymer is given in terms of the hole theory, in which the probe is supposed to diffuse through the polymer matrix as well as a polymer segment. The correctness and applicability of this theory was experimentally tested using probes differing in size for natural rubber, poly(vinyl acetate), and polystyrene. This theory was applied to estimate the segment size of typical amorphous polymers such as vulcanized natural and acrylonitrile–butadiene rubbers, poly(vinyl acetate), polystyrene, and polypropylene. The estimated segment size of vulcanized natural rubber is 45–98 backbone atoms, and that for vulcanized acrylonitrile–butadiene rubber is 65–215 backbone atoms, the values depending upon the degree of vulcanization. The segment size for poly(vinyl acetate) and polypropylene was found to be 7–8 monomeric units.

A linear spin-probe was used to study the molecular motion of oriented polyethylene. The temperature variation of the anisotropy of the ESR spectra revealed uniaxial rotational motion. The activation energy for motion of the probe was found to increase with increasing elongation. The temperature dependence of the order parameter was rather complex, reflecting stepwise changes in the motional mode and molecular rearrangement of host polymers.

INTRODUCTION

Many studies have been made on synthetic polymers using paramagnetic spin probes embedded in polymer matrices to study the molecular motions of the host polymers. Experimental evidence has shown that the molecular motions of the probe are closely related to those of the host polymer. Stryukov[1] found that motions of the probe are dominated by those of the polymer matrix and Rabold[2] reported that this method was useful for characterizing polymers.

Two techniques have been used to represent the motion of the spin probe in these experiments. One employs the calculation of the correlation frequency of the tumbling probe, which can be estimated using the theory of Kivelson[3] in the frequency range 10^9–10^{11} Hz. The equations of Freed[4] and spectral simulation[5,6] have been used where the correlation frequencies are below 10^9 Hz.

The second technique observes the change in the extrema separation of the triplet spectra of the nitroxide radical with temperature. This line separation narrows with increasing motion of the probe due to the averaging of the anisotropic hyperfine interaction and the anisotropy of the g-value. In most cases this narrowing takes place rapidly, and the narrowing temperature is defined as T_n which characterizes the polymer matrix.

T_n can be correlated to the glass transition temperature of polymers. The ESR spectra are those of the nitroxide probe and not of the host polymer. It is conceivable that the change in the ESR spectra with temperature may be a characteristic of the probe used. In fact, Rabold[2] found that T_n was shifted to higher temperatures when the size of the probe was increased and observed a linear relation between T_n and T_g for polyethylene, polypropylene, and poly(vinyl chloride). Gross[7] found that T_n was shifted to higher temperatures when a larger probe was used for five polyester films. Kumler[8] compared T_n with T_g using seven different probes in three different polymer systems. Kusumoto reported similar probe size effects on T_n.[9,10,11]

We are seeking a quantitative interpretation of the relationship between T_n and T_g by studying the interaction of a probe with the polymer matrix. We hope to apply this interpretation to a better understanding of the fundamentals of the segmental motion of the host polymer.

In addition to the variation of T_n with probe size, the polarity and the flexibility of the probe may affect the correlation between T_n and T_g. These problems have not been completely clarified as yet.

Another factor which may be important in applying the spin probe technique to synthetic polymers is the intrinsic shape of the probe and its motional mode with respect to the polymer matrix. To date, many probes differing in shape have been synthesized; of these probes, a linear one would be expected to rotate about its molecular axis when embedded in an oriented polymer system. Such probes have been used in biological membranes to study the molecular order of lipid molecules. An example of the application of a linear probe to a drawn synthetic polymer film is described in this paper.

TEMPERATURE VARIATION OF EXTREMA SEPARATION AND THE MEANING OF NARROWING TEMPERATURE T_n (T_{50G})[12]

Before discussing the relation between T_n and T_g it is desirable to explain the temperature variation of the extrema separation and the meaning of T_n in view of the motional mode of the probe. Figure 1 shows the temperature variation of the ESR spectra for a sample of an acrylonitrile–butadiene elastomer (NBR)

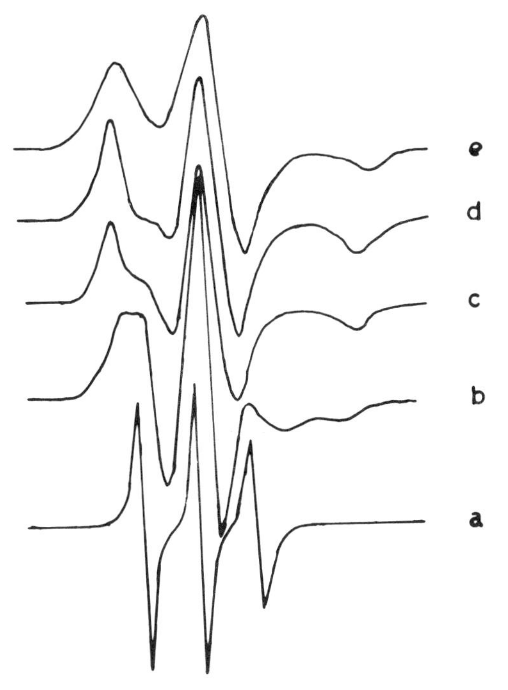

FIGURE 1 ESR spectra of NBR at various temperatures: (a) 75°C, (b) 25°C, (c) −20°C, (d) −90°C, (e) −180°C.

when 2,2,6,6-tetramethylpiperidinol-1-oxyl was used as the paramagnetic probe. At liquid nitrogen temperature it shows a broad asymmetric triplet spectrum indicating probe in a rigid matrix. This broad spectrum becomes sharper as the line separation narrows with increasing temperature, indicating increased mobility of the probe at higher temperatures. A small subsplitting appears at around −90°C, increasing in intensity as temperature increases. The broad spectral component narrows as the intensity of the narrow component increases, and the spectral shape finally becomes a sharp triplet.

This multiplicity of spectra has been interpreted in two ways.[13] One possibility is that the superposition of broad and narrow spectra may arise from the inhomogeneity of the fine structure of the matrix. An intrinsic explanation may be the partial averaging of the anisotropic hyperfine interaction and g-value in the intermediate frequency range.[5] The present data support the former case, in that an effect of the fine structure of the polymer on the spectra was observed; namely, the higher the content of combined sulphur in the samples, the clearer the separation of the two peaks in the intermediate temperature range.

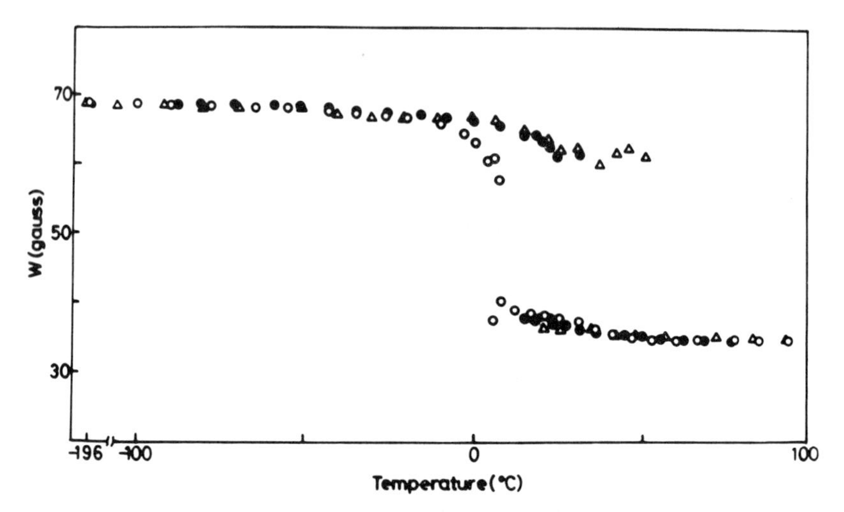

FIGURE 2 Temperature dependence of extrema separation for NBR: ○, $S_c = 2\%$; ●, $S_c = 11.9\%$; △, $S_c = 23.4\%$. S_c, combined sulphur.

It has been proposed that the extrema separation of the triplet spectra, W, can be used as a measure of the motion of the probe.[2] Figure 2 shows examples of the variation of W with temperature for samples of NBR. The extrema separation is 69 gauss at liquid nitrogen temperature and shows rapid narrowing above the glass transition. Since the spectra show a complex pattern, the figure shows the temperature dependence of the line separation of both components. The temperature difference between the two narrowing temperatures is found to increase as the content of combined sulphur is increased.

For convenience, the temperature at which the intensity of the two peaks becomes equal is taken as the narrowing temperature, T_n, and is comparable to T_{50G}. T_{50G} is the temperature at which the hyperfine separation equals 50 gauss. The relations between T_n and T_g for the vulcanizate series of natural rubber (NR) and NBR are shown in Figure 3, indicating a possible correlation between T_n and T_g. T_n was found to increase with increasing T_g. From these curves one observes that T_n coincides with T_g, at 34°C for NR, and at 28°C for NBR. The relation between T_n and T_g will be discussed later for these two polymer systems.

When the temperature is high enough to cause sufficient sharpening of the line width, the correlation time for isotropic tumbling motion of the probe can be calculated using the theory of Kivelson.[3]

$$\tau_c = W_0\{(h_0/h_1)^{1/2} - (h_0/h_{-1})^{1/2}\}C \quad (1)$$

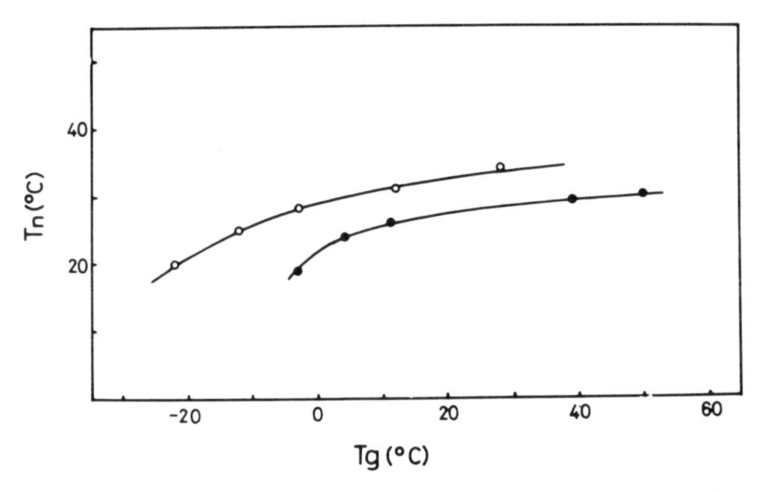

FIGURE 3 Relations between T_n and T_g, vulcanizate series: ○, NR; ●, NBR.

$$\tau_c = W_0\{(h_0/h_1)^{1/2} + (h_0/h_{-1})^{1/2} - 2\}C' \qquad (2)$$

W_0 is the maximum slope width of the central peak of the three line spectrum, h is the peak height, whose subscript represents the nuclear quantum number of the nitrogen atom, C and C' are constants determined by experimental conditions, e.g., the hyperfine interaction characteristic of the probe and the magnitude of a static magnetic field. These two equations are mathematically equivalent. Stryukov[1] has discussed the accuracy in using them and certain rate conditions[12] must be fulfilled.

Figure 4 shows the plots of the logarithmic correlation frequency $(1/\tau_c)$ versus the reciprocal of temperature. The correlation frequency increases with increasing combined sulphur content over the whole temperature range, indicating that the polymer matrix is immobilized by an increase in combined sulphur. A slight upward departure in the frequency from a linear plot was found for NR samples with a low sulphur content at low temperatures, and a progressive decrease in the frequency was observed for samples with a high sulphur content at higher temperatures.

On the other hand, for NBR an upward departure at low temperature is not observed, but a progressive decrease at higher temperature is noted for all samples; the decrease becoming larger with increasing sulphur content.

The activation energy of the motion of the probe is calculated by the equation:

$$E = \mathrm{R}d\ln(1/\tau_c)/d(1/T) \qquad (3)$$

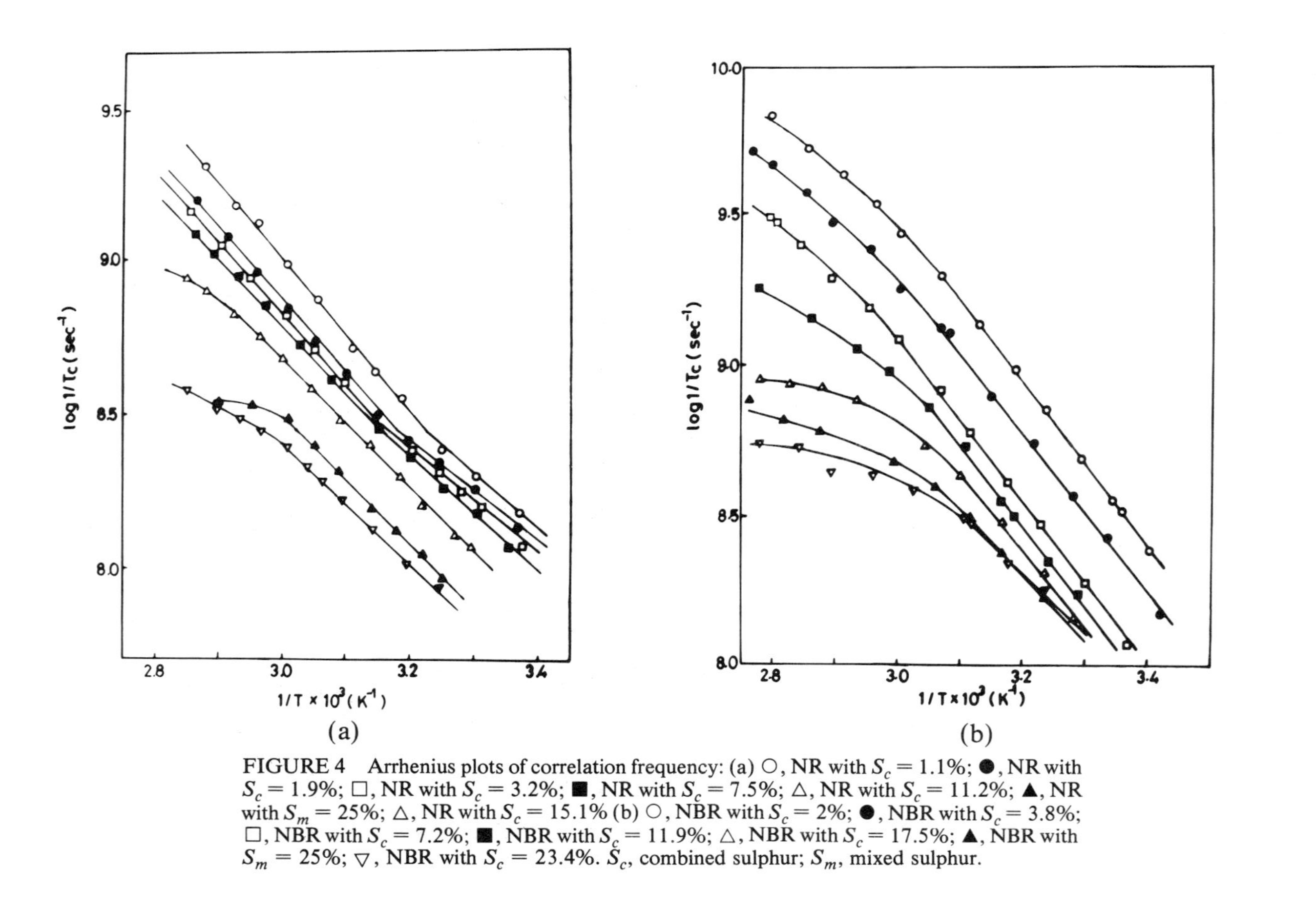

FIGURE 4 Arrhenius plots of correlation frequency: (a) ○, NR with $S_c = 1.1\%$; ●, NR with $S_c = 1.9\%$; □, NR with $S_c = 3.2\%$; ■, NR with $S_c = 7.5\%$; △, NR with $S_c = 11.2\%$; ▲, NR with $S_m = 25\%$; △, NR with $S_c = 15.1\%$ (b) ○, NBR with $S_c = 2\%$; ●, NBR with $S_c = 3.8\%$; □, NBR with $S_c = 7.2\%$; ■, NBR with $S_c = 11.9\%$; △, NBR with $S_c = 17.5\%$; ▲, NBR with $S_m = 25\%$; ▽, NBR with $S_c = 23.4\%$. S_c, combined sulphur; S_m, mixed sulphur.

TABLE I

Activation energy of the tumbling motion of the probe.

	NR				NBR		
No.	S_m (%)	S_c (%)	E_1 (kcal/mol)	E_2 (kcal/mol)	S_m (%)	S_c (%)	E_1 (kcal/mol)
1	2	1.1	7.8	11	2	2	12
2	4	1.9	7.4	11	4	3.8	12
3	8	3.2	7.2	10	8	7.2	12
4	14	7.5	7.6	10	14	11.9	12
5	20	11.2	—	10	20	17.5	12
6	25	—	—	10	25	—	10
7	32	15.1	—	9	32	23.4	10

S_m, mixed sulphur; S_c, combined sulphur; curing was performed at 140°C.

Activation energies calculated from the curves of Figure 4 are listed in Table I. For the NR samples, showing two linear regions, the activation energy at higher, E_1, and lower temperatures, E_2, were estimated. The values of E_1 and E_2 are comparable to those of bulk polyethylene,[13] although much higher than those estimated from Arrhenius plots of polyethylene single crystals[13] in the lower temperature region.

It has been suggested,[13] that this relatively large activation energy, E_2, may be associated with the jumping mode of the probe into adjacent holes. Andrew[15] has shown by measurements of the second moment of wideline NMR that cyclohexane molecules begin to diffuse through its crystal lattice between −53 and −33°C. In this case, the activation energy of molecular motion was estimated to be 8 kcal/mol, which is supposedly two-thirds of the lattice energy, suggesting the occurence of self-diffusion of cyclohexane molecules. This explanation suggests the existence of self-diffusion of the probe in rubbers. The value of 9–11 kcal/mol found in the present study and the fact that the activation energy is nearly independent of the sulphur content, implies that the probe is not restricted in a local potential barrier, the so-called cavity model, but can diffuse through the polymer matrix.

At low temperatures the probe is in a rigid state having only restricted rotational oscillation, as the temperature increases some free rotation is allowed. The rotational oscillation and the free rotation may possibly average out the anisotropic hyperfine interaction and g-value, causing the narrowing of the extrema separation until self-diffusion can take place. However, the appearance of the narrowing depends on the rotational frequency of the probe, being recognized only when the frequency exceeds the ESR frequency of 10^{7-8} Hz. Since the motional mode of the probe and its frequency depends on the nature of the probe, i.e., size, shape, and possible interaction with the polymer, the cause of the narrowing is not always the same.

When narrowing occurs by a rotational mode, the activation energy will be below 10 kcal/mol. The values obtained for the low temperature region of polyethylene,[13,14] below 6kcal/mol, may be associated with this mode.

The appearance of the break point around 50°C in the Arrhenius plot in Figure 4(a) suggests that the rotational mode of the probe is mixed, leading to a change in the slope of the curves in the low temperature region.

THEORETICAL TREATMENT OF THE RELATIONS BETWEEN T_n, T_g, AND THE RATIO OF THE MOLAR VOLUME OF THE SEGMENT TO THE PROBE[12]

We have discussed that the motional mode of the probe in NR in the highest temperature range can be self-diffusion of the probe in the polymer matrix. Further, the correlation between T_n and T_g implies that this self-diffusion takes place in a manner similar to that of the segmental diffusion of the polymer above its glass transition temperature.

In terms of the hole theory, a segment of a polymer molecule can begin to jump into adjacent holes formed by the displacement of other segments when sufficient thermal expansion of the polymer matrix is realized, usually around T_g. Similarly, if a probe is of comparable size to a segment it also may jump into an adjacent hole in competition with the polymer segment. According to the theory by Bueche,[16] consider a polymeric system consisting of polymer molecules with n segments and probes and with free volume, V_f, as shown in Figure 5. V_f is the summation of all packets having the volume, v_f:

$$V_f = n \times v_f \tag{4}$$

If q packets are gathered together to form a hole of volume v_f' in the vicinity of a segment or probe, the segment or probe can jump into this hole when v_f' exceeds the activation volume for jumping, v_m^* or v_p^*, of the segment or probe, respectively. In this system, the probability that q packets are gathered around a certain segment is given by:

$$p(q) = (1-1/n)^{n-q}(1/n)^q\{n!/q!\,(n-q)!\} \tag{5}$$

This calculation can be compared to the case where a person throws n balls toward n boxes and has the possibility of getting q balls into a certain box. The ball corresponds to the hole and the box to the segment. Let us consider the case where a few of the boxes are replaced by ones differing in size and having q balls in one of them. This corresponds to the case where q balls are gathered around the probe in a polymer matrix. Here, the difference between the

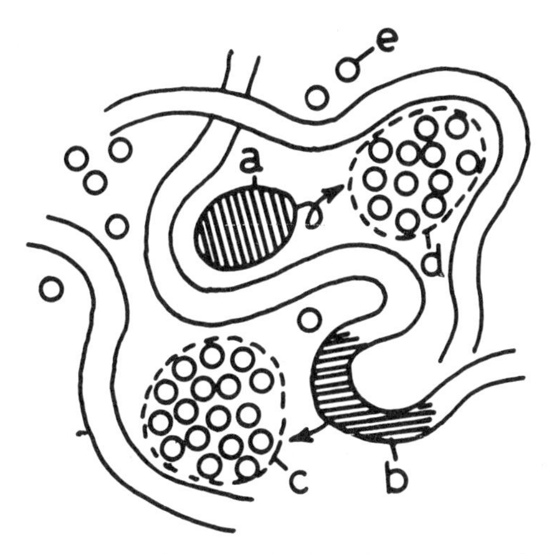

FIGURE 5 Schematic model of jumping of probe and segment: a, probe (volume v_p); b, segment (volume v_m); c, hole (volume $v_f' > v_m^*$); d, hole (volume $v_f' > v_p^*$); e, packet (volume v_f).

interaction of the probe or the segment with the polymer matrix is ignored. When the number of probes is negligibly small compared to the number of segments, the probability of having q holes around the probe may be approximated by rewriting equation (5) as:

$$p(q) = (1-v_p/nv_m)^{n-q}(v_p/nv_m)^q\{n!/q!\,(n-q)!\} \tag{6}$$

where v_p and v_m denote the molar volume of the probe and the segment, respectively. In equation (5) the probability of a hole near a segment is $1/n$; this probability is modified in equation (6) as being proportional to the volume ratio of the probe to the segment.

Assuming $n \gg q$ and using:

$$\lim_{n\to\infty}(1-v_p/nv_m)^n = \exp\{-v_p/nv_m\}$$

and

$$n!/q!(n-q)! = n(n-1)\cdots\cdots\cdots(n-q+1)/q! \simeq n^q/q!$$

equation (6) may be rewritten as:

$$p(q) = \exp\{-v_p/v_m\} \times (v_p/v_m)^q(1/q!) \tag{7}$$

While the gamma function $\Gamma(2)$ is: $\Gamma(n+1) = n!$ and from the Stirling formula:

$$\lim_{n\to\infty} \Gamma(n+1)/(2n\pi)^{1/2}(n/e)^n = 1.$$

Therefore, when q is not so small:

$$1/q! = (2\pi q)^{-1/2} \times (q/e)^{-q} \tag{8}$$

Substituting equation (7) into equation (8):

$$p(q) = (2\pi q)^{-1/2}(v_m q/v_p)^{-q} \times \exp(q - v_p/v_m) \tag{9}$$

where $q = v_f'/v_f$. Therefore, the probability that the probe encompasses the free volume, v_f':

$$p(v_f') = (2\pi v_f'/v_f)^{-1/2}\exp\{-\{\beta + 1n(v_m/v_p)\}v_f'/v_f - v_p/v_m\} \tag{10}$$

where $\beta = \ln(v_f'/v_f) - 1$.

The probe jumps into the hole when $v_f' \geq v_p^*$, the probability of the jumping of the probe is given by:

$$\int_{v_p^*}^{\infty} p(v_f')dv_f' \simeq (\text{const.})\exp\{-\{\beta^* + 1n(v_m/v_p)\}v_p^*/v_f - v_p/v_m\} \tag{11}$$

Accordingly, the jumping frequency of the probe depends upon the probability

$$\Phi = \Phi_0\exp\{-\{\beta^* + 1n(v_m/v_p)\}v_p^*/v_f - v_p/v_m\} \tag{12}$$

The microscopic free volume v_f changes with temperature above T_g as:

$$v_f = v_{fg} + v_{mg}\Delta\alpha(T - T_g) \tag{13}$$

where, v_{fg} and v_{mg} are the free volume, and the occupied volume of a segment at T_g, respectively. $\Delta\alpha$ is the difference between the thermal volume expansion coefficients above and below T_g.

Substituting equation (13) into equation (12) and rearranging we obtain:

$$T - T_g = \left\{ \frac{\{\ln(v_m/v_p) + \beta^*\} v_p^* v_{fg} v_m^*}{\ln(\Phi_0/\Phi) \times v_{fg} \times v_m^* - v_p v_m^* v_{fg}/v_m} \right\} /v_{mg}\Delta\alpha \tag{14}$$

According to Bueche,[16] when $v_{fg}/v_{mg}\Delta\alpha = 52$, $v_m^*/v_{fg} = 40$, and $\beta^* \simeq 1$ are substituted into equation (14) and assuming $\ln(\Phi/\Phi_0) \gg$ f, we obtain:

$$T - T_g = 52[40\ f\{1n(1/f) + 1\}/1n(\phi_0/\phi) - 1] \tag{15}$$

where $f = v_p{*}/v_m{*} = v_p/v_m$.

$\Phi_0 = 10^{14}$ is assumed, according to Bueche.[16] The narrowing of the line separation may occur at $\Phi = 10^8$; this supposition is satisfactory since the extrapolation of most of the curves in Figure 4 to the narrowing temperatures is of the order of 10^8. Thus, we obtain the simple formula:

$$T_n - T_g = 52[2.9\ f\{1n(1/f) + 1\} - 1] \tag{16}$$

When T_n and T_g are determined experimentally and the molar size of the probe is known the segment size of the polymer can be estimated using equation (16).

Figure 6 shows the value of $T_n - T_g$ as a function of f, calculated from equation (16). $T_n - T_g$ has a maximum at $f = 1$. This may mean that the probe is not physically able to jump at T_g, when the size of the probe is larger than that of the segment. The maximum value of $T_n - T_g$ is 98.8°C. On the other hand, $T_n - T_g$ may be negative, as mentioned in the previous section. This may occur when the probe is considerably smaller than the segment and the jumping mode occurs by local motion of the segment. For example, this can occur in a system

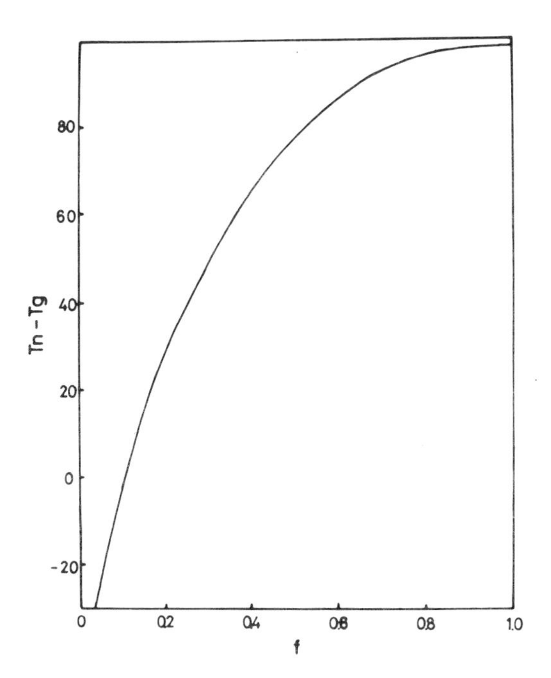

FIGURE 6 Theoretical relation between $T_n - T_g$ and f.

having a wide distribution of segmental sizes because of the inhomogeneity of the matrix. The deduction that the inhomogeneity of the matrix becomes larger with increasing content of combined sulphur may explain this phenomenon. This could also be explained by the local mode relaxation. In this case, macroscopic volume expansion and a total increase in free volume, is not necessary for the displacement of the probe. The diffusion of the probe due to the displacement of local free volume may afford the narrowing of the ESR pattern. The case where $T_n - T_g < 0$ is seen in polymer systems having large side groups, such as polystyrene. It was found[10] that $T_n - T_g$ is $-15°C$ for polystyrene when di-*t*-butyl nitroxide $(CH_3)_3C(NO\cdot)C(CH_3)_3$ is used as the probe; while $T_n - T_g$ is 35°C for the probe used in this study. The molar volume of di-*t*-butyl nitroxide is 88% of that of the probe used here, accordingly the narrowing occurs at lower temperatures. The effect of probe size is well interpreted by equation (16); details of the data are described in the next section.

EXPERIMENTAL VERIFICATION OF THE THEORY[10]

In the previous section we proposed a theory relating T_n and T_g. If we use a certain probe to estimate the size of the polymer segment, we can find the value of f in eq. (16) for any probe; then, measuring T_n we can obtain a relation between $T_n - T_g$ and f for any probe used. Attempts were made to verify the

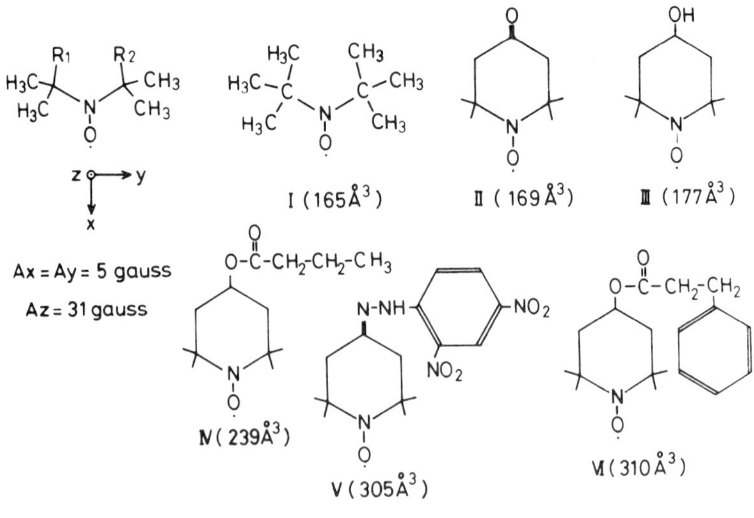

FIGURE 7 Paramagnetic probes.

theory experimentally, using probes differing in molar volume in three amorphous polymers: vulcanized natural rubber, poly(vinyl acetate) (PVAc), and polystyrene (PS). The paramagnetic probes used in this experiment are shown in Figure 7. The molar volume of each probe was estimated by Kitaigorodskii's method[17] and is given in the parentheses.

Figure 8 shows the temperature dependence of extrema separation, W, for each polymer. The separation narrows rapidly at a given temperature, and is markedly influenced by probe size. Table II lists the narrowing temperatures observed. T_n was found to increase with increasing probe size. For PS and PVAc the dependence on probe size is greater than that for NR. This may be due to the differences in the interaction of the probes with the polymer chains. In NR the narrowing temperatures were found to be located above the glass transition temperature for all six probes. In PVAc, similarly, all probes exhibited a T_n above T_g except probe I. However, in PS, all T_n values were considerably below T_g, except that for probe V, having a value above T_g.

Figure 9 shows the Arrhenius plots of correlation frequency for each probe estimated above its T_n value for each polymer. It should be noted that the plots are linear for NR, but have breaking points for PVAc and PS. The activation energy of motion of the probe can be estimated from the linear part of the lines; these values are listed in Table III.

For PVAc and PS the activation energies below and above the breaking point are denoted as E_1 and E_2, respectively. As suggested previously the motional mode of the probe depends upon the magnitude of the activation energy. Below the breaking point the probe may be experiencing hindered rotation in a microvoid among polymer chains (cavity model); and above the breaking point the probe may be allowed to diffuse through the polymer matrix. A relatively large scattering in the magnitude of the energies below the

TABLE II

Narrowing temperature.

Probe	T_n (°C)		
	NR	PVAc	PS
I	12.5	22	−15
II	18	62	43
III	21	49	35
IV	35	71	61
V	50	108	120
VI	34	86	65
T_g	−26	33	79

T_g was determined by dilatometry.

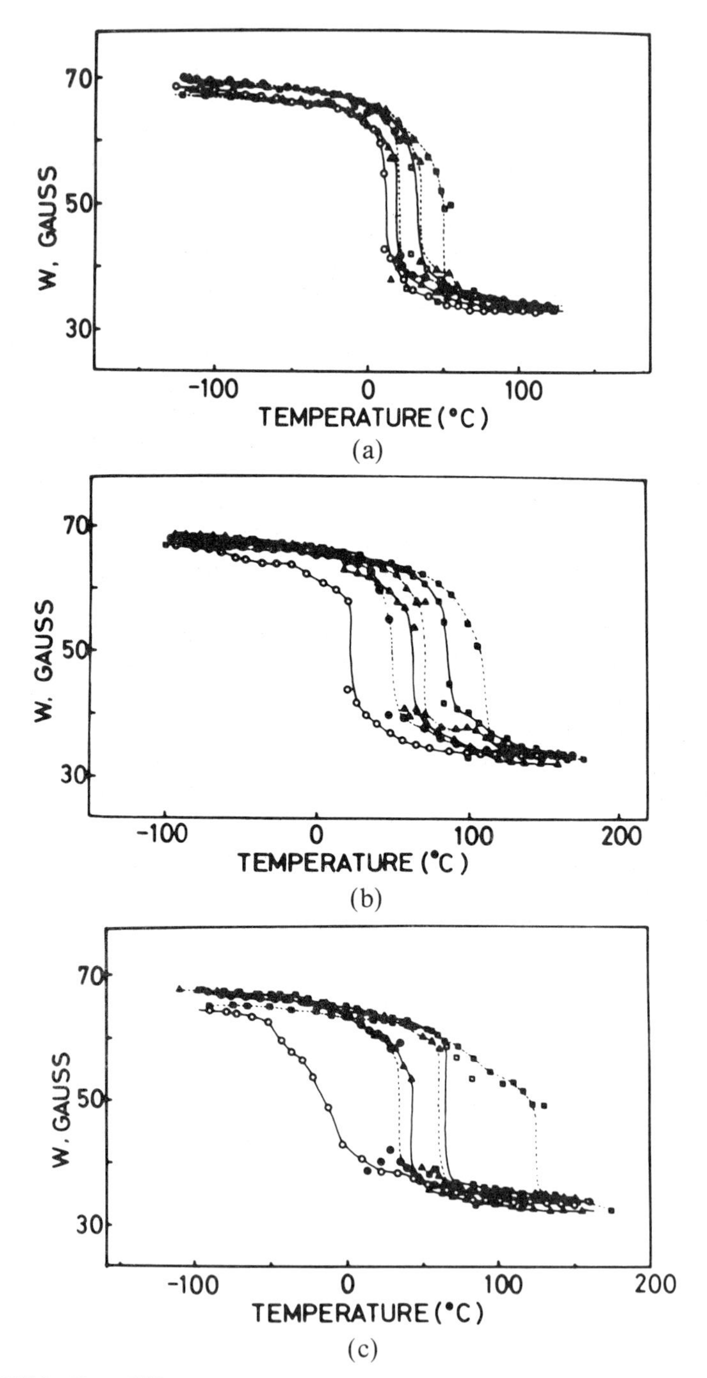

FIGURE 8 Plots of W *vs.* temperature: ○, Probe I; △, Probe II; ●, Probe III; ▲, Probe IV; ■, Probe V; □, Probe VI; (a) NR, (b) PVAc, (c) PS.

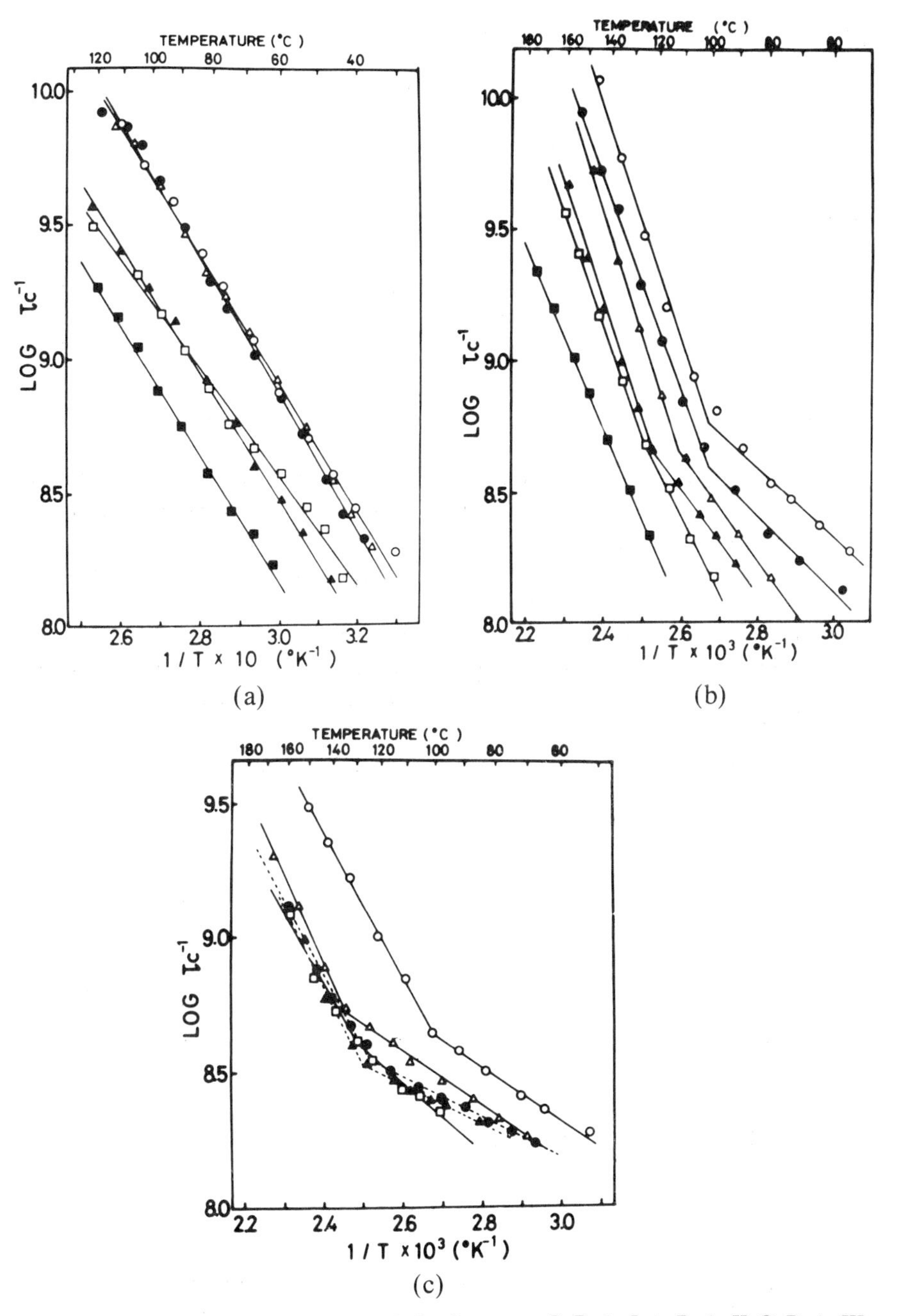

FIGURE 9 Arrhenius plots of the correlation frequency: ○, Probe I; △, Probe II; ●, Probe III; ▲, Probe IV; ■, Probe V; □, Probe VI; (a) NR, (b) PVAc, (c) PS.

TABLE III
Activation energy (kcal/mol).

	NR	PVAc		PS	
Probe	E	E_1	E_2	E_1	E_2
I	11.4	6.2	20.5	4.4	13.0
II	11.6	9.0	21.9	4.7	14.8
III	11.6	7.1	19.2	3.6	13.0
IV	11.2	8.9	19.6	5.5	15.0
V	11.2	16.5	16.5	—	—
VI	9.2	13.9	19.5	3.5	15.0

breaking point is observed, in contrast to those above the breaking point. This supports the idea that the motional behavior of the probe in a cavity is strongly influenced by the specific shape and size of the probe.

In the temperature region above the breaking point there is little probe size dependence on the activation energy and the order of magnitude of the energy is of the same order as the cohesive energy density of the polymer (NR, 62 cal/cc; PVAc, 92 cal/cc; PS, 73 cal/cc). Therefore, one can say that there is an averaged potential field, characteristic of the host polymer in active motion and the field is overrun by diffusion of the probe.

Figure 10 shows the relation between T_n and f for NR. The solid line is the

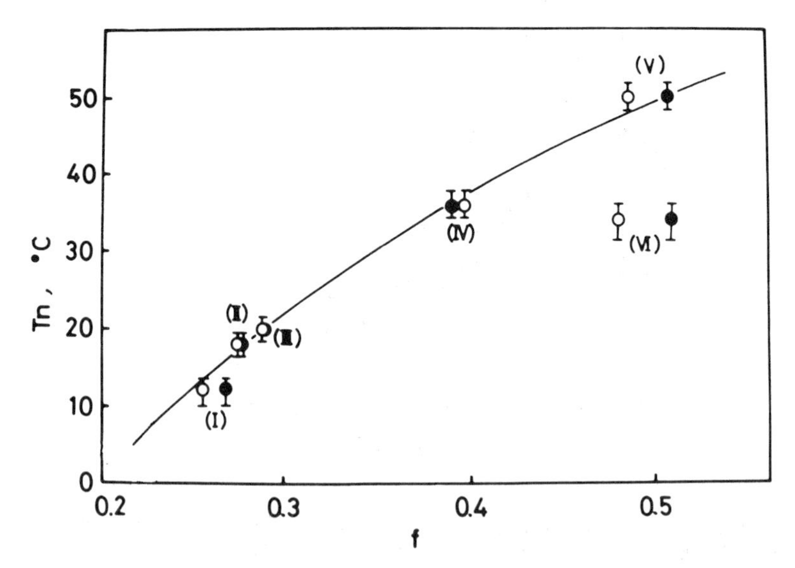

FIGURE 10 Relation between T_n and f for NR: solid line, theoretical; ○, experimental (Kitaigorodskii's method); ●, experimental (molar volume at boiling point).

theoretical curve calculated from eq. (16). The value of f was determined initially for probe III, from its coincidence with the theoretical curve. From this, the value of f for all the other probes was estimated. All probes fit the T_n *vs.* f curve quite well except probe VI. In this case it is thought that the phenyl group may have some independent motion, thereby decreasing the effective molar volume of this probe.

For PVAc and PS eq. (16) is not applicable since the assumption that the probe diffuses through the polymer matrix at T_n does not hold. However, if the Arrhenius plots for PVAc in the temperature region above the breaking point are extrapolated to the temperature at which the correlation frequency is 10^8 Hz, and this temperature (T_n') is used instead of T_n, then eq. (16) is found to be satisfactory as shown in Figure 11.

For PS considerable deviation from the theory is observed. In this case T_n is always below T_g except for probe V and side chain motion may possibly activate probe motion up to 10^8 Hz.

Similar experiments were carried out on two semicrystalline polymers, polyethylene and polypropylene. The probes used for this study are shown in Figure 12; the molar volumes calculated by Kitaigorodskii's method are also noted.

Figures 13(a) and (b) show the variation of W with temperature for high density polyethylene (HDPE) and isotactic polypropylene (PP), respectively. Curves having rapid narrowing are observed similar to those for the amorphous polymers. The values of T_n for HDPE and PP are listed in Table IV. The

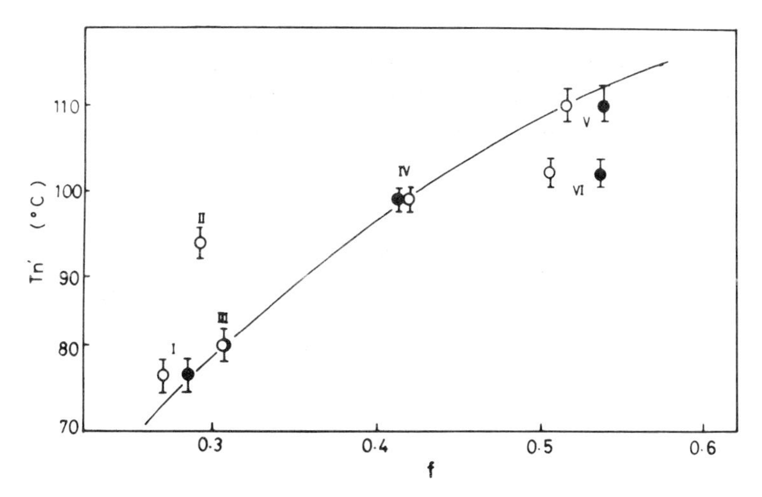

FIGURE 11 Relation between T_n' and f for PVAc: solid line, theoretical; ○, experimental (Kitaigorodskii's method); ●, experimental (molar volume at boiling point).

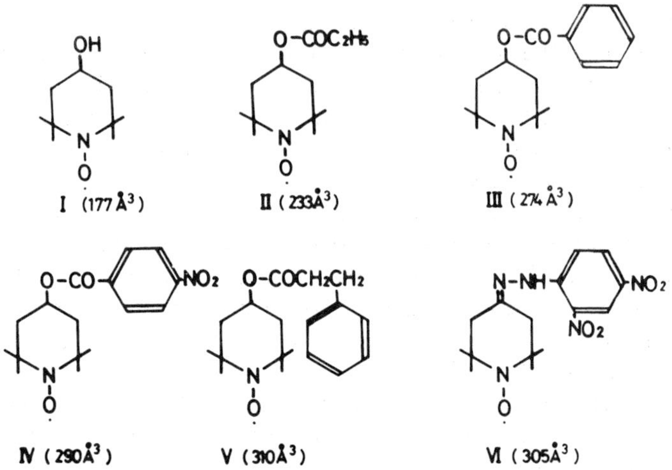

FIGURE 12 Nitroxide probes.

narrowing temperature, T_n, increases with probe size except for probes III and V in HDPE, and probe III in PP.

Figure 14 shows the Arrhenius plot of correlation frequency for probe II in HDPE; two distinct breaks in the cruve are observed. In the frequency range $10^9 > 1/\tau_c > 10^6$ an equation by Freed[18] was used.

$$\tau_c = a(1 - S)^b \quad (17)$$

where S is the ratio of one-half the extrema separation at a given temperature to the z component of the hyperfine interaction tensor, a and b are constants which depend on the diffusion model used. In this paper the values of a and b

TABLE IV

Activation energies in kcal/mol.

	HDPE			PP		
Probe	Ea_1	Ea_2	Ea_3	Ea_1	Ea_2	Ea_3
I	0.7	5.7	11	1.7	5.3	13
II	2.0	5.2	11	1.9	6.5	13
III	2.6	5.4	11	1.8	6.5	14
IV	3.0	5.0	13	2.8	6.5	13
V	1.3	5.0	11	1.2	6.6	14
VI	1.3	4.9	12	1.4	5.7	14

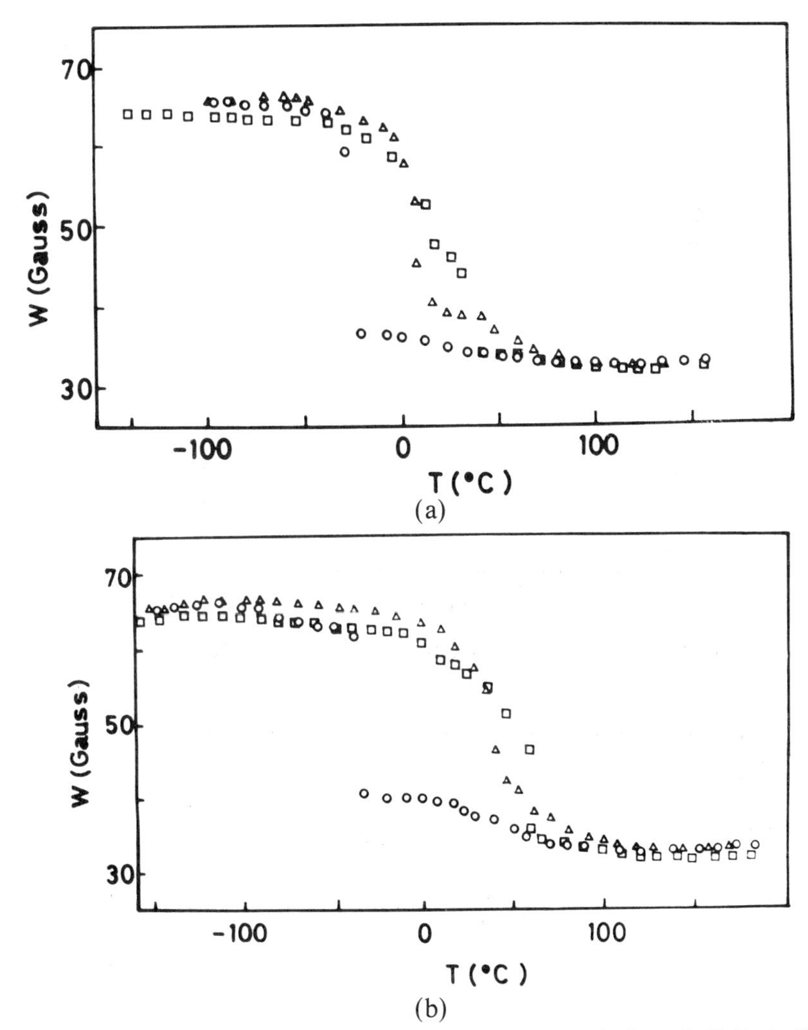

FIGURE 13 Temperature dependence of extrema separation: ○, Probe I; □, Probe IV; △, Probe V; (a) HDPE, (b) PP.

for the "moderate jump" diffusion model were used. The temperatures at which the two breaking points occur are dependent on the polymer species and independent of the probe species. The activation energies at lower (Ea_1), medium (Ea_2), and higher temperature (Ea_3), were estimated and are listed in Table IV. Note that the value of Ea_3 is nearly independent of the probe species. The magnitude of these activation energies suggests that they may possibly be associated with the oscillational, rotational, and jumping modes of the probes, respectively.

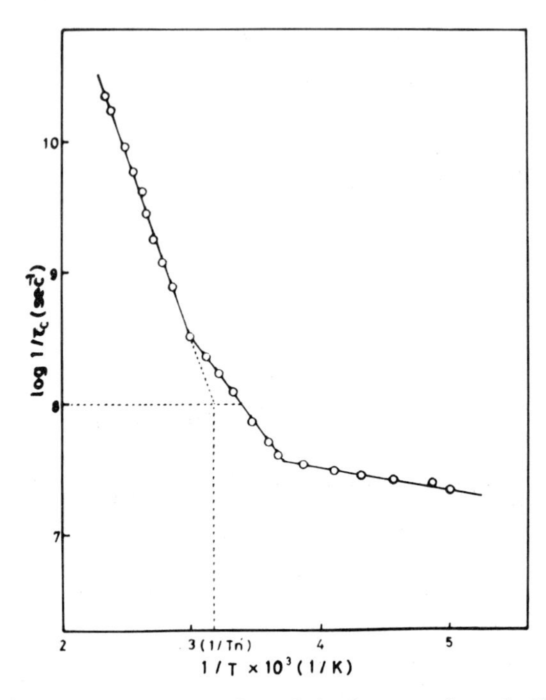

FIGURE 14 Arrhenius plot of correlation frequency for probe II in HDPE.

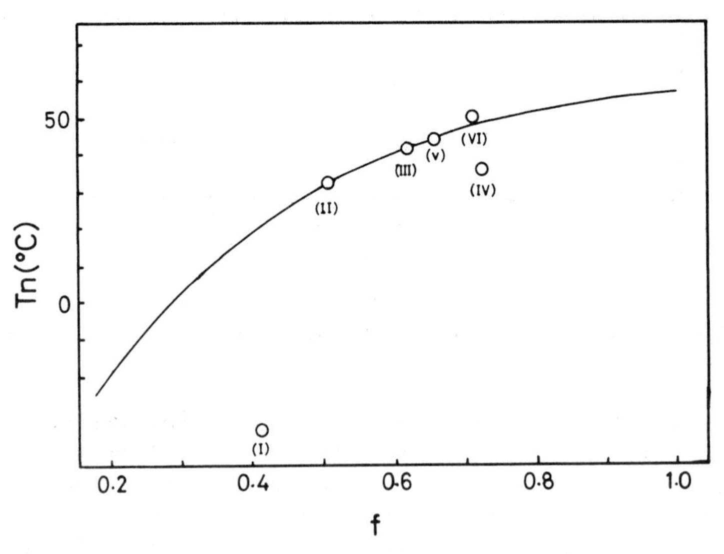

FIGURE 15 Relation between T_n and f for PP: solid line, theoretical; ○, experimental (Kitaigorodskii's method).

Figure 15 shows the plot of T_n *vs.* f for PP, normalizing probe II to the theoretical curve. Probe I deviates considerably from the curve in this figure. Thus, a supposed narrowing temperature, T_n', is employed. As depicted in Figure 14, T_n' is defined as the temperature at which the Arrhenius relation giving Ea_3 intersects the point of the $\log 1/\tau_c = 8$. Table V lists the values of T_n' for the probes I–VI. Figure 16 shows the plot of T_n' *vs.* f for PP. This plot was found to be in better agreement with the theoretical curve.

TABLE V

Values of T_n' for PP.

Probe	T_n' (°C)
I	34
II	48
III	52
IV	62
V	58
VI	66

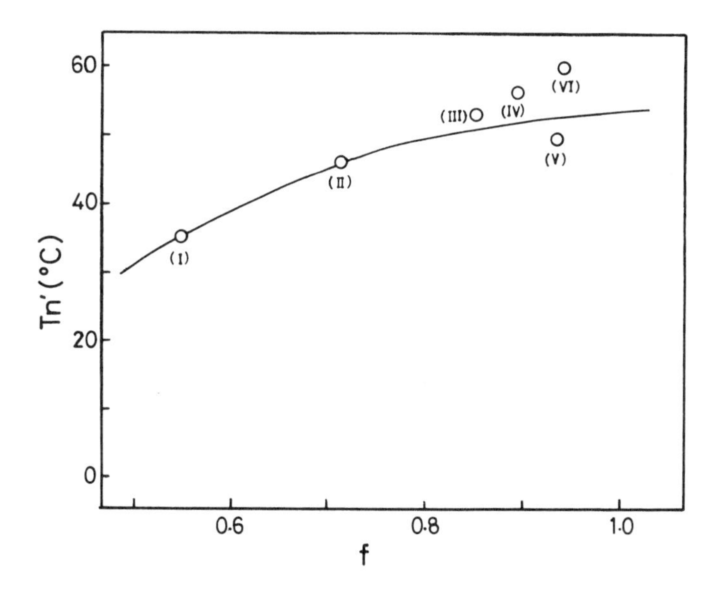

FIGURE 16 Relation between T_n' and f for PP: solid line, theoretical; ○, experimental (Kitaigorodskii's method).

APPLICATION OF THE THEORY[10,12]

From Figure 6 we can estimate the segmental size of a polymer above its glass transition by measuring T_n and T_g since we know the molar volume of the probe. Table VI lists the estimated values of f, the molar volume of the segment, the number of backbone atoms in the segment, and the diameter of the segment, calculated assuming a spherical shape for the segment.

For the calculation of v_m and v_m', the molar volume at boiling point[19] and Kitaigorodskii's method[17] were employed. The latter case includes only the excluded volume of the segment. The number of backbone atoms in a segment was estimated by multiplying the number of backbone atoms in the repeating unit of the polymer by the excluded volume ratio of the repeating unit to the segment. Thus, 45–98 and 65–215 backbone atoms are included in a segment of NR and NBR, respectively, depending on the content of combined sulphur. Sulphur was not taken into consideration for these calculations. The values, however, are nearly equivalent to those estimated by dividing the number of backbone atoms of the polymer[12] by f, within $\pm 2\%$, thus the number estimated may be valuable to a first approximation.

Figure 17 shows the number of backbone atoms as a function of T_g. The values for NBR are found to be located above the values for NR. When the segment size of NBR is the same as that of NR, T_g is about 5–15°C lower than that of NR.

TABLE VI

Calculated size of the segment.

No.	T_n (°C)	T_g (°C)	ΔT (°C)	f	v_m cc/mol	d_m Å	v_m' Å^3	d_m' Å	N
NR									
3	20	−22	42	0.268	876	14.0	660	10.8	45
4	25	−12	37	0.246	955	14.5	719	11.1	50
5	28	− 3	31	0.219	1073	15.0	808	11.5	56
6	31	12	19	0.175	1609	15.1	1011	12.4	70
7	33	27	6	0.125	1880	15.3	1416	13.9	98
NBR									
3	19	− 3	22	0.182	1291	16.0	973	12.3	65
4	24	4	20	0.175	1343	16.2	1011	12.4	68
5	26	11	15	0.156	1505	16.8	1135	13.3	76
6	29	36	− 7	0.090	2611	17.1	1966	15.5	142
7	30	41	−11	0.055	4273	23.8	3218	18.3	215

$\Delta T = T_n - T_g$, v_m and d_m are from molar volume at boiling point,[19] v_m' and d_m' are calculated from Kitaigorodskii's method,[17] N is the number of backbone atoms in the segment, T_g was measured by dilatometry.

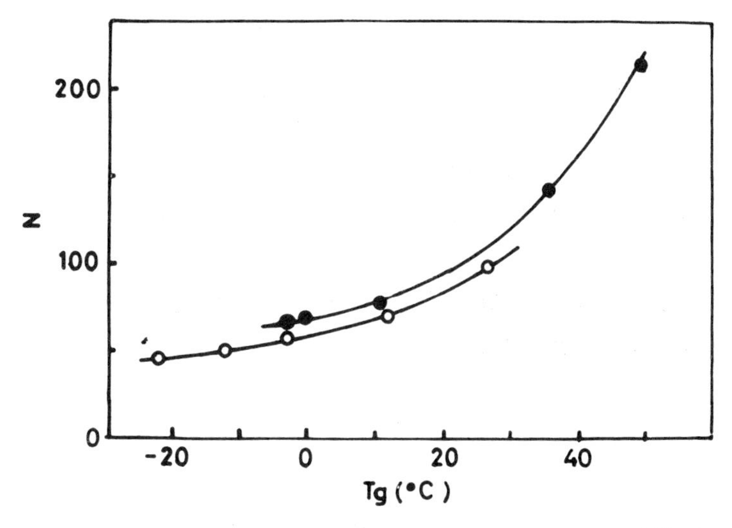

FIGURE 17 Relation between number of backbone atoms of segment and T_g: ○, NR; ●, NBR; data from Table VI.

If the supposed T_n (T_n') is used, the segment size for PVAc and PP is estimated to be ~7–8 monomer units. The value for PP ($T_g = -10°C$) is smaller than that from viscoelastic data of 30 monomer units.[20]

The glass transition temperature of high density polyethylene was estimated from eq. (16) using two probes of different size. T_n' was found to be −16°C and −2°C for probe I and III, respectively. Substituting these values into eq. (16), setting up two equations, and solving them graphically, T_g and the molar volume of the segment, v_m', were estimated to be −108°C and 297 $Å^3$, respectively. This value of T_g is somewhat lower than that reported by Kumler and Boyer[21] in the range of −70 to −80°C.

STUDY OF ORIENTED POLYMERS USING A LINEAR PROBE[22]

In biological research the spin label method has been widely used. A linear spin probe is especially effective to study biological membranes whereby the probe aligns itself with the host lipid molecules and reflects the motions and the molecular arrangement of its surroundings. This technique may be similarly applied to the study of oriented synthetic polymer systems.

Figure 18 shows the linear spin probe used in these experiments. The axis of the nitrogen p-orbital in the oxazolidine ring is parallel to the alkyl chain.

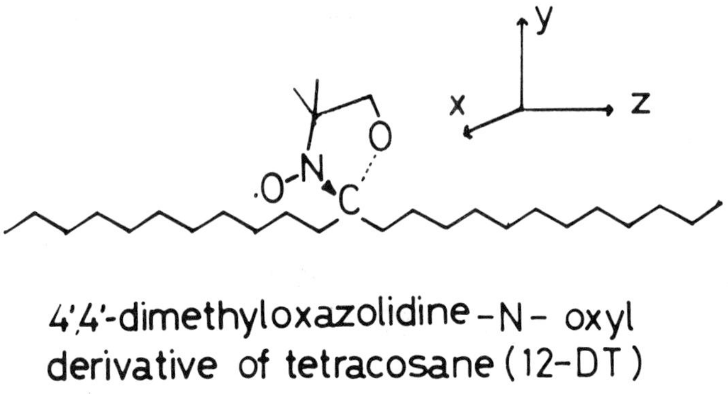

4',4'-dimethyloxazolidine-N-oxyl derivative of tetracosane (12-DT)

FIGURE 18 Chemical structure of linear probe.

Figure 19(a) shows the ESR spectra for the undrawn sample of polyethylene at various temperatures. At −196°C the spectral line shape shows a typical rigid spectrum very similar to that observed with the spherical probe (2,2,6,6-tetra-methyl-4-piperidinol-1-oxyl). From −70 to 50°C small subsplittings appear which are not observed for the spherical probe. Since the probe used in these experiments has long alkyl chains, it is expected that the probe may experience anisotropic motion, namely, rotational motion about the chain axis. This subsplitting may be due to the averaging of the x and y components of the hyperfine tensor.

Figures 19(b) and (c) show the ESR spectra of the drawn samples in which the drawing directions are parallel and perpendicular to the static magnetic field, respectively. If the polyethylene chains and the probe are completely aligned along the draw direction, then only the z-component of the hyperfine tensor determines the spectral line shape, when the draw axis is parallel to the static magnetic field. When the draw axis is perpendicular to the static magnetic field, the spectral line shape is determined by the x and y components of the hyperfine tensor. The extrema separation in the parallel spectrum taken at −196°C is larger than that in the perpendicular spectrum. Spectra taken at −90°C and at −10°C show a similar tendency. These spectra show subsplittings due to the x and y components of the hyperfine tensor, suggesting uniaxial rotation of the probe. At around 75°C the parallel and perpendicular spectra show very similar line shapes suggesting the beginning of true isotropic motion. Above 150°C the spectra are observed to be sharp triplets, indicating rapid isotropic motion of the probe. Assuming that these probe motions reflect similar motions in the host polymer, then, stepwise changes in the motional modes of polyethylene is suggested.

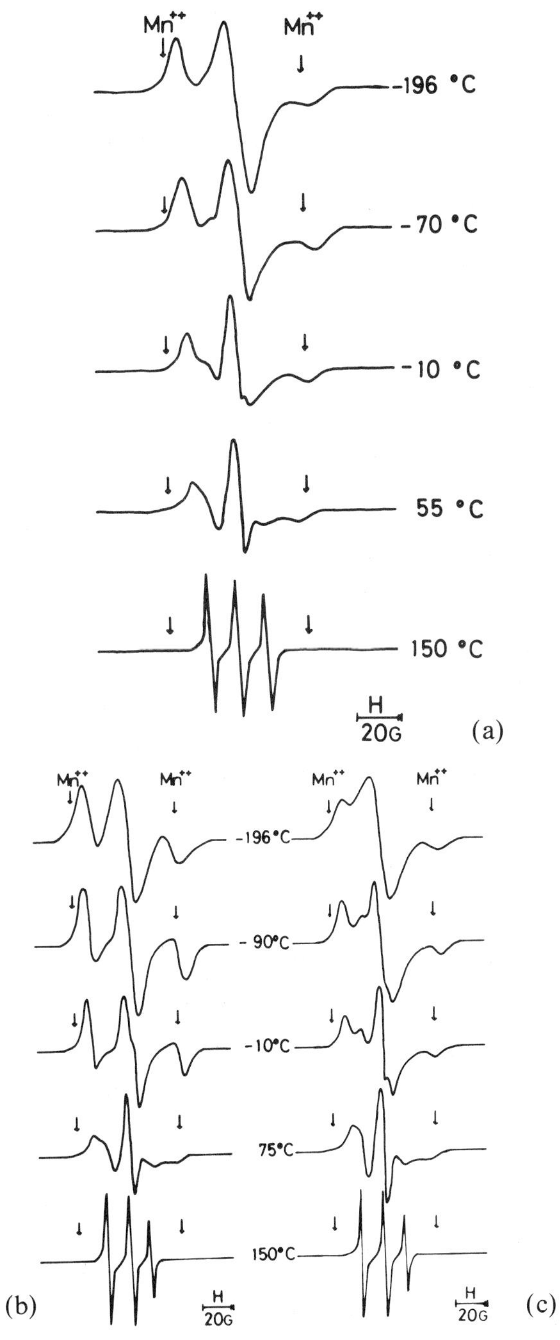

FIGURE 19 ESR spectra of polyethylene using linear probe: (a) undrawn sample, (b) drawn sample, draw axis is parallel to H_0, (c) drawn sample, draw axis is perpendicular to H_0; peak positions for Mn^{++} standard shown by arrows.

Figure 20 shows the change of extrema separation with temperature for the undrawn and 900% drawn samples. It should be noted that the value of T_n, around 90°C, is considerably higher than that found using a spherical probe such as 2,2,6,6-tetramethylpiperidinol-1-oxyl. The highly restricted motion of the oxazolidine ring may be due to the incorporation of the alkyl chains of the probe with the surrounding host polymer chains. For the drawn samples the extrema separation in the parallel sample is always larger than that in the perpendicular sample. Around −20°C a small narrowing of the separation is observed in the perpendicular sample, or in a bulk sample, corresponding to the appearance of the spectral subsplittings shown in Figure 19. This may be due to the initiation of rotational motion of the polymer chain. In the temperature range from 30 to 90°C, the extrema separation in the perpendicular sample is larger than that in the bulk sample. This is attributable to a depression of molecular motions by an increase in crystallinity by drawing. T_n increases with elongation in both the parallel and perpendicular samples leveling off above 600%, as shown in Figure 21.

Above 100°C the ESR spectra show very sharp triplets with a line separation of 34 gauss, indicating isotropic motion of the probe. Figure 22 shows the Arrhenius plots of correlation frequency of the linear probe in the temperature range from 100 to 150°C. The plots show good linearity in the temperature range measured and activation energies were estimated. Figure 23 shows the plot of activation energy versus elongation. The activation energy for the undrawn sample agrees well with that of jumping diffusion of a

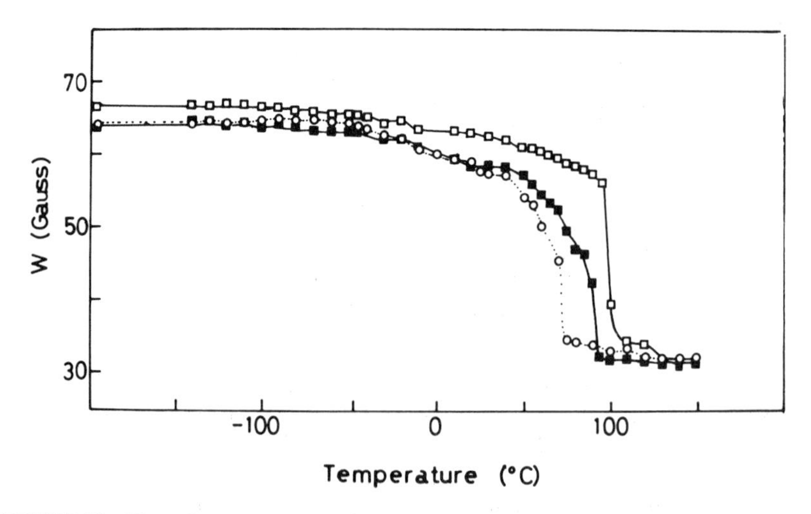

FIGURE 20 Plots of extrema separation *vs.* temperature for PE: ○, undrawn; □, elongation 900%, draw axis is parallel to H_0; ■, elongation 900%, draw axis is perpendicular to H_0.

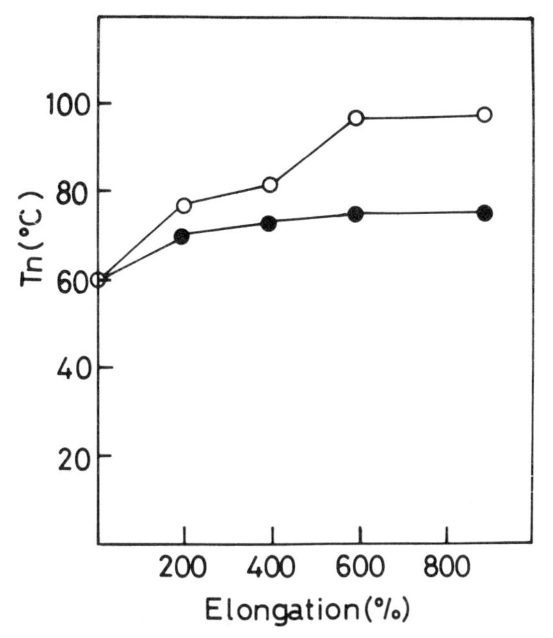

FIGURE 21 Plots of T_n *vs.* elongation for drawn PE. ○, draw axis is parallel to the static magnetic field; ●, draw axis is perpendicular to the static magnetic field.

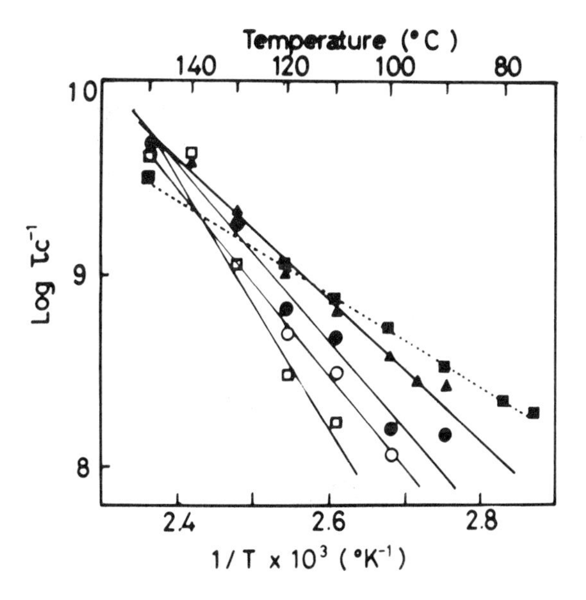

FIGURE 22 Arrhenius plots of correlation frequency for PE: ■, undrawn; ▲, 200% elongation; ●, 400%; ○, 600%; □, 900%.

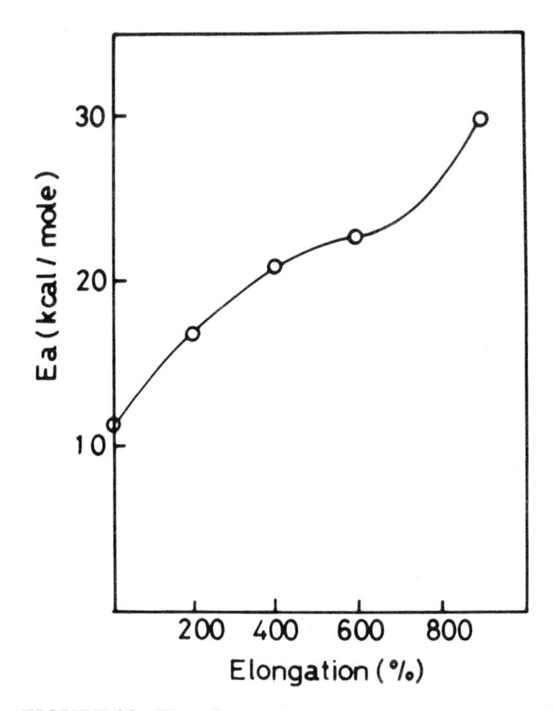

FIGURE 23 Plot of activation energy *vs.* elongation for PE.

spherical probe used in the rubbers. In the undrawn sample the oxazolidine ring of the probe may be located in the amorphous region and experiences micro-Brownian motion in this temperature range. The energy markedly increases with elongation, indicating immobilization of the chains by drawing.

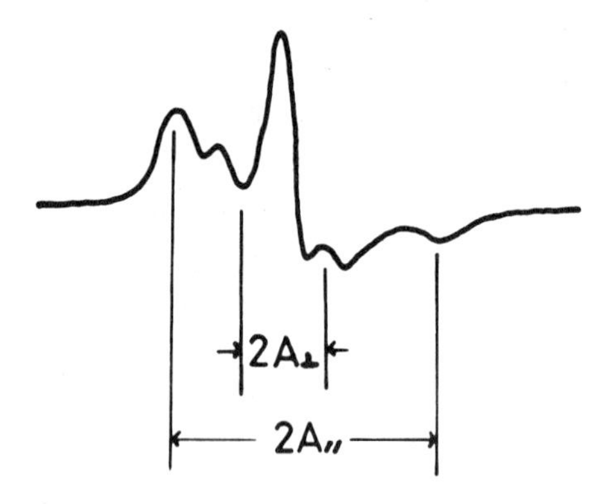

FIGURE 24 Hyperfine separations determining the order parameter S.

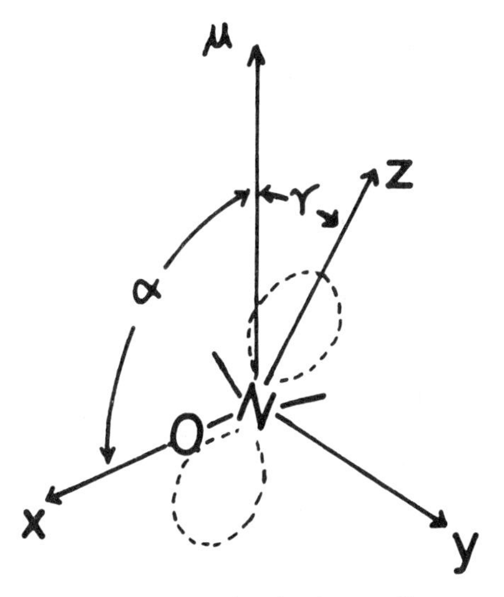

FIGURE 25 Schematic representation of molecular coordinate system and rotating axis of probe.

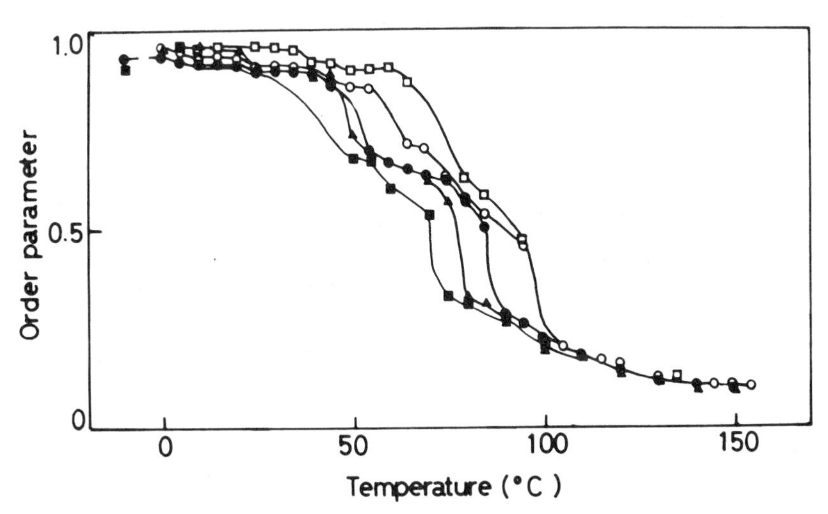

FIGURE 26 Plots of order parameter *vs.* temperature for PE: ■, undrawn; ▲, 200% elongation; ●, 400%; ○, 600%; □, 900%.

The order parameter S is given by the following equation:[23]

$$S = \frac{A_{\parallel} - A_{\perp}}{A_{zz} - A_{xx}} \frac{a_N}{a'_N} \tag{18}$$

where $A_{\parallel}$ and $A_{\perp}$ are the hyperfine separations shown in Figure 24, A_{zz} and A_{xx} are the z and x components of the hyperfine tensor, $a_N = 1/3(A_{zz} + 2A_{xx})$, and $a_N' = 1/3(A_{\parallel} + 2A_{\perp})$. The order parameter represents the alignment of the nitrogen p-orbital with the rotating axis of the probe, μ, as shown in Figure 25. Figure 26 shows the change of the order parameter with temperature. The order parameter seems to change in four distinct steps: a first slight decrease around 0°C, a relatively rapid decrease around 40°C, the most rapid decrease around 70°C, and a final gradual decrease above 80°C, for the bulk sample. These temperatures are shifted upwards and the temperature range narrows with elongation. The detailed origin of these stepwise decreases in the order parameter is unclear at present. They may be related to a stepwise change in the degree of uniaxial oscillational and/or rotational modes and arrangements of the host molecules. Above 110°C all samples have the same order parameter value indicating that no effect of elongation on molecular motion is present above this temperature.

REFERENCES

1. V.B. Stryukov and E.G. Rozantsev, *Vysokomol. Soedin., A*, **10**, 626 (1968).
2. G.P. Rabold, *J. Polymer Sci., A-1*, **7**, 1203 (1969).
3. D. Kivelson, *J. Chem. Phys.*, **33**, 1094 (1960).
4. S.A. Goldman, G.V. Bruno, and J.H. Freed, *J. Chem. Phys.*, **76**, 1858 (1972).
5. M.S. Itzkowitz, *J. Chem. Phys.*, **46**, 3048 (1967).
6. M. Shiotani and J. Sohma, *Polymer J.*, **9**, 283 (1977).
7. S.C. Gross, *J. Polymer Sci., A-1*, **9**, 3327 (1971).
8. P.L. Kumler, S.E. Keinath, and R.F. Boyer, *Polymer Prepr.*, 28 (1976).
9. N. Kusumoto and H. Mukoyama, *Rep. Progr. Polymer Phys., Japan*, **16**, 555 (1973).
10. N. Kusumoto, S. Sano, and T. Kijima, *ibid.*, **20**, 519 (1977).
11. K. Honda, K. Hamada, and N. Kusumoto, *ibid.*, **20**, 515 (1977).
12. N. Kusumoto, S. Sano, and N. Zaitsu, *Preprints of 14th ESR Symposium*, Tokyo, Japan, 1975; N. Kusumoto, S. Sano, N. Zaitsu, and Y. Motozato, *Polymer*, **17**, 448 (1976).
13. N. Kusumoto, M. Yonezawa, and Y. Motozato, *Polymer*, **15**, 793 (1974).
14. N. Kusumoto and H. Mukoyama, *Rep. Progr. Polymer Phys., Japan*, **16**, 551 (1973).
15. E.R. Andrew, "Nuclear Magnetic Resonance", University Press, Cambridge, 1955.
16. F. Bueche, "Physical Properties of Polymers", Wiley, New York, 1962.
17. A.I. Kitaigorodskii, *Organicheskaya Kristallokhimiya, Izd. Acad. Nauk., USSR* (1955).
18. S.A. Goldman, G.V. Bruno, and J.H. Freed, *J. Phys. Chem.*, **76**, 1858 (1972).
19. J.H. Perry, "Chemical Engineer's Handbook", McGraw-Hill, New York, 1950, p. 538.
20. L.T. Muus, *SPE J.*, **15**, 368 (1959).
21. P.L. Kumler and R.F. Boyer, *Macromolecules*, **9**, 903 (1976).
22. N. Kusumoto and T. Ogata, *Rep. Progr. Polymer Phys. Japan*, **21**, 463 (1978).
23. W.L. Hubbell and H.M. McConnell, *J. Amer. Chem. Soc.*, **93**, 314 (1971).

DISCUSSION

P. L. Kumler (State University of New York, Fredonia, New York): Could you explain how you determine the T_n values?

N. Kusumoto: We often see doubled peaks in the ESR triplet spectra of nitroxide radicals near the narrowing temperature region. This makes determination of T_{50G} difficult in the narrowing region. Therefore, we define T_n as the temperature at which the intensity of the two peaks becomes equal. Usually, T_n nearly coincides with T_{50G}.

P. Törmälä (Tampere University of Technology, Tampere, Finland): Your model gave values of N (the number of backbone atoms of the polymer segment) as a function of the degree of crosslinking of natural rubber (NR). Do you also have results of non-crosslinked NR?

N. Kusumoto: The value of N is not dependent on the probe size but on the T_g of each sample. I do not have results for non-crosslinked rubber. I made attempts to duplicate the sample preparation conditions to that of work by wide-line NMR, which has been done and reported in *J. Polymer Sci.* by H. Kusumoto.

J. Sohma (Hokkaido University, Sapporo, Japan): You presented a model by which the size effect could be explained. But it seems to me that the important factor for averaging is not the free volume but the rate of motion. You implicitly assume that a spin-labelled molecule in a void gets sufficiently rapid motion. Is that correct?

N. Kusumoto: Yes. This theory predicts the relation between T_n and T_g. T_n appears when the probe gets enough free volume for jumping diffusion and a jumping frequency of the order of 10^8 Hz. These problems are provided for in our theory.

Spin Probe Studies of Polymers at Temperatures > T_g

PETER M. SMITH

*Midland Macromolecular Institute, Midland, Michigan 48640**

This was a relatively simple experiment which gave anything but simple results. The use of nitroxide spin-probes in synthetic polymers has concentrated on the T_{50G} *vs* T_g correlation as described by P. L. Kumler. This work extended the studies into temperature regions > T_g where the polymers were rubbery or molten. The three line motionally narrowed ESR spectrum of the radical at these temperatures gave, by the usual line-width analysis outlined by A. T. Bullock in his talk, the rotational correlation time τ_c for isotropic Brownian diffusion. A plot of log τ_c *vs.* reciprocal temperature gave Arrhenius plots and activation energies of the relaxation processes involved. The radical was kept at concentrations < 0.01% in the polymers. The systems studied were (a) anionic polystyrene (PS) (fractions of $\bar{M}_n$ = 4,000, 17,500 and 111,000) plasticized with 50% or (in one case) 30% m-bis(phenoxy-phenoxy)benzene (to bring transitions down to regions < 160°C where the probe was stable) (b) atactic, low m.w. polypropylene (PP) (c) crystalline, commercial polybutene-1 (PB-1). The probes were BzONO (I), TEMPOL (II) (in PP) and the rigid steroid probe III (in PP and PB-1).

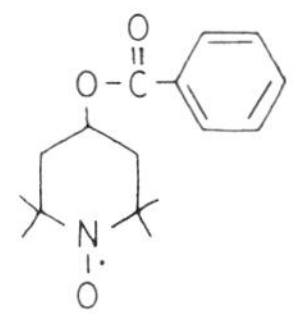

I BzONO

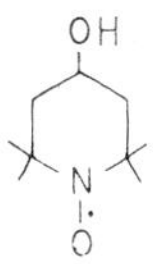

II TEMPOL

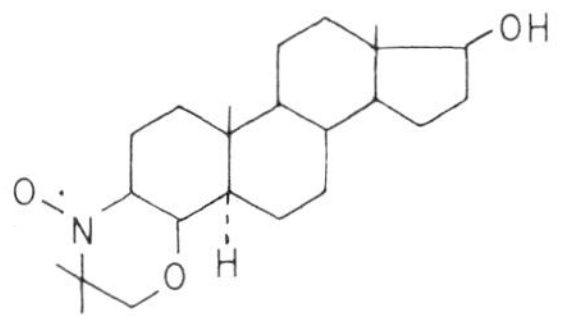

III

*Present address: American Cyanamid Company, Stamford, Connecticut 06904

It was found that there was a marked break in the Arrhenius plots for the probes in amorphous polymers, but not in the crystalline PB-1. (The Arrhenius plot was linear over two orders of magnitude of τ_c for probe I in PB-1.) Accompanying this break was a marked change in the shape of the ESR spectrum.

Figure 1 shows the Arrhenius plots for probe I in the plasticized PS samples. Table I summarizes the results and also in Table I are results estimated from the torsional braid analysis (TBA) data of Gillham *et al.* on one sample of plasticized PS. (See ref. 1 for details and literature references).

The break in the Arrhenius plot has been identified as the controversial T_{ll} transition observed in amorphous polymers by a variety of techniques (see ref. 1) including Gillham's work. This is considered to be a third-order transition when the polymer chains change from hindered motion into more true liquid like motion. It is believed that this ESR work provides strong evidence for the existence of the T_{ll} transition.

The problem of anisotropic motion must be considered. The equations due to Kivelson for calculating τ_c assume isotropic rotation, but the probe was

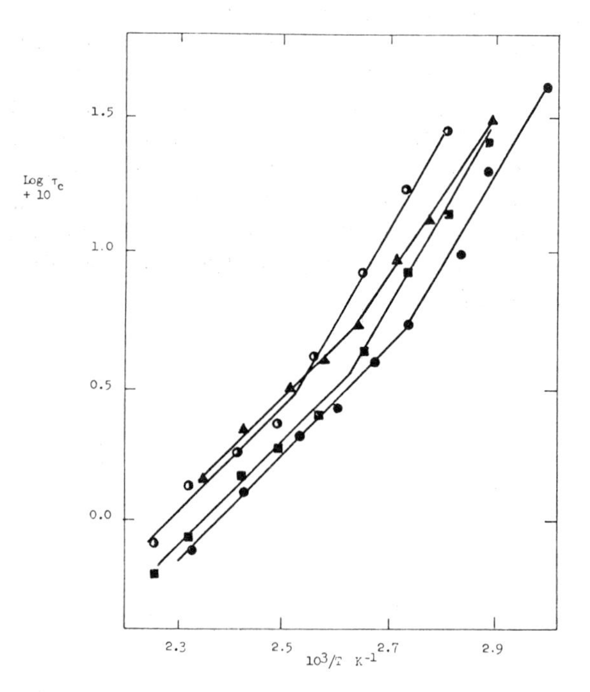

FIGURE 1 Arrhenius plots for probe I in plasticized polystyrene. ●, PS 4,000 + 50% plas.; ■, PS 17,500 + 50% plas.; ▲, PS 110,000 + 50% plas.; ◑, PS 4,000 + 30% plas. Reprinted with permission from ref. 1. Copyright by the American Chemical Society.

TABLE I

T_{ll} and T_g values by ESR spin-probes and TBA. Temps. in K, activation energies in kJ mol^{-1}

		Technique	ESR			TBA			ESR		
Probe	Polymer	% Plas.	T_{ll}	T_{50G}	T_{ll}/T_{50G}	T_{ll}[a]	T_g[a]	T_{ll}/T_g	$E_1 < T_{ll}$[b]	$E_2 < T_{ll}$[b]	E_1/E_2
I	PS 110,000	50	377	323	1.17	333	290	1.15	57	38	1.5
—	PS 97,000[c]	50	342	—	—	—	—	—	125.4	62.7	2.0
I	PS 17,000	50	381	327	1.17	324	288	1.14	65	38	1.7
I	PS 4,000	50	363	328	1.11	301	269	1.12	67	39	1.7
I	PS 4,000	30	398	338	1.18	334	302	1.11	67	36	1.9
I	PP atactic	0	325	289	1.12	298[d]	265[d]	1.12	61	28	2.2
II	PP atactic	0	314	250	1.26	298[d]	265[d]	1.12	63	58	1.09
III	PP actactic	0	372	301	1.23	298[d]	265[d]	1.12	49	30	1.63
I	PB-1 isotactic	0	—	300	—	—	—	—	55	—	—
III	PB-1 isotactic	0	—	336	—	—	—	—	55	—	—

a. Estimated from the data of J. K. Gillhan, J. A. Benci, and R. F. Boyer, *Polym. Eng. and Sci.*, **16**, 357 (1977).

b. ±4 kJ mole^{-1} maximum error.

c. Data from R. S. Colburn, *J. Macromol. Sci.-Phys.* **B1**, 517 (1967), for Polystyrene $\bar{M}_w = 212{,}000$, $\bar{M}_n = 97{,}000$, plas. di(2-ethyl hexyl)phthalate, from melt viscosities by the falling ball method.

d. R. F. Boyer, *Encyclopedia of Polymer Science & Technology,* (Wiley Interscience) Supp. Vol. 2, 745 (Fig. 8), (1977).

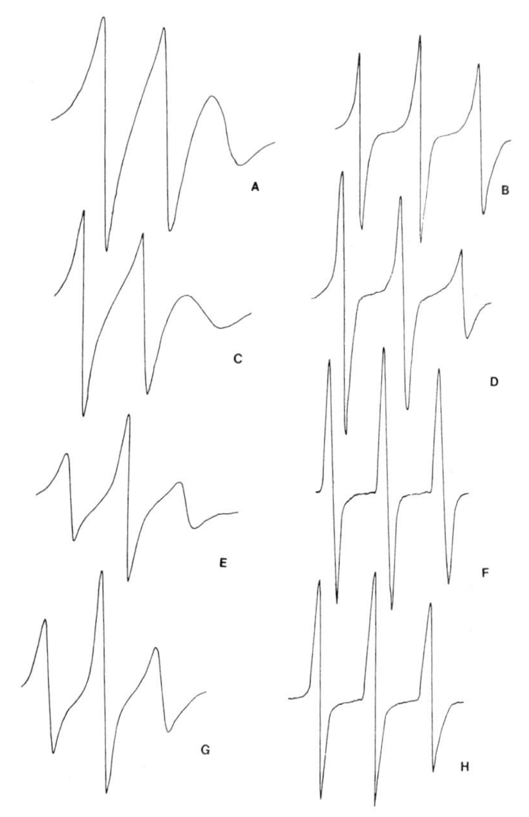

FIGURE 2 Representative spectra. A,C,E, and G for $\tau_c = 2 \times 10^{-9}$ s, B,D,F, and H for $\tau_c = 2 \times 10^{-10}$ s. (A) PP + I at 312 K, (B) PP + I at 355 K, (C) PB + I at 346 K, (D) PB + I at 378 K, (E) PP + III at 331 K, (F) PP + III at 395 K, (G) PB + III at 384 K, and (H) PB + III at 434 K. (From ref. 2). (Reproduced by permission of the Pergamon Press).

clearly moving anisotropically. This is seen in Figure 2 (Spectra C and D). The detailed analysis of anisotropic rotation due to Freed was not applied here, but the work of the Russian group (Kovarskii, Buchachenko *et al.*, details, ref. 2), was utilized to discuss this anisotropic motion. According to Kovarskii, the correlation time for isotropic diffusion and an average τ_c obtained from anisotropic rotation are close in value, and the anisotropic motion determines the shape of the spectrum.

These workers define the degree of anisotropy by the parameter

$$\varepsilon = \frac{B + C}{B - C} = \left\{\frac{(R_+ - 1)}{(R_- - 1)}\right\}_{\text{exp}}$$

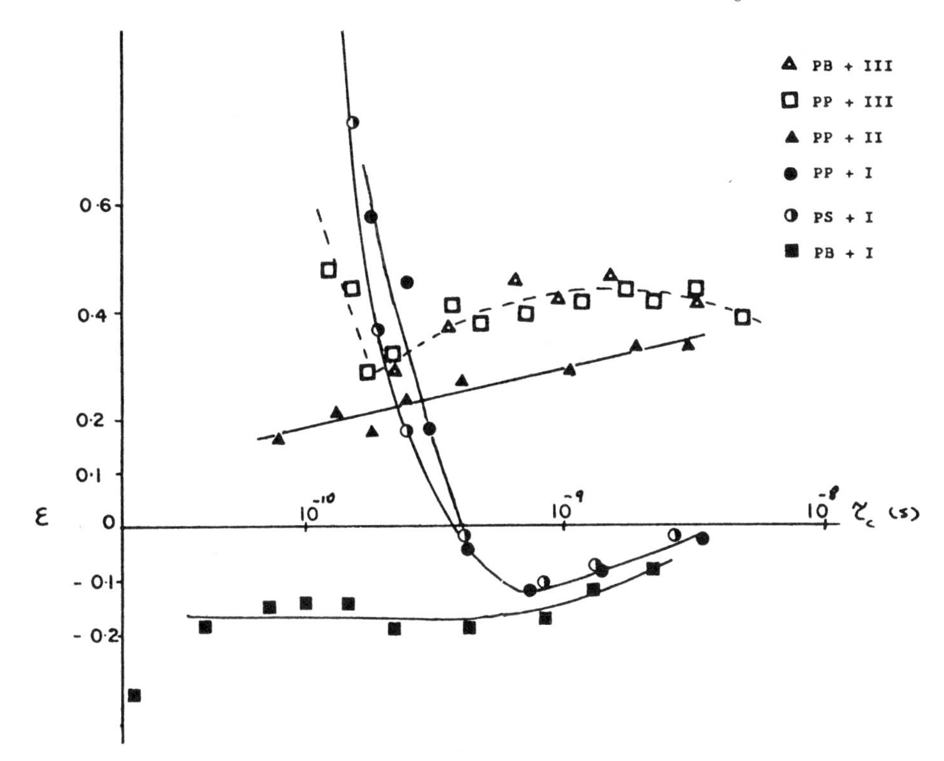

FIGURE 3 Degree of anisotropy ε *vs.* τ_c (isotropic) for: △, PB-1 + III; □, PP + III; ▲, PP + II; ●, pp + I; ◑, PS 17,500 (+ 50% plas.) + I; and ■, PB-1 + I. (From ref. 2). (Reproduced by permission of the Pergamon Press).

where B and C are the coefficients of the m_I terms of the basic line width equation, and R_+ and R_- are the experimental root-ratios of line intensities (see ref. 2). A calculation of ε for the present systems, plotted against τ_c, is shown in Figure 3. It was observed that the motion of probe I in PB-1 remained anisotropic with $\varepsilon < 0$ for all τ_c values, and the same probe in PP and plasticized PS had similar motion for $\tau_c > 10^{-9}$ s. Then there was a dramatic change in motion (at the same value of $\tau_c \sim 4 \times 10^{-10}$ s for PP and PS) and then the probe moved with reversed diffusion axes ($\varepsilon > 0$). This change in the ε value occurred at T_{ll} as shown by the break in the Arrhenius plot. Whatever the details of the probe's motion (and computer simulations of the spectra might clarify this) it did not seem to change its rotational character at T_{ll} in the amorphous systems.

ACKNOWLEDGMENTS

The author wishes to acknowledge the advice and encouragement of R. F. Boyer, who, along with P. L. Kumler made the original suggestion to study T_{ll} by the spin-probe technique.

REFERENCES

1. P.M. Smith, R.F. Boyer, and P.L. Kumler, *Macromolecules*, **12**, 61 (1979).
2. P.M. Smith, *Europ. Polym. J.*, **15**, 147 (1979).

DISCUSSION

A. Yelon (Ecole Polytechnique, Montreal, Canada): What happens if the τ *vs.* T behavior is analysed as a WLF process rather than as an Arrhenius process?

R. F. Boyer (Midland Macromolecular Institute, Midland, Michigan): Computer analysis in the T_{ll} region shows that we clearly have two straight lines rather than a quadratic which you would expect if we had WLF type behavior.

D. J. Meier (Midland Macromolecular Institute, Midland, Michigan): Your results show that the temperature dependence of the correlation times can be fitted with an Arrhenius-type equation with a constant energy of activation. This seems surprising since the correlation times should be directly related to the melt viscosities—which, in your temperature region of interest, should follow a WLF type of equation—which gives "activation energies" which are not constant and a strong function of temperature.

P. M. Smith: We're really above the WLF range in most of our systems, we start our measurements 30–40° above T_{50G} which may be 50° above T_g already.

R. F. Boyer: I have analyzed by computer a great deal of the published data for the WLF equation and we find that in the small range around the T_{ll} transition the Arrhenius plot is in fact two linear portions intersecting at one temperature.

W. G. Miller (University of Minnesota, Minneapolis, Minnesota): I am confused! The correlation times were calculated assuming isotropic rotation, yet you have shown that you have anisotropic motion. What meaning, then, do breaks in the Arrhenius plots have?

P. M. Smith: Russian workers have shown that τ_c for isotropic rotation, and an average τ_c for anisotropic rotation, are very close. The change in ε for probe I in amorphous polymers which quantifies the spectral shape change corresponds to a marked change in rotational motion, and occurs at T_{ll}. Therefore, there is a real transition at this temperature and this transition is observed by other techniques. If the break in the Arrhenius plots were an artifact of anisotropic rotation, the linearity of the plot for probes I and III in PB-1 would be inexplicable. I gratefully acknowledge W. G. Miller bringing to my attention a paper by Polnaszek and co-workers (*Arch. Biochem. and Biophys.*, **167**, 505 (1975)) which discusses the problem of false breaks in the Arrhenius plots. Although caution is necessary, the present evidence for T_{ll} seems too coherent to dismiss the transition as an artifact due to anisotropic motion of the probes.

A. T. Bullock (University of Aberdeen, Old Aberdeen, Scotland): I would like to suggest that the reason for E_η (ESR) $< E_\eta$ (bulk viscosity) in the case of 50% plasticized polystyrene is that the probe is preferentially solvated by the plasticizer. In this case, where two components are present, the temperature dependence of the probe correlation times will give an activation energy appropriate to the microviscosity, not the bulk viscosity.

P. M. Smith: I would agree with this, the absolute values of the activation energies in the plasticized system below and above T_{ll} will have less significance than their ratio. However, the agreement with literature values for bulk viscosity when we calculate E_η using correlation times for the non-plasticized systems (PB-1 in particular) is good support for the validity of our measurements.

B. Rånby (Royal Institute of Technology, Stockholm, Sweden): The T_{ll} transitions are clearly indicated for the amorphous polymer–plasticizer systems described. These transitions signal the onset of a new mode of molecular motion. The T_{ll}/T_g ratios are expected to be affected by the presence of plasticizer, the size of the spin probe, etc. The temperature range between T_g and T_{ll} for an amorphous polymer may cover the "rubbery" zone for the polymer. For crystalline polymers no rubbery zone is found because the crystalline phase gives the polymer solid-state properties between T_g and T_m.

D. J. Meier: Why don't you see T_{ll} in PB-1 which was only about 60% crystalline?

P. M. Smith: T_{ll} is a relaxation of the amorphous region, and also the noncrystalline phase of PB-1 is quite highly structured, even up into the melt.

R. F. Boyer: Some Russians have observed that T_{ll} disappears as the degree of vulcanization increases in polybutadiene. We see T_{ll} in quenched isotactic polystyrene but don't see T_{ll} in the crystalline state. In all these cases we see a restriction of the motions of these large segments which come into play at T_{ll}.

A. M. Bobst (University of Cincinnati, Cincinnati, Ohio): Your tumbling is anisotropic but you are using the isotropic formalism to calculate Arrhenius plots. Why aren't you using $\tau = 1/6R$ of Freed's formalism?

P. M. Smith: The use of Freed's spectral simulation program would be an obvious next step in this analysis. However, I doubt if it would alter the experimental evidence for a change in probe motion at T_{ll}.

NEW TECHNIQUES

Applications of Electron-Nuclear Double Resonance to Polymeric Systems

R. B. CLARKSON

Varian Associates, Inc., Palo Alto, California 94303

I. INTRODUCTION

Electron paramagnetic resonance (EPR) has proven to be an extremely useful spectroscopic technique in studying the structure and mechanical properties of polymeric systems. Utilizing native radicals characteristic of certain systems, or introducing radicals by means of radiation damage or chemical additions to a polymer in the form of spin labels, experimenters have been able to probe the chemistry and modes of motion of this important class of matter.

In cases of solids, where the radicals are native to the polymer system, such as the radicals found in coals and the products of coal liquefaction, or where they are introduced by high-energy radiation or fracture processes, EPR alone may not reveal details of structure or motion desired by the experimenter. This most often is the result of inadequate resolution of the EPR spectrum due to phenomena such as broadening of the resonance lines by fast spin-lattice relaxation, or the averaging of anisotropic g or hyperfine tensors of the paramagnetic center over the entire range of angles formed between principal magnetic axes of the center and the external magnetic field, H_0, resulting in the broad, "powder pattern" spectrum typical of disordered, and even highly ordered but randomly oriented solids.

Under such circumstances, it may be advantageous to exploit the coupling that often exists between the paramagnetic electron and neighboring nuclei with nuclear spins $I \neq 0$, in order to gain insights into unresolved hyperfine structure, electron-nuclear dipole-dipole interactions, and motions of the electron spin relative to lattice nuclei. This technique, known as electron-nuclear double resonance (ENDOR) spectroscopy, was first proposed and demonstrated by Feher in 1956.[1] This paper will survey the general theory of the ENDOR technique, briefly examine the instrumental requirements for performing the experiment, and then discuss its recent application to polymers. A bibliography is provided at the conclusion of the article for those readers wishing to gain more extensive, first-hand information about this powerful double-resonance approach to the study of polymeric systems.

II. THEORY OF THE ENDOR TECHNIQUE

Put most simply, ENDOR spectroscopy obtains useful information about a paramagnetic center by at least partially saturating the electronic Zeeman transitions of the system with microwave radiation at a fixed external magnetic field, H_0, and then inducing electron spin relaxation by causing rapid nuclear spin transitions among nuclei coupled to the electron. Nuclear spin transitions are induced by means of a second, radio-frequency field simultaneously impressed on the sample, hence the *double* resonance name generic to this class of experiments. Induced transitions in the nuclear magnetic resonance region provide alternate relaxation pathways for the saturated electron spins; the overall effect is observed by monitoring variations in the EPR signal that is being "pumped" with microwave and rf radiation, and noting the rf frequencies at which EPR signal changes occur. A diagram of the pump (NMR) and observe (EPR) portions of the ENDOR experiment are shown in Figure 1 for a system with $S = ½$, $I = ½$, in which the electron and nucleus are coupled by some unspecified mechanism. It should be noted here that the electron and nucleus under discussion may be part of the same atom, or may be associated with different, usually neighboring atoms. In more complex cases, the electron may find itself coupled to many different nuclei with I≠0, and each different coupling could provide a separate relaxation pathway for the electronic spin.

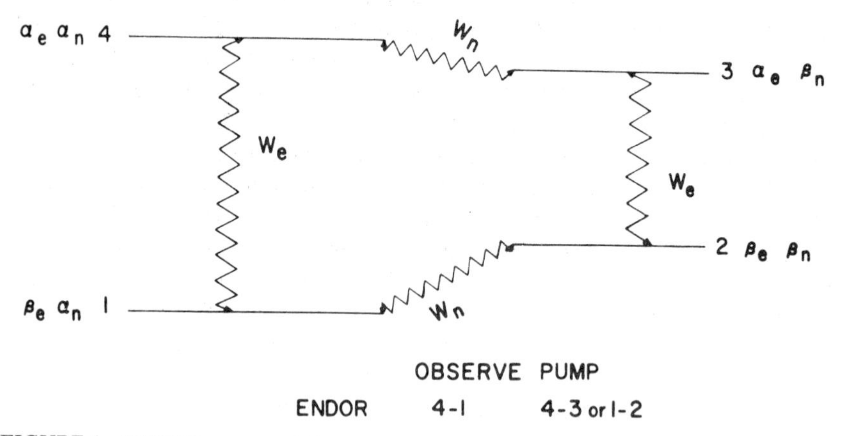

FIGURE 1 ENDOR pump and observe scheme for $I = ½$, $S = ½$ system. W_e is the electron spin transition probability; W_n is the nuclear magnetic transition probability. W_e is the observe and W_n the pump. $\alpha = +½$, $\beta = -½$.

For the present, however, we will restrict our discussion to one electron and one nucleus.

In order to understand what is measured in the ENDOR experiment, let us continue with our $S = ½, I = ½$ system. Most simply, such a system would have a spin Hamiltonian of the form

$$\mathcal{H} = g\beta\mathbf{H}\cdot\mathbf{S} - g_n\beta_n\mathbf{H}\cdot\mathbf{I} + h\mathbf{S}\cdot\mathbf{A}\cdot\mathbf{I}. \quad (1)$$

To compute the magnetic energy levels of this system, let us assume that the electronic Zeeman term dominates Equation (1) and that the hyperfine interaction is isotropic. The first assumption enables us to take $\mathbf{H}$ as the z-direction of the spin system and makes m_s and m_I good quantum numbers (strong-field approximation). The second assumption enables us to write the hyperfine term in Equation (1) as h A $\mathbf{I}\cdot\mathbf{S}$.

The magnetic energy levels of the system may then be written as

$$E(m_s,m_I) = g\beta Hm_s - g_n\beta_nHm_I + hAm_sm_I. \quad (2)$$

Substituting via Planck's equation, $E = hv$, we obtain

$$\nu_{\text{res}} = \frac{E(m_s,\ m_I)}{h} = \nu_e\ m_s - \nu_n\ m_I + A\ m_s\ m_I \quad (3)$$

where ν_{res} is the observed EPR frequency, the purely electron and nuclear resonant frequencies of the system are given by

$$\nu_e = \frac{g\beta H}{h} \quad \text{and} \ \nu_n = \frac{g_n\beta_nH}{h} \quad ,$$

and where A, the isotropic hyperfine coupling constant, is given in frequency units.

Now we are able to distinguish between two extreme cases in the energy level scheme of this sysem. Case I, as illustrated in Figure 2, assumes $\nu_n > |A|/2$; Case II assumes $\nu_n <|A|/2$, as shown in Figure 3. In both figures, the first splitting, that of the electronic Zeeman term, is very large and not drawn to scale, and the sign of A is assumed negative.

With the two energy level diagrams derived from limiting cases of Equation 3 in hand, we are now ready to develop hypothetical ENDOR spectra. If we sit at the maximum of the EPR line, $\nu_{\text{res}} = g\beta H_{\text{res}}/h$, and pump our spin system so as to saturate that transition, then sweep over the range of possible nuclear resonances with a strong rf field, we will encounter, in each of our two cases, two values of ν_{rf} that will excite the nuclear spins coupled to the paramagnetic electron. At those frequencies, some of the saturation of the electronic spin will be alleviated due to the strongly induced nuclear transitions, and the EPR

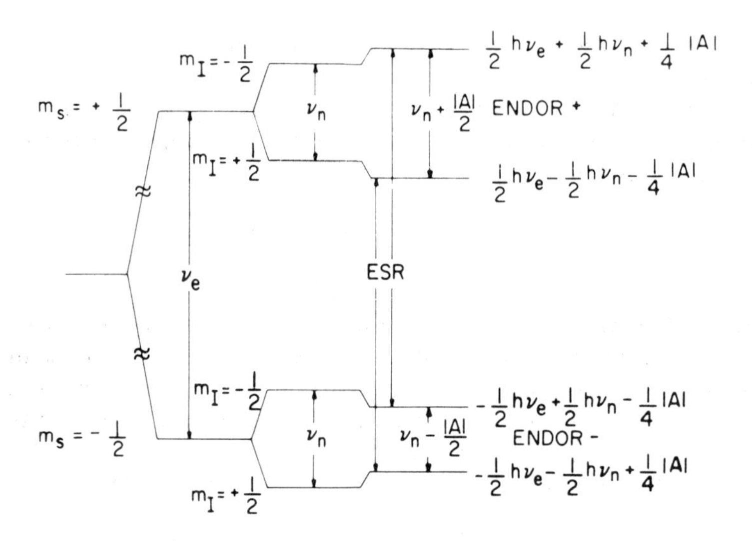

FIGURE 2 Case I, where $\nu_n > |A|/2$ and $A < 0$.

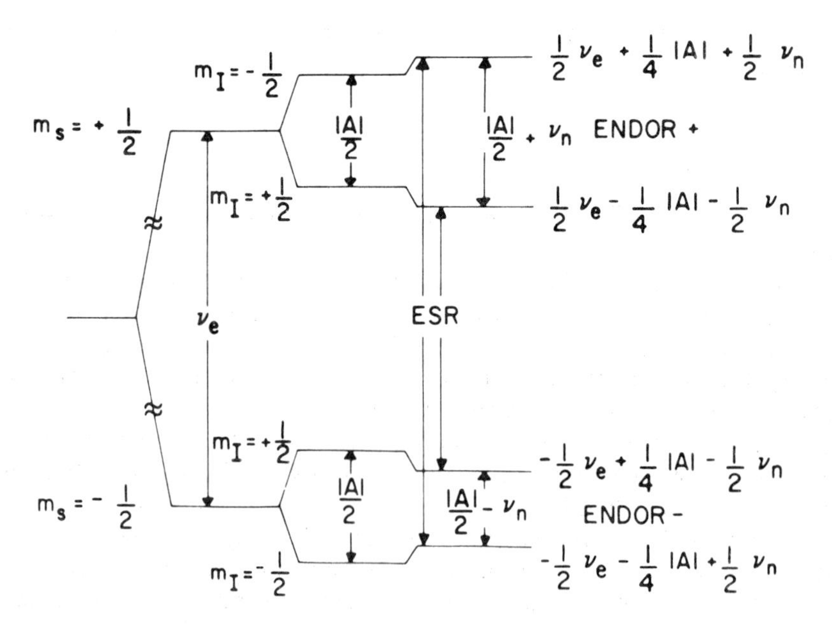

FIGURE 3 Case II, where $\nu_n < |A|/2$.

signal will be observed to increase. Figure 4 illustrates the change in EPR signal intensity noted as we sweep through the coupled electron-nuclear transition frequencies. These peaks are the resonances noted in an ENDOR spectrum and, for our particular example, correspond to the *rf* frequencies

$$\begin{aligned} \nu_{\text{ENDOR}} &= \nu_n \pm \frac{|A|}{2} \text{ (Case I)} \\ &= \frac{|A|}{2} \pm \nu_n \text{ (Case II)} \end{aligned} \tag{4}$$

Thus, the ENDOR experiment for our system results in a spectrum of resonances whose frequencies correspond to $\nu_{\text{ENDOR}} = |\nu_n \pm A/2|$. Note that because of the relationship between ν_n and g_n, the ENDOR technique actually permits the measurement of both $|g_n|$ and $|A|$, thus usually unambiguously identifying the nuclear species whose electron-nuclear interaction is being observed. Of course, the value of g in the electronic Zeeman term is also obtained, making the identification of the paramagnetic species possible in many cases.

Although only the absolute values of g_n and A are generally available, the ENDOR experiment also affords the possibility of the determination of signs. To accomplish this, however, requires that certain additional experimental criteria be fulfilled. Also, it should be noted that while we assumed an isotropic hyperfine interaction, A, and neglected nuclear quadrupole effects completely in Equation (1), these assumptions and omissions are not necessary. In fact,

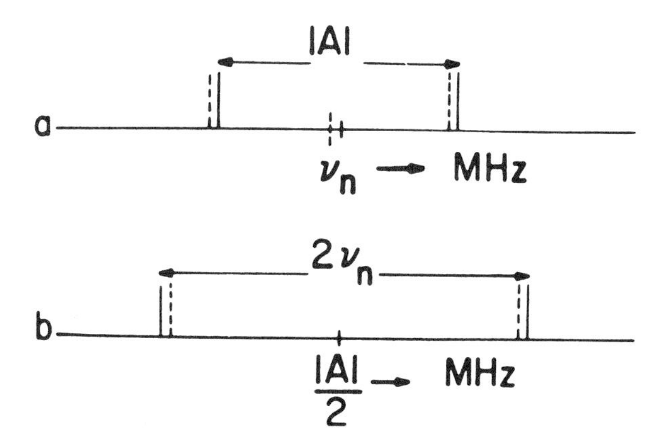

FIGURE 4 ENDOR resonances: (a) $\nu_n > |A|/2$, (b) $\nu_n < |A|/2$. The dashed lines indicate positions when the low-field EPR line is being observed.

the anisotropic hyperfine tensor may be investigated by ENDOR in many ordered solids, and quadrupole couplings also may be measured. The reader is referred to the bibliography following this article for more general treatments of the interpretation of ENDOR spectra.

Thus far we have considered only the information that is available from our simple ENDOR experiment. We shall return later to details of spectral interpretation when we consider real polymer systems. Before moving on, however, we must consider ENDOR mechanisms in greater detail than that illustrated in Figure 1. Only when these mechanisms are understood will we be in a position to estimate what experimental criteria need to be met in order to observe the ENDOR effect.

Figure 5 illustrates a typical ENDOR mechanism for our $s = ½, I = ½$ system. Here, the electronic Zeeman transition is being pumped by microwave radiation of energy $h\nu_e$. In Figure 5(a) and 5(b), the allowed transitions $|-½, +½\rangle \leftrightarrow |+½, +½\rangle$ and $|-½, -½\rangle \leftrightarrow |+½, -½\rangle$, respectively, are being driven. In each diagram is also shown an allowed ($\Delta m_s + \Delta m_I = \Delta m = \pm 1$) spin-lattice relaxation pathway, denoted by T_{1e}. Further, a "forbidden" transition ($\Delta m = 0$) is denoted by the pathway T_{x1}. This transition is the result of the isotropic hyperfine interaction $A\mathbf{I}\cdot\mathbf{S}$, which produces terms in the Hamiltonian of the form (S_+I_-) and (S_-I_+): operators that simultaneously induce electron and nuclear spin flips. The nuclear relaxation path, T_{1n}, also is shown. For this discussion, it is assumed that $T_{1e} \sim T_{x1} \ll T_{1n}$.

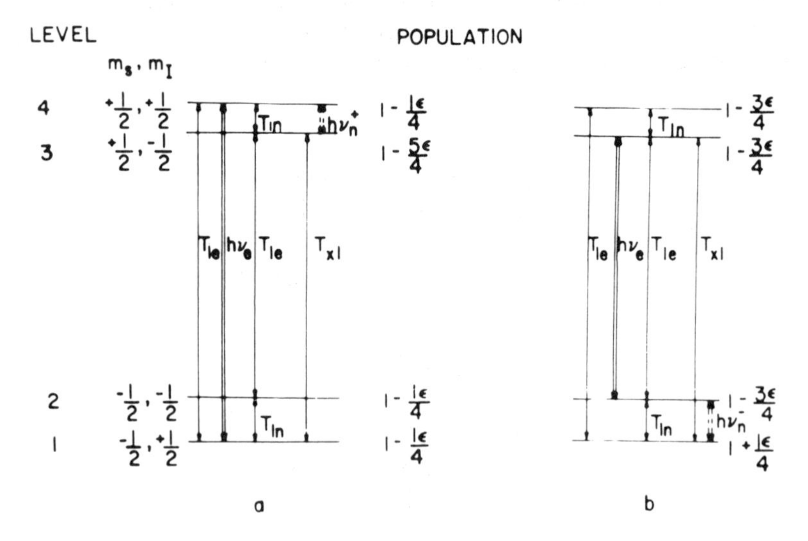

FIGURE 5 (a) ENDOR mechanism for $|A|/2 > \nu_n; h\nu_n^+$ driven. (b) ENDOR mechanism for $|A|/2 > \nu_n$; $h\nu_n^-$ driven. In each case: $\varepsilon = g\beta H/kT$.

Initially, a steady state spin population not in thermal equilibrium with the lattice is achieved by the $h\nu_e$ pump. T_{1e} is operative, but is really effective as a relaxation path only for a single, $\Delta m = \pm 1$ route. Now as the *rf* frequency is swept in the ENDOR experiment, the coupled nuclear transition corresponding to $\nu_n \pm = (|A|/2) \pm \nu_n$ is excited. Suddenly, the induced transition $h\nu_n \pm$ rapidly begins equalizing populations between the $|+½, +½\rangle$ and $|+½, -½\rangle$ states in Figure 5(a) and between the $|-½, -½\rangle$ and $|-½, +½\rangle$ states in Figure 5(b). As a result of this transition, two additional electron spin-lattice relaxation pathways become viable (namely, the second T_{1e} and T_{x1}), and the saturation of the pumped Zeeman transition is alleviated in some measure, resulting in an enhancement of the EPR signal being monitored. It should be noted here that in Figure 5(a), only the nuclear frequency ν_n^+ will be effective in inducing more rapid electron spin relaxation, while in Figure 5(b) only ν_n^- will accomplish that result.

Figure 6 shows an energy level scheme in which both cross or "forbidden" transitions are operative. This is the case when the coupling between electron and nucleus is dipole-dipole in nature, or when the hyperfine interaction is anisotropic (under certain circumstances, such as when there is strong thermal

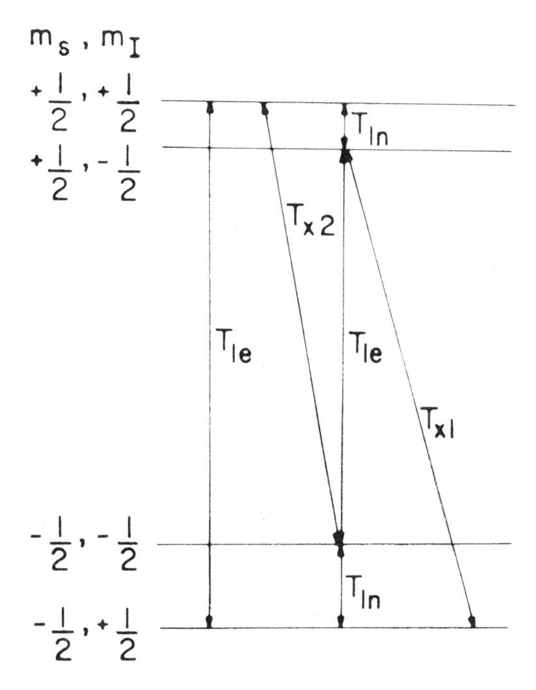

FIGURE 6 Cross-relaxation mechanism in ENDOR with anisotropic hyperfine interaction.

modulation of the hyperfine interaction, this is also allowed in the isotropic case[2]).Then both the (S_+I_-, S_-I_+), and the (S_+I_+, S_-I_-) terms are present in the Hamiltonian, making both ENDOR transitions, $h\nu_n\pm$, effective in producing relaxation.

Up until now, we have considered only an isolated electron interacting with a single $I = 1/2$ nucleus. If the electron in our analysis communicates with other paramagnetic electrons by means of a spin-spin interaction, it can transfer saturation through the $\mathbf{S}_1 \cdot \mathbf{S}_2$ interaction. In cases of extremely rapid spin-spin interaction, the T_{1e} process becomes saturated, and no ENDOR effect is observed. For intermediate cases, spin-spin interactions only weaken the ENDOR response by providing alternate relaxation pathways to that afforded by the ENDOR experiment.[3] Thus, increasing the concentration of paramagnetic centers in order to achieve stronger EPR signal intensities reaches a limit when $\mathbf{S}_1 \cdot \mathbf{S}_2$ interactions become too dominant. Hyde found, for example, that the ENDOR response in a system of tetracene cations dissolved in sulfuric acid at 273 K *decreased* by a factor of 2.5 when the tetracene concentration was raised from 10^{-3} to 10^{-2} M.[4]

Mechanisms other than those affecting the rate of electron spin relaxation to the lattice also are possible ENDOR processes: packet-shifting,[5] line-shifting,[6] and "distant ENDOR".[7] As yet, these mechanisms have not been found to be important to the ENDOR of polymer systems, and so we omit further discussion of them.

In summary then, we see that to achieve the ENDOR effect we need to meet several experimental criteria:

a) The electron spin transition must be capable of saturation (often possible only at very low temperatures);
b) The ENDOR transition must be driven by an *rf* field of sufficient intensity to significantly alter the rate of electron spin relaxation to the lattice;
c) Competing relaxation mechanisms, such as spin-spin interactions, should be kept to a minimum; and
d) The EPR transition being pumped must be continuously monitored in a stable fashion while the *rf* frequency is being swept.

If conditions a)–d) are met, then ENDOR spectroscopy offers the experimenter enhanced resolution of hyperfine interactions, a probe of electron delocalization via an interpretation of electron-nuclear dipole-dipole mechanisms, and a tool for the study of molecular motion, to name but three applications we will consider in later sections of this paper.

III. INSTRUMENTATION

For a fundamental ENDOR experiment, one needs to have a stable EPR spectrometer capable of monitoring a selected resonance for rather long periods of time without drift, and a source of variable frequency *rf* radiation able to impinge on the sample within the microwave structure. For optimum sensitivity, the microwave H_1 field and the *rf* H_1 field should be mutually orthogonal. Both must be orthogonal to the external magnetic field H_0. This basic arrangement, with a rather remarkable number of variations, has been achieved by a number of workers. The reader is referred to the bibliography for details of individual spectrometer systems.

Work described in this paper which was done at Varian Associates, Inc., was performed on an E-1700 ENDOR spectrometer, in conjunction with an E-112 EPR spectrometer operating at a nominal frequency of 9.5 GHz. A block diagram of the instrument is shown in Figure 7. The E-1700 was operated in a pulsed mode, giving a better signal-to-noise ratio for polymer

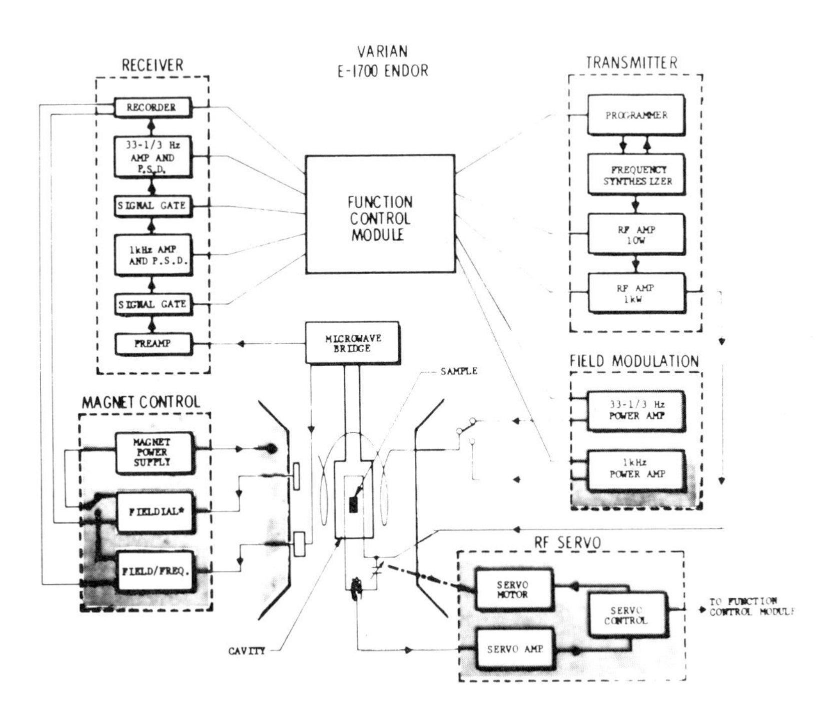

FIGURE 7 Block diagram of the Varian E-1700 ENDOR Spectrometer.

systems that were difficult to saturate. Continuous wave ENDOR experiments on polymer systems also were carried out on the E-1700 system. EPR stability was insured by using a field-frequency lock, and temperatures were controlled either by a Varian E-257 variable temperature accessory (77 K–573 K) or by an Air Products Heli-Tran system (5 K–273 K).

IV. ENDOR IN POLYMERIC SYSTEMS

We are now in a position to consider those features of ENDOR spectra most frequently encountered in polymeric systems. This will be accomplished in two stages: first we shall deal specifically with the origin and interpretation of the so-called Matrix ENDOR line frequently seen in randomly oriented solids; second, we will discuss two examples of ENDOR in polymers currently under investigation in our laboratory. This route, it is hoped, will acquaint the reader with the advantages of ENDOR spectroscopy applied to polymeric systems as well as with problems in the interpretation of spectra that frequently are encountered.

A. Matrix ENDOR

In the ENDOR spectrum of a polycrystalline solid containing paramagnetic centers surrounded by protons interacting with the paramagnets via magnetic dipole-dipole interactions, one usually sees an ENDOR line at the resonance frequency of free protons (14.4 MHz in a field of 3390 G). The origin of this line, called the Matrix ENDOR line, was first considered by Hyde, *et al.*,[8] and has subsequently been the object of much theoretical and experimental study.[9–15] The Matrix peak is believed to arise exclusively from electron-nuclear dipolar (END) interactions in a rigid lattice. Electron spin relaxation to the lattice is accelerated when the *rf* field excites protons that have an essentially unshifted value of ν_n (or γ_n). The END mechanism gives rise to terms in the spin Hamiltonian of the form:

$$\mathscr{H}_{\text{dipole}} = \frac{\gamma_e \gamma_n \hbar^2}{r^3} (A + B + C + D + E + F) \tag{5}$$

where:

$A = I_z S_z (1-3\cos^2\theta)$ (secular term)

$B = -\frac{1}{4}(1-3\cos^2\theta)(S_+ I_- - S_- I_+)$ (nonsecular)

$C = -3/2 \sin\theta \cos\theta \, e^{-i\phi} (S_z I_+ + S_+ I_z)$ (pseudosecular)

$D = -3/2 \sin\theta \cos\theta \, e^{i\phi} (S_z I_- + S_- I_z)$

$E = -3/4 \sin^2 \theta \, e^{-2i\phi} S_+ I_+$

$F = -3/4 \sin^2 \theta \, e^{2i\phi} S_- I_-$ (nonsecular)

Of these terms, C and D are singularly important for our problem, since for an $S = ½$, $I = ½$ system like that discussed previously in Figure 5, these are the terms permitting nuclear spin relaxation between the $|+½, -½> \leftrightarrow |+½, +½>$ or $|-½, -½> \leftrightarrow |-½, +½>$ level pair.[16] Terms B, E, and F, which induce the cross transitions ($\Delta m = 0$ and $\Delta m = 2$), become important Matrix ENDOR pathways when the matrix of protons surrounding the paramagnetic electron exhibit considerable motion, as Leniart, *et al.*[11,12] showed, and as we shall discuss presently. For the rigid lattice, however, it is the pseudosecular terms C and D of the END Hamiltonian that provides a "second" T_{1e} pathway via T_{1n}, thus creating a more rapid spin-lattice relaxation for electrons and alleviating the saturation. In the context of Figure 5, it is as if T_{x1} were very slow ($T_{1e} \ll T_{x1}$) and relaxation to produce the ENDOR effect proceeded accordingly.

The usefulness of Matrix ENDOR lies precisely in its sensitivity to motion, environment, and paramagnetic electron delocalization, thus allowing the experimenter to probe these parameters, to which the EPR signal may be virtually insensitive. For example, as protons of the matrix begin to execute forms of motion (rotation, vibration, etc.), Hyde[8] and later Leniart[11,12] showed that this produced significant changes in the Matrix ENDOR lineshape, sometimes causing it to disappear completely! Lineshape was sensitive to the model used for nuclear relaxation, as Leniart[11] showed by simulating Matrix ENDOR lineshapes for the case of constant and angle-independent nuclear relaxation and variable and angle dependent T_{1n} dominated by the pseudo-secular (C and D) terms in the dipolar Hamiltonian, as shown in Figure 8. He also showed that *all* the terms of Equation (5) may contribute to the lineshape if motion of the spins becomes so rapid that the correlation time for their motion, τ_c, becomes comparable with the EPR resonant frequency ($\omega_e \tau_c \sim 1$). Sensitivity to motion is useful, not only to those studying indigenous radicals in polymers, but also to those working with nitroxide spin-labeled compounds. Here, although the nitrogen hyperfine structure in the EPR spectrum may be obscured by line broadening, the nitrogen ENDOR signal can be analyzed as a motional probe.[17]

Because the origin of Matrix ENDOR involves interaction of the paramagnetic electron with many nuclei comprising its environment, the signal can yield important information on the degree of localization of the electron, the size of its wave function and, in conjugated polymers, an estimate of the number of double bonds comprising a continuous chain. Kevan and co-workers[14] have made calculations on the chain lengths of γ-irradiated

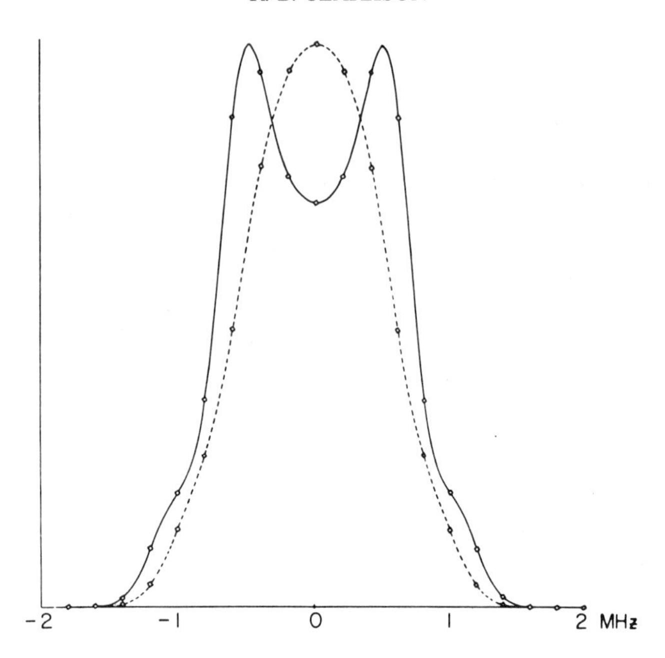

FIGURE 8 A comparison of the ENDOR lineshapes resulting from two different types of nuclear relaxation. Reprinted with permission from (D. S. Leniart, J. S. Hyde, and J. C. Vedrine, *J. Phys. Chem.*, **76**, 2097 (1972)). Copyright by the American Chemical Society.

poly(vinyl chloride), poly(methyl methacrylate), poly(vinyl fluoride), and poly(vinylidene fluoride). They also offer proof that the radical formed in the radiation damage is of the polyenyl type, based on the narrow width of the Matrix ENDOR signals observed.[13] In the case of PVF and PVF_2, a matrix line was observed for the fluorine nucleus as well as for protons.

In summary, then, Matrix ENDOR provides a probe of the nature of radicals in solid polymer systems that can be of great value—particularly in elucidating the environment of unpaired electrons, the degree of spatial delocalization, and the motion of nuclei neighboring the radical.

B. Two Applications of ENDOR to Polymer Systems

For the past year, our laboratory has been actively studying the ENDOR of coals and coal derivatives, and of DuPont's Kevlar 49 and 29. Both of these systems exhibit native paramagnetic resonances, and thus provide convenient, if challenging applications of EPR and ENDOR in the investigation of structure and the characterization of the radicals found in the polymeric systems. We will conclude this article with a brief description of the work thus far completed on these two very different ENDOR problems, as well as suggestions of the direction future work will take.

The discovery of EPR signals in natural carbons was made independently by Uebersfeld[18] and Ingram.[19] Since then many EPR studies on coals have been published, and the literature prior to 1966 has been reviewed.[20,21] Retcofsky and co-workers have done much recent EPR work,[22,23] and representative EPR spectra of American coals taken by them are shown in Figure 9. While there there is some variation in linewidth and lineshape in the spectra as one goes from peat to a very hard, graphitic coal like Dorrance anthracite, there is very little in the EPR spectrum to suggest what some of the characteristics of the radical species in these samples might be. In col-

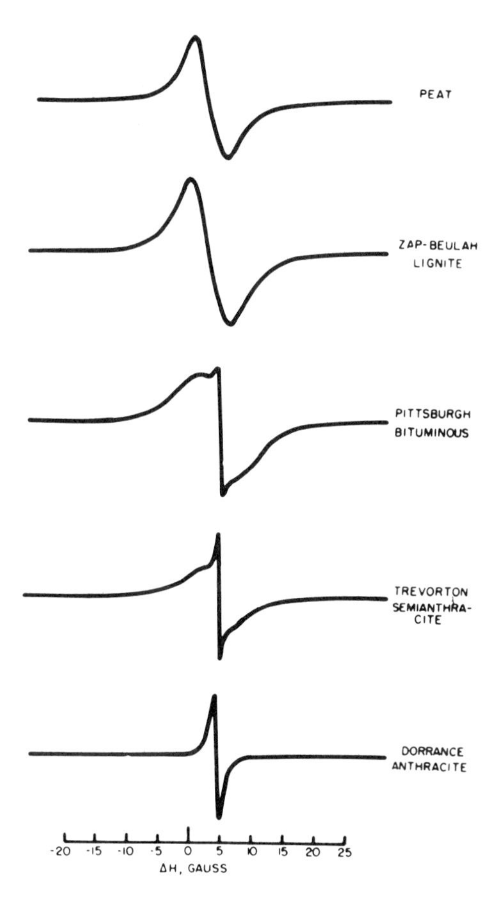

FIGURE 9 Representative EPR spectra of selected American coals. Reprinted with permission from (H. L. Retcofky, J. M. Stark, and R. A. Friedel, *Anal. Chem.*, **40**, 1699 (1968)). Copyright by the American Chemical Society.

laboration with H. Retcofsky, we began a systematic ENDOR study of bituminous and sub-anthracite coals in order to study their structure, and the relationship between radicals found in the solid with those produced by coal liquefaction processes.

In all cases, the coal samples were pulverized, then placed in tubes and evacuated at 383 K for 48 hours at 10^{-6} Torr (1 Torr = 133.3 Nm^{-2}). Sealed tubes were placed in an E-1700 Varian ENDOR spectrometer, and spectra were taken at temperatures from 295 K to 113 K. Figure 10(a) shows the ENDOR spectrum of Bruceton bituminous coal powder at 113 K immediately after very hard grinding in a ball mill and agate mortar. Similar, well-resolved ENDOR spectra were obtained for other coals of this hardness (and harder) and the hyperfine coupling constants obtained from these spectra are given in Table I, together with the measured hyperfine constants for protons in some aromatic compounds believed to be polymerized into large sheets in ordered coals of this type. General trends in the hyperfine constants support the views of Retcofsky[23] that the coals are highly aromatic (> 80%) and that EPR linewidths are a function of unresolved hyperfine structure. The absence of a Matrix ENDOR line for the Bruceton sample shown in Figure 10(a) is particularly surprising, since it suggests that the paramagnetic electron encounters virtually no END interactions.

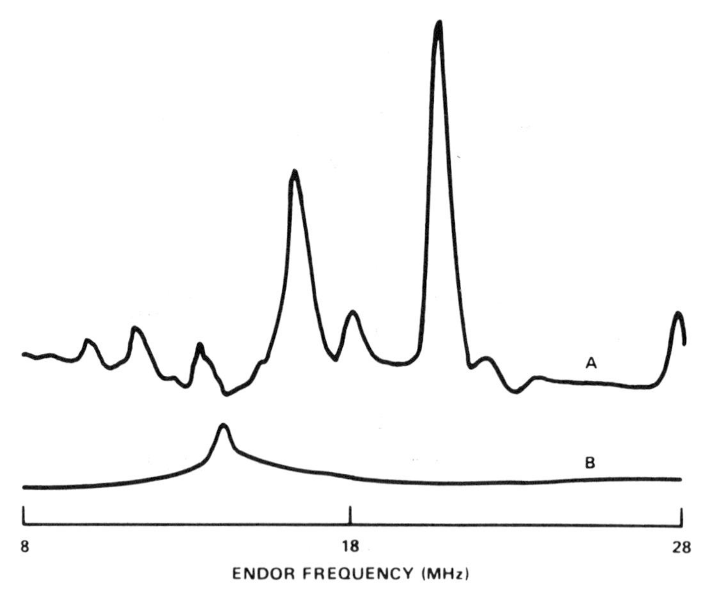

FIGURE 10 (a) ENDOR of freshly crushed Bruceton bituminous coal after heating to 383 K and evacuation to 10^{-6} Torr. (b) ENDOR of Bruceton coal 10 months after crushing, and hydrated by air.

TABLE I
ENDOR of Coals

Origin	A (MHz)	A (Gauss)
Adaville (sub A)	13.6	0.36
	14.1 (free proton)	
	16.4	1.64 (−)
	16.8 (shoulder)	1.92
	18.3	3.0 (−)
	20.5	4.57
	23.8	6.9
	27.7	9.7
Bruceton	13.5	0.43 (−)
	14.1 (free proton not seen)	
	16.4	1.64
	18.1	2.86
	20.6	4.64
	22.3	5.86
	27.8	9.79
Pocahontas #4	13.6	0.35
	14.1 (free proton)	—
	16.4	1.64
	18.3	3.0
	20.7	4.7
	22.1	5.71
Now for comparison:		
perylene	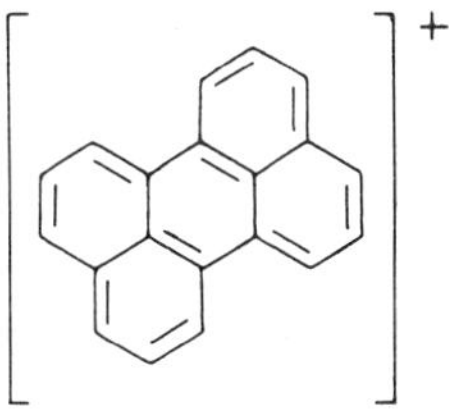 [perylene cation structure, $[\ ]^+$]	$A_\alpha = 4.11$ G $A_\beta = 3.09$ $A_\gamma = 0.46$
anthracene +	$A_\alpha = 6.53$ $A_\beta = 3.06$ $A_\gamma = 1.38$	

If the samples described in Table I and shown in Figure 10(a) are allowed to stand under moist air for several months, the ENDOR signal observed when the samples are re-run will be as illustrated in Figure 10(b). This resonance, at 14.15 MHz, and with a half-width at half-height of ~ 0.8 MHz, is very similar

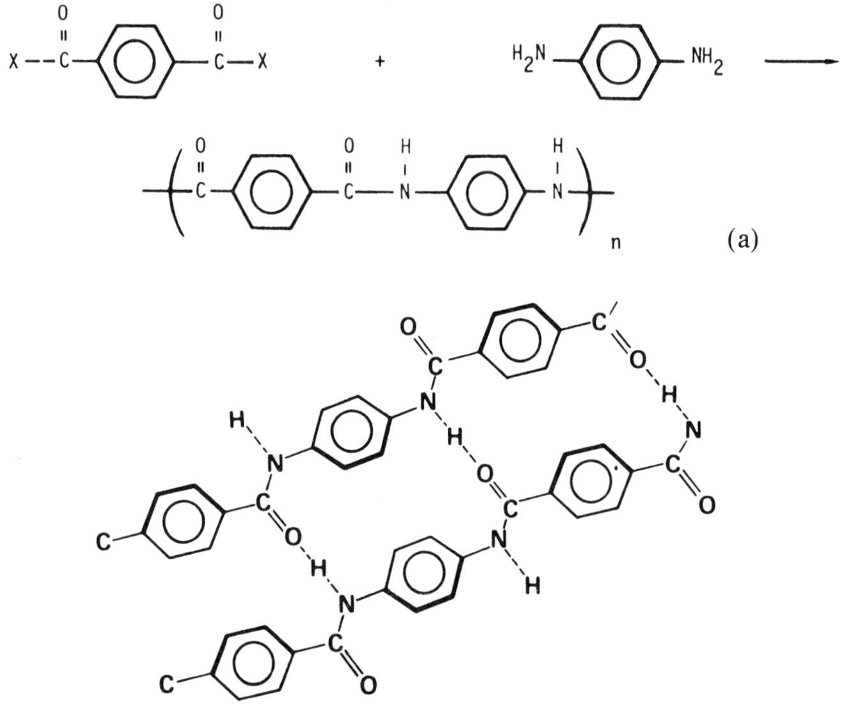

(a)

1. H-bonds for stability
2. Aromatic rings for stability and stiffness
3. Extended chains for high modulus

(b)

FIGURE 11 (a) PPT polymer. (b) Presumed cross-linked PPT in Kevlar.

to that recently reported by Kevan, *et al.*[15] If the sample is re-heated and evacuated, some structure on the Matrix ENDOR line is resolved, but the new splittings of 1.6 MHz and 6.2 MHz do not closely correspond to the splittings previously obtained. Not only does the ENDOR spectrum change, but the EPR spectrum goes from a very narrow line ~ 2 G to a line ~ 10 G peak-to-peak.

Preliminary results indicate the original signals can be obtained from these samples only if they again are subjected to great pressure in grinding. At the moment, we can only speculate that grinding at very high pressures produces a metastable, glassy radical species in the coal, and that time and heat cycling causes this to undergo a phase transformation to the normal, polycrystalline state. This would account for the changes observed between the spectra seen in

Figure 10(a) and 10(b). Work continues on this problem in our laboratory, together with the ENDOR of products of coal liquefaction.

Turning now to a man-made polymer system, we report preliminary spectra of DuPont's Kevlar 49, a fiber made from the polymer poly(p-phenylene-terephthalamide), PPT. The chemical structure of PPT is shown in Figure 11(a), together with the cross-linking chain structure, Figure 11(b). Kevlar 49, as received from DuPont, consists of 1000 filaments, each approximately 11 μm in diameter. X-ray diffraction studies by Northolt[24] show the material to be highly crystalline in character, with a unit cell structure shown in Figure 12. Areas of crystallinity form microcrystals in the fiber, with dimensions on the order of 100 Å by 500 Å.[25]

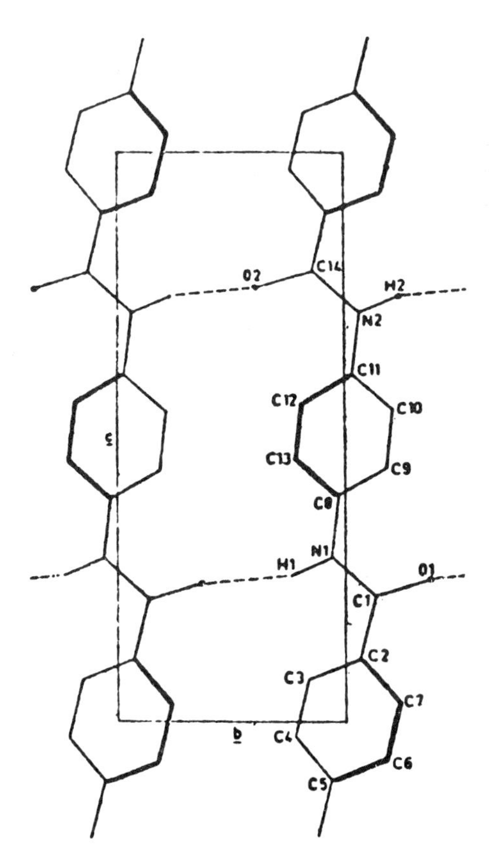

FIGURE 12 Unit cell of PPT. Reprinted with permission from (M. G. Northolt, *Europ. Polym. J.*, **10**, 799 (1974)), copyright by Pergamon Press, Ltd.

Kevlar 49 shows an intense EPR spectrum, illustrated in Figure 13. While the small, sharp line at $g \simeq 2$ may be intensified by heating the fiber to ~ 400 K and evacuating (Kevlar is very hydroscopic), the basic spectral features remain unchanged, even when the material is washed in good organic solvents and dilute aqueous acid.[26] The sharp $g \simeq 2$ resonance is virtually orientation independent, as our work at both 9.5 GHz and 35 GHz has shown. The broad resonances show marked dependence on the angle made by the fiber axis and the external magnetic field, H_0, however, as shown in Fig. 14, and this suggests a highly ordered character to the paramagnetic center responsible for the resonance. The nature of these centers is currently under investigation, with attention being given to the suggestion of G. C. Stevens, *et al.*[27] that the broad resonances may arise from transition metal impurities, notably Cu(II).

ENDOR of Kevlar 49 at 19 K in a totally random orientation is shown in Figure 15. A prominant Matrix ENDOR peak is seen, as well as at least one other resonance at about 27.5 MHz. The half-width at half-height of the Matrix peak is ~ 0.65 MHz; the line seems too narrow to rule out significant delocalization of the electron responsible for it (see References 13 and 14 for the effects of delocalized electrons on Matrix lineshapes). The resonance at 27.5 MHz has not yet been assigned, and ENDOR of the oriented fiber system, similar to that used in the EPR rotational studies, is currently being undertaken. Much greater ENDOR resolution may be possible in the oriented system. Future work will focus on the suggestion that strong exchange interactions between PPT chains exist, giving the polymer its unique mechanical properties, and on the radicals formed during chain scission.

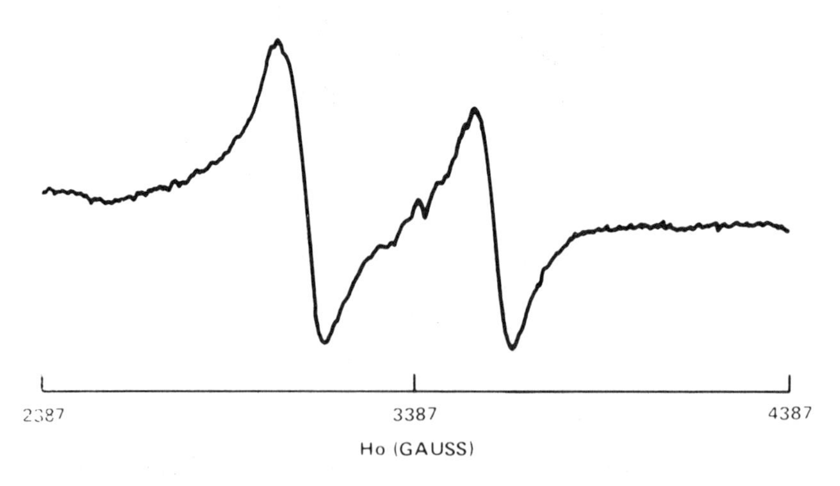

FIGURE 13 EPR spectrum of oriented Kevlar 49.

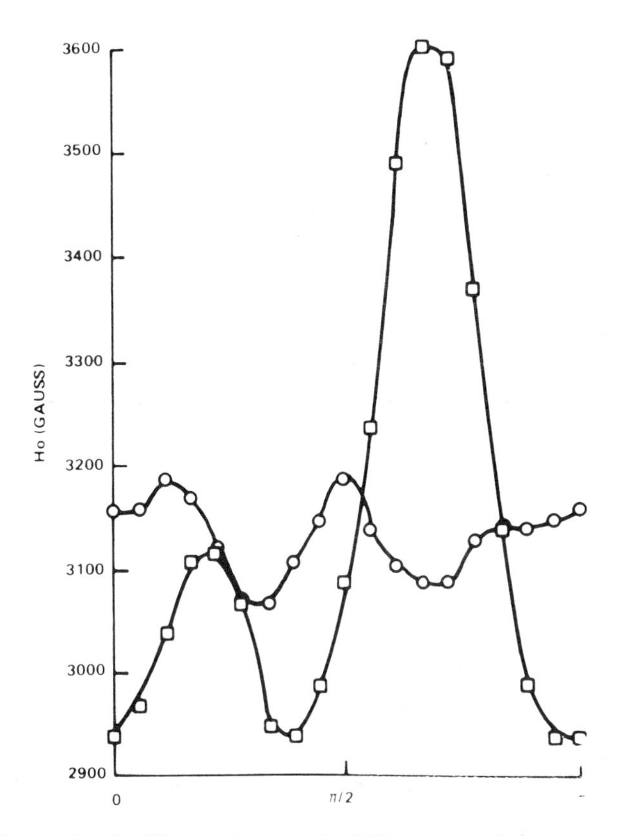

FIGURE 14 Angular (θ) dependence on the EPR resonance lines of oriented Kevlar 49.

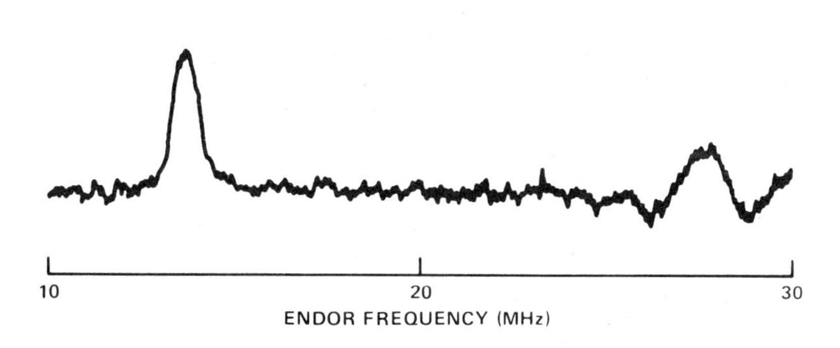

FIGURE 15 ENDOR of unoriented Kevlar fiber at 19 K.

V. SUMMARY

We have traced the origin of the ENDOR effect to induced electron spin-lattice relaxation brought about by radio-frequency radiation. We have seen that existing relaxation mechanisms in the system under investigation play a dominant role in the ENDOR lineshapes and intensities that are to be expected. Matrix ENDOR was seen to be a particularly useful example of this fact, proving sensitive to paramagnetic electron environment, degree of delocalization and motion. These features make ENDOR spectroscopy a particularly useful tool in polymer research. As we learn more about the application of the technique, important new information on the structure of macromolecular and polymeric systems surely will be forthcoming.

VI. REFERENCES

1. G. Feher, *Phys. Rev.*, **103**, 834 (1956).
2. A. Abragam, *Phys. Rev.*, **98**, 1729 (1955),
   C.D. Jeffries, *Phys. Rev.*, **117**, 1056 (1960),
   A. Abragam and B. Bleaney, "*Electron Paramagnetic Resonance of Transition Ions*", Oxford, 1970, pg. 248.
3. L. Kevan and L.D. Kispert, "*Electron Spin Double Resonance Spectroscopy*", John Wiley and Sons, 1976, pg. 18.
4. J.S. Hyde, *J. Chem. Phys.*, **43**, 1806 (1965).
5. G. Feher, *Phys. Rev.*, **114**, 1219 (1959).
6. G. Feher and R.A. Isaacson, *J. Mag. Res.*, **7**, 111 (1972).
7. J. Lambe, N. Laurance, E.C. McIrvine, and R.W. Terhune, *Phys. Rev.*, **122**, 1161 (1961).
8. J.S. Hyde, G.H. Rist, and L.E.G. Eriksson, *J. Phys. Chem.*, **72**, 4269 (1968).
9. L.E.G. Eriksson, J.S. Hyde, and A. Ehrenberg, *Biochim. Biophys. Acta*, **192**, 211 (1969).
10. G.H. Rist and J.S. Hyde, *J. Chem. Phys.*, **52**, 4633 (1970).
11. D.S. Leniart, J.S. Hyde, and J.C. Vedrine, *J. Phys. Chem.*, **76**, 2079 (1972).
12. D.S. Leniart, J.S. Hyde, and J.C. Vedrine, *J. Phys. Chem.*, **76**, 2087 (1972).
13. J. Helbert, B. Bales, and L. Kevan, *J. Chem. Phys.*, **57**, 723 (1972).
14. J.N. Helbert, B.E. Wagner, E.H. Poindexter, and L. Kevan, *J. Polymer Sci. (Phys.)*, **13**, 825 (1975).
15. S. Schlick, P.A. Narayana, and L. Kevan, *JACS*, **100**, 3322 (1978).
16. A. Abragam, "*Principles of Nuclear Magnetism*", Oxford, 1961, pg. 380.
17. D.S. Leniart, J.C. Vedrine, and J.S. Hyde, *Chem. Phys. Lett.*, **6**, 637 (1970).
18. J. Uebersfeld, A. Étienne, and J. Combrisson, *Nature*, **174**, 614 (1954).
19. D.J.E. Ingram, J.G. Tapley, R. Jackson, R.L. Bond, and A.R. Murnahgan, *Nature*, **174**, 797 (1954).
20. H. Tschamler and E. De Ruiter, "*Chemistry of Coal Utilization*", Supp. Vol. (H.H. Lowry, Ed.), Wiley, 1963, pg. 78.
21. W.R. Ladner and R. Wheatley, *Brit. Coal Util. Res. Assoc. Monthly Bull.*, **29**, 202 (1965).
22. H.L. Retcofsky, J.M. Stark, and R.A. Friedel, *Analytical Chem.*, **40**, 1699 (1968).
23. H.L. Retcofsky, G.P. Thompson, R. Raymond, and R.A. Friedel, *Fuel*, **54**, 126 (1975).
24. M.G. Northolt, *Europ. Polym. J.*, **10**, 799 (1974).
25. M.G. Dobb, D.J. Johnson, and B.P. Saville, *J. Polym. Sci. (Phys.)*, **15**, 2201 (1977).
26. K.L. DeVries and B.A. Lloyd, Final Report, Contract No. DAAG46-73-C-0251, Army Materials and Mechanics Research Center, 1975.
27. G.C. Stevens, D.J. Ando, D. Bloor, and J.S. Ghotra, *Polymer*, **17**, 623 (1976).

VII. BIBLIOGRAPHY

The following is a short bibliography of books and general articles on aspects of ENDOR spectroscopy.

ENDOR of Polymers

B. Rånby and J.F. Rabek, "*ESR Spectroscopy in Polymer Research*", Springer, 1977.
J.N. Helbert, B.E. Wagner, E.H. Poindexter, and L. Kevan, *J. Polymer Sci. (Phys.)*, **13**, 825 (1975).
S. Schlick, P.A. Narayana, and L. Kevan, *JACS*, **100**, 3322 (1978).

ENDOR Technique

A. Abragam and B. Bleaney, "*Electron Paramagnetic Resonance of Transition Ions*", Oxford, 1970.
N.M. Atherton, *Electron Spin Resonance,* **1**, 32 (1972).
L. Dalton, *Mag. Resonance Rev.,* **1**, 301 (1972).
J.S. Hyde, "*Free Radicals in Chemistry*" (Ed. L.A. Blumenfeld), Nauka, 1972, p. 24 (in Russian).
J.S. Hyde in "*Magnetic Resonance in Biological Systems*" (A. Ehrenberg, B.G. Malmstrom, and T. Vaangard, Ed.), Pergamon, 1967, p. 81.
L. Kevan and L.D. Kispert, "*Electron Spin Double Resonance Spectroscopy*", Wiley, 1976.

ENDOR Instrumentation

C.P. Poole, Jr., "*Electron Spin Resonance: A Comprehensive Treatise on Experimental Techniques*", Wiley, 1967.
R.J. Cook, *J. Sci. Instr.*, **43**, 548 (1966).
J.S. Hyde, *J. Chem. Phys.*, **43**, 1806 (1965).
E.R. Davies, *Phys. Lett.*, **47A**, 1 (1974).

DISCUSSION

B. Rånby (Royal Institute of Technology, Stockholm, Sweden): During the 1960's there was considerable work published on free radicals in lignin samples of different origin, with speculations on the nature of these radicals and their possible relations to the synthesis of lignin *in vivo*. Various lignin specimens of widely different origin and treatment often showed the same radical concentration and similar ESR spectral shapes (singlet spectra). At this time (1968), I was a visiting professor at North Carolina State University in Raleigh. We decided to study the effect of light on the formation of lignin radicals. Therefore, one dark night, we dug up a root of a pine tree and studied its radical content in the dark. There was a very low radical content in the pine lignin of this sample, never exposed to light. As soon as the pine root wood with its lignin was exposed to light (daylight or electric fluorescent light), its radical content increased rapidly to the levels reported in the literature. Storing the wood samples in the dark decreased their radical content to the level of the original "dark" value. We concluded that the lignin radicals were largely photochemically induced, presumably related to phenol–hydroquinone–quinone systems which are present in the lignin structure. These results were publishd as a short communication [B. Rånby, K. Kringstad, E. B. Cowling, and S. Y. Lin, *ACTA Chem. Scand.*, **23**, 3257 (1969)]. As far as I know there have been no more papers on free radicals in native lignin published in the literature since then.

B. Rånby: Some radicals in coal samples may be due to trapped electrons in the graphite structure. But, is it possible that some of the ESR spectra recorded for coal samples are related to photochemical radicals formed during grinding and subsequent treatments?

R. B. Clarkson: While it is possible that some EPR signals contributing to the observed line seen in coals are photochemical in origin, a significant portion of the spectral line is light-independent, as evidenced by the existence of signals in solid, uncrushed coal samples. Also, the crushed samples showed EPR and ENDOR spectra which *changed* with time after crushing and evacuation, but which did not disappear, although they were stored in the dark.

A. Yelon (Ecole Polytechnique, Montreal, Canada): Has this work on coal been published?

R. B. Clarkson: The present ENDOR work on coal has not yet been published. The best reviews on EPR of native coals may be found in recent articles by H. L. Retcofsky.

A. Yelon: Is there a rule of thumb on linewidths which might be resolved by ENDOR?

R. B. Clarkson: While it is difficult to make any totally general statements concerning ENDOR linewidths, if we assume that the line-broadening mechanism for the ENDOR line is principally T_{1e}, then the half-width at half-height of the ENDOR line, $\Delta ½$, for proton ENDOR, will be:

$$\Delta ½ = T_{1e} - \frac{1}{2.8 \text{ MHz/gauss}}$$

if T_{1e} is given in seconds.

W. G. Miller (University of Minnesota, Minneapolis, Minnesota): Kevlar can be dissolved in suitable solvents to give a locally oriented liquid crystal phase in which, judging from rheological measurements, the polymer chains are not intermolecularly connected. Have you looked at liquid crystal Kevlar solutions to see if they have EPR activity? This could distinguish between interchain versus intrachain delocalization.

R. B. Clarkson: While we have not attempted the experiment described, it is an excellent suggestion for future work.

Saturation-Transfer Spectroscopy

JAMES S. HYDE†

National Biomedical ESR Center, Department of Radiology, The Medical College of Wisconsin, Milwaukee, Wisconsin, U.S.A.

Nitroxide radical spin-labels have been used to gather motional information in three ranges, as follows:

Fast	$10^{-11} - 10^{-9}$ sec
Slow	$10^{-9} - 10^{-7}$ sec
Very slow	$10^{-7} - 10^{-3}$ sec

In the fast tumbling domain, very simple theories such as that of Stone *et al.*[1] permit determination of the correlation time. In the slow tumbling domain, more sophisticated theoretical approaches, mostly due to Freed,[2] are used. The time scale of the "fast" and "slow" categories is established by the inverse of the anisotropies of the magnetic interactions.

Saturation-Transfer Spectroscopy is concerned with the *very slow* domain. As motion slows, the usual ESR spectra asymptotically approach a line shape that is characteristic of an immobilized unoriented powder. To a good approximation, this limit is reached at rotational correlation times of 10^{-7} sec and longer.

If an intense saturating microwave field is incident on a sample, then rotational diffusion gives rise to spectral diffusion of saturation because of the anisotropy of the magnetic interactions. It is this *spectral diffusion of saturation* that gives rise to the experimental observables that permit measurements to be made of rotational diffusion in the very slow tumbling domain.

Microwave power saturation phenomena depend on T_1, the spin-lattice relaxation time of the nitroxide. We have discovered that T_1 of nitroxide radical spin-labels is about 10^{-5} sec in the very slow tumbling domain and is independent of motion. The effects depend on the extent of spectral diffusion

† Supported by NIH Grants 5 R01 GM22923 and 5 P41 RR01008.

that can occur in the basic memory time of the spin system, which is of course T_1.

While several experimental methods based on the saturation transfer concept can be employed to make motional measurements in the 10^{-3} to 10^{-7} range, the V_2' second harmonic absorption out-of-phase display has turned out to be the most practical approach. Here one uses magnetic field modulation of 50 kHz and detects at 100 kHz—the second harmonic. The out-of-phase condition is defined with respect to the in-phase setting of the reference phase of the phase sensitive detector where the ordinary second derivative is detected. The time scale of the very slow tumbling experiments is established by a combination of T_1, the field modulation frequency, and the inverse of the anisotropies of the magnetic interactions.

Theoretical simulations of saturation transfer spectra have been carried out with excellent agreement. In general, however, motional assignments are made by comparison with spectra from model systems. Maleimide labeled hemoglobin in glycerol-water solution has been chiefly used as a model system.

Saturation-Transfer Spectroscopy has been reviewed in two recent articles.[3,4] The basis of the method is presented in the standard reference by Thomas, Dalton, and Hyde.[5]

The range of applications is increasing rapidly. Listed below are motional problems studied by saturation transfer spectroscopy that may have analogies in the field of organic polymers:

Isotropic rotational diffusion. [Hemoglobin in glycerol water.][5]

Anisotropic rotational diffusion of a rigid rod. [S–1 myosin fragment.][6,7]

Segmental flexibility of one part of a protein of molecular weight 500,000 with respect to the rest of the protein. [S–1 head group of myosin.][6,7]

Flexibility of exterior active groups or crossbridges with respect to a macromolecular array of 5×10^7 molecular weight. [S–1 myosin head group in thick filaments.][6,7]

Cooperative stiffening of a polymeric array during titration of a reagent. [Actin thin filaments; myosin as titrating reagent.][8]

Polymeric growth. [Sickle cell hemoglobin.][4]

Tumbling of nitroxide *probes* as in clathrate cages.[4]

The author is not trained in the field of organic polymers, but it appears to him that the technique of Saturation-Transfer Spectroscopy will be of very significant utility in future years in this area.

REFERENCES

1. T.J. Stone, T. Buckman, P.L. Nordio, and H.M. McConnell, *Proc. Natl. Acad. Sci. USA*, **54**, 1010 (1965).

2. J.H. Freed in "Electron Spin Relaxation in Liquids" (L.T. Muus and P.W. Atkins, eds.), Plenum Press, New York (1972).
3. J.S. Hyde, Saturation-Transfer Spectroscopy in "Methods in Enzymology", (C.H.W. Hirs and S.N. Timasheff, eds.), **49G**, No. 19, pp. 480–511, Academic Press, New York (1978).
4. J.S. Hyde and L.R. Dalton, Saturation-Transfer Spectroscopy in "Spin-Labeling: Theory and Applications, Vol. 2", (L.J. Berliner, ed.), Academic Press, New York, pp. 1–70, (1979).
5. D.D. Thomas, L.R. Dalton, and J.S. Hyde, *J. Chem. Phys.*, **65**, 3006 (1976).
6. D.D. Thomas, J.C. Seidel, J.S. Hyde, and J. Gergely, *Proc. Natl. Acad. Sci. USA*, **72**, 1729 (1975).
7. D.D. Thomas, J.C. Seidel, J. Gergely, and J.S. Hyde, *J. Supra. Mol. Struct.*, **3**, 376 (1975).
8. D.D. Thomas, J.C. Seidel, and J. Gergely, *J. Mol. Biol.*, (in press).

DISCUSSION

P. Törmälä (Tampere University of Technology, Tampere, Finland): Do you think that it is possible to obtain quantitative information about label motions in polymer solids and melts by means of nonlinear methods?

J. S. Hyde: Yes. One approaches the "quantitative" aspect of your question at several levels of approximation: One can easily determine the correct order of magnitude under the assumption of isotropic rotational diffusion. More detailed analysis of saturation transfer spectra is now possible assuming anisotropic Brownian rotational diffusion with co-linear magnetic and diffusion axes. At a still higher level of motional detail, current research is directed towards understanding the effects of anisotropic restoring potentials. It should be emphasized that such detailed motional information cannot be obtained in the usual spin-label technique where motional narrowing occurs.

L. J. Berliner (Ohio State University, Columbus, Ohio): What is the practical level of decrease in the sensitivity of saturation transfer spectra vs. normal immobilized X-band spectra?

J. S. Hyde: The question of relative sensitivity cannot be simply answered, but consider these aspects:

1. The ordinary immobilized spectrum is obviously decreased in peak-to-peak amplitude relative to a fast tumbling spectrum.
2. The saturation transfer spectra are obtained under saturating conditions (~ 60 mW), whereas normal in-phase spectra are usually obtained at lower powers (~ 5 mW).
3. The out-of-phase dispersion displays (U_1') are similar in amplitude to the in-phase (U_1), both measured at 60 mW.
4. The second harmonic displays are generally about 5 times lower in amplitude than the first harmonic displays.

5. The practical limit determined by D. D. Thomas is 10^{-5} M in spin label concentration in aqueous media. Sensitivity would be much higher with samples of lower dielectric loss.

6. If one is interested in detecting a special anisotropic motion, one has much higher sensitivity in the very slow tumbling domain, since higher frequency motion tends to obscure such details.

7. If one is satisfied with paramaterization of the saturation transfer spectra by measuring signal amplitudes at just two spectral positions, then the available time can be used more efficiently and the effective signal-to-noise ratio is higher.

R. B. Clarkson (Varian Associates, Palo Alto, California): Have you observed changes in τ_c via 35 GHz 2nd harmonic out-of-phase Saturation Transfer EPR at the phase transition temperatures of spin-labeled membrane systems, and with what success, relative to X-band ST-EPR?

J. S. Hyde: Yes. Motion in spin-labeled membranes is highly anisotropic and Q-band is a better frequency for investigating anisotropic motions. The so-called "pretransition" of dipalmitoyl phosphatidyl choline gives a dramatic change in the C'/C ratio corresponding to a change in $\tau_{\|}$ between 1 and 2 orders of magnitude occurring over just a few degrees centigrade.

A. M. Bobst (University of Cincinnati, Cincinnati, Ohio): Why can pre-transitions be more readily seen with the immobilization parameters H''/H, C'/C, and L''/L?

J. S. Hyde: The C'/C ratio is primarily determined by motion *about* the p_z axis (so-called x-y averaging) of the nitroxide, while the H''/H and L''/L ratios are primarily determined by rotational diffusion of the p_z axis about x or y. This is of course just a rough approximation. The spin-labeler may have the option of using labels with different orientations of the magnetic axes with respect to the diffusional axes.

E.S.R. Study of Free Radicals in Electrical Trees in Polyethylene

J. DIB, O. DORLANNE, M. R. WERTHEIMER, A. YELON

Départment de génie physique, Ecloe Polytechnique, P.O. Box 6079, Station A, Montreal, Quebec H3C 3A7, Canada

G. BACQUET

Université Paul Sabatier, Laboratoire de Physique des Solides, 118 route de Narbonne, 31077 Toulouse Cedex, France

and

J. R. DENSLEY

National Research Council, Montreal Road, Building M-50, Ottawa, Ontario K1A 0R8, Canada

The electron spin resonance (ESR) spectrum of polyethylene samples containing electrical trees (or, where appropriate, breakdown channels) is found to be rich in structure. Much of this structure is correlated with the conditions under which the tree was produced, thus providing valuable information on the treeing phenomenon. The most reproducible feature, a sharp singlet near the free electron g value (2.0023) is due to char formation.

Details of this work may be found in *IEEE Trans. Electr. Insul.*, **13**, 157 (1978), and in IEEE Conf. Record of 1978 IEEE Int. Symp. on Electr. Insul., p. 134.

DISCUSSION

D. B. Losee (Philip Morris Research Center, Richmond, Virginia): Has the iron content of the sample been determined? The presence of the 4.24 resonance is certainly suggestive of Fe(III) in some appropriately distorted site, while the broad resonance might be clustered Fe(III).

A. Yelon: The "high purity" material contains exceedingly little iron. We cannot exclude the possibility of an Fe(III) signal, but it seems very unlikely, especially if we suppose that it must be at or near the surface of a tree channel, giving us extremely little material.

A. T. Bullock (University of Aberdeen, Old Aberdeen, Scotland): Following Losee's suggestion that the $g = 4$ signal comes from iron, I would like to suggest a probable source of this, namely evaporation from your probe needle under discharge conditions.

A. Yelon: This seems very unlikely, as we would expect the signal to get stronger and stronger as the treeing continues. This is not the case. We saw $g = 4.24$ only at the earliest stage of treeing. We cannot, however, exclude the possibility that some metal from the needle enters the polymer. We have recently done neutron activation analysis experiments which show that metal diffusion in polymers is much greater than one might expect.

B. Rånby (Royal Institute of Technology, Stockholm, Sweden): I have several comments to make. Firstly, it seems possible that some of the ESR spectra recorded for crosslinked polyethylene are due to peroxy radicals. Crosslinked polyethylene is known to contain peroxy groups if the crosslinking agent used is a peroxide or hydroperoxide. Secondly, the sharp singlet ESR spectrum observed may be due to trapped electrons, e.g., as shown by Sohma in sudies of mechanically degraded and γ-irradiated polyethylene and polypropylene. Finally, ESR spectra with g-values of ~ 4 are assigned to radical pairs with a separation of a few Å units. These spectra are due to the forbidden transition, $M_s = 2$.

J. Sohma (Hokkaido University, Sapporo, Japan): Your attribution of the sharp singlet to char seems to me doubtful, it is well known that the singlet from chars made from sugars is hardly observed in an air atmosphere. Don't you think it is rather reasonable to assign this singlet to electrons trapped in the polymer matrix?

A. Yelon: We considered the possibility that the signals with $2.0022 \leq g \leq 2.0049$ are due to trapped electrons; we also considered the possibility of polyenyl radicals. However, we feel that the evidence we presented in an earlier report [*IEEE Trans. Electr. Insul.,* **13**, 157 (1978)] for char is very strong. The relationship to tree color, the long life, the range of g values, the linewidths, and the saturation behavior all point to char.

R. F. Boyer (Midland Macromolecular Institute, Midland, Michigan): Could you learn something about the nature of the radicals by exposing the tree to a nitroxide trapping molecule, e.g., di-*t*-butyl nitroxide?

A. Yelon: We had not thought about this before this week, but it seems that this might work.

L. J. Berliner (Ohio State University, Columbus, Ohio): Could you dope the polyethylene with spin traps in their preparation?

A. Yelon: It ought to be possible and apparently this has been done by Kusumoto.

A. T. Bullock: It should be possible to produce your trees, then immerse the cable in a solution of a spin trap. Similar experiments have been done by dissolving irradiated crystals in solutions of spin traps.

L. J. Berliner: The Russians have used nitroxides as antioxidants. If nitroxides were diffused into these polyethylene cables and if peroxy radicals or oxidation products are responsible for your ESR signals, the lines should diminish or disappear.

A. Yelon: We take precautions to try to avoid oxygen. We melt the polyethylene under vacuum, and have done our treeing of pure material under argon. This, of course, does not mean that we have eliminated all oxygen, and large quantities of antioxidant might change the situation. However, Törmälä has told me that he found that nitroxides do not inhibit the ozonation of rubber.

GENERAL DISCUSSIONS

Spin Labels vs. Spin Probes vs. NMR (^{1}H, ^{13}C, ^{19}F)

Session Chairmen: A. T. BULLOCK and J. SOHMA

A. T. Bullock (University of Aberdeen, Old Aberdeen, Scotland): In looking at a comparison of NMR and ESR techniques for studying polymers, I would like to start by pointing out that for sensitivity reasons, NMR can rarely tell us anything about chain end relaxations. Although, having worked in the spin-labeling area for eight or nine years now, I must say that for studying segmental relaxation, especially in solution, ^{13}C relaxation techniques have many benefits not shared by the spin-labeling technique.

While we were working on spin labeled polystyrene,[1] it wasn't clear until we finished the study that we were not perturbing the chain but were in fact looking at isotropic motion of the label. Allerhand and Hailstone[2] published results of their ^{13}C relaxation work at about the same time. They were able to look at every position in the molecule and calculate correlation times from T_1 measurements. The correlation times for every carbon atom were the same with any variation between the ring and main chain positions being of a purely random nature. Their results agree quite well with our correlation times even though we haven't allowed for the different viscosities of the two samples.

We can forget about proton relaxations in solution; even in fully deuterated solvents intermolecular interactions can make a substantial contribution to the overall relaxation rate and make it difficult to extract information about intramolecular modes of motion.

R. F. Boyer (Midland Macromolecular Institute, Midland, Michigan): Sillescu[3] and co-workers have deuterated the ring and the main chain of PS and have followed the motion of each group by NMR as a function of temperature and concentration in several solvents. From this work they have deduced that around 180–200°C both ring and main chain motions are equally facile.

A. T. Bullock: As mentioned, hydrogen doesn't look too good but it may be the only way to get T_1's in the solid state by NMR. Fluorine is a bit complicated in that you are dealing with several correlation functions. There's a dipole-dipole interaction between fluorine and every other magnetic nucleus; we also have the anisotropic chemical shielding tensor. Over the past few years we have seen a considerable upsurge of ^{13}C NMR work; but it still cannot give us any information about chain end relaxations. The one great advantage of ^{13}C is that it is totally dominated by the dipole-dipole interaction of the ^{13}C and the protons bonded to just that carbon.

L. J. Berliner (Ohio State University, Columbus, Ohio): Berendsen's group[4] in Groningen have been looking at protons presumably on protein surfaces in aqueous media. They have encountered enormous problems with spin diffusion, or cross relaxation, where all of the protons on the macromolecule appear to have the same relaxation time because they are relaxing out to the surface and through the solvent. The way in which this can be detected is to look at the very, very early stage of the decay in a relaxation experiment, in the microsecond and tenths of microsecond range. I can't go into any great detail but just wanted to alert people to this work.

J. Sohma (Hokkaido University, Sapporo, Japan): Because proton resonance is my old field, I feel some obligation to say something about its usefulness. There was some work done on poly(vinyl carbazole); this system has very bulky side groups and forms a very tight helix with the carbazole units stacked above each other. Correlation times from ^{13}C relaxation data gave information about the motion of the chain itself. From proton NMR using deuterated solvents we can get an additional relaxation time arising from dipole-dipole interactions between protons on adjacent carbazole moieties.[5] If the chain is in a random coil state this interaction is quite negligible. By adjacent, I don't mean in the next repeating unit but rather the closest approach position of two carbazole units in the helix.

G. G. Cameron (University of Aberdeen, Old Aberdeen, Scotland): I would like to point out that if you normalize the ^{13}C and ESR data for polystyrene to the same viscosity they agree very well.[1,2] I would agree with Bullock, that for studying segmental motion, ^{13}C is probably superior, but to examine motions at specific parts of the chain, not necessarily just end groups, then the ESR method may have something to offer.

L. J. Berliner: There are some fancy pulsed techniques in proton NMR for selecting out differnt ranges of relaxation time in multinuclear samples, nuclei say which have different domains of T_1's or T_2's. These techniques have been developed by DeMarco and Wütrich[6] and others recently with proteins. I think there may be some hope yet for looking at segmental motions by proton NMR, utilizing some good computer handling techniques.

R. F. Boyer: I'd like to make a few comments here. I've had the experience in talking about our ESR work that the NMR spectroscopists always seem to jump on us saying that we perturb the system by adding a probe or label, whereas NMR is looking at an unperturbed system by observing parts of the molecule naturally present. And then, Smith tells me we are foolish to use probes, as only labels can give a true understanding of the molecular motions. But, the other day we heard Törmälä say that there is essentially no difference between a probe and a label. Sohma came here talking about a comparison of ^{13}C NMR and ESR indicating that no serious perturbation was present in his system. Finally, we have Bullock's comment that ^{13}C may be alright but ^{1}H and ^{19}F spectroscopy may not be useful.

L. J. Berliner: I'd like to mention a new technique we've been using recently with proteins where sensitivity is a problem. It's a surface probe technique such as at a liquid–solid or solid–solid interface. The specific example is the side chain motion of the amino acid tyrosine which may be similar to say a phenolic type polymer. If this unit is rotating rapidly we would have several pairs of equivalent protons. If the unit is not rotating, as is sometimes the case, the protons would be non-equivalent, if we could sufficiently resolve them.

The new technique is called laser photo CIDNP—chemically induced dynamic nuclear polarization, and requires that one have essentially a transient radical pair. This requires that the repeating unit studied in the macromolecule be capable of forming a radical by an extraction mechanism.

In our particular case we used a flavin dye, which was photoexcited by the laser using a flat bottom NMR tube and a system of mirrors.The flavin is excited to the singlet state and goes to the triplet state by intersystem crossing. The excited triplet presumably abstracts a proton from the tyrosine unit to give us two radicals which summarily simply recombine after a certain time; this is essentially the CIDNP process.

The person responsible for the development of this general technique is Kaptein.[7]

REFERENCES

1. A.T. Bullock, G.G. Cameron, and P.M. Smith, *J. Phys. Chem.*, **77**, 1635 (1973).
2. A. Allerhand and R.K. Hailstone, *J. Chem. Phys.*, **56**, 3718 (1972).

3. H. Sillescu, *et al.*, *Makromol. Chem.*, **178**, 1445, 2401 (1977).
4. A. Kalk and H.J.C. Berendsen, *J. Magn. Res.*, **24**, 343 (1976).
5. N. Tsuchihashi, M. Hatano, and J. Sohma, *Makromol. Chem.*, **177**, 2739 (1976).
6. A. DeMarco and K. Wütrich, *J. Magn. Res.*, **24**, 201 (1976).
7. R. Kaptein in "11th Jerusalem Symposium on Quantum Chemistry and Biochemistry", Reidel-Dordrecht, Holland, in press (1978).

Spin Labeling Methods for Synthetic Polymers, Characterization of Molecular Motion, Industrial Applications of ESR and Quality Control, and Cost/Benefit Considerations

Session Chairman: L. J. BERLINER

L. J. Berliner (Ohio State University, Columbus, Ohio): Today we would like to discuss methods of labeling polymers, presumably evaluating what's been done to date and future prospects. We should think about some of the new nitroxides and methods available to incorporate the label either onto preformed polymer or through copolymerization schemes.

An evaluation of present labeling methods might be best reflected in what we would like to have in the future both from a chemical and physical viewpoint for a better means of studying structural and motional aspects of polymers. One point I think is important would be to insert the nitroxide moiety directly into the main chain of the polymer, instead of what appears to be more side chain labeling up to now. More direct information about the main chain rather than the spin label itself should be obtained in this way.

W. G. Miller (University of Minnesota, Minneapolis, Minnesota): Incorporating the label into the chain backbone I feel would have an opposite effect. It will certainly allow measurement of motion right near the nitroxide unit, but it will be less able to report on what the rest of the molecule is doing, and thus will not be as representative of the whole polymer as a tethered label would be.

L. J. Berliner: What if you have several labels distributed along the polymer backbone and maybe several different kinds as well?

W. G. Miller: You will only be measuring backbone motion unless the labels are separated from the chain by at least a few bonds.

R. F. Boyer (Midland Macromolecular Institute, Midland, Michigan): Miller has published work on a styrene-methacrylate copolymer with the methacrylate unit labeled.[1] Could we have a comment on your results?

W. G. Miller: In solvents we were to a first approximation monitoring only the motion of the methacrylate unit and were not getting any information about the styrene units nearby. In the solid state, it might be something different though.

A. T. Bullock (University of Aberdeen, Old Aberdeen, Scotland): We may be perturbing the chain by inserting a label directly. One should take a flexible approach to this problem and put the group in several positions and perhaps change the pendant group as well. You obviously need to check your experiment for perturbation possibilities.

L. J. Berliner: As a rule of thumb we always use several kinds of labels in terms of pendant groups and never only one. I would say that if three pendant groups agree, then you are probably observing motions of their particular environment and not the motions of the label itself.

A. T. Bullock: I don't think one can be entirely dogmatic about this. It obviously depends on the polymer system you are dealing with and what kind of chemistry can be done on it. In the case of polystyrene by labeling various positions in the ring one can elucidate motions ascribed to main chain, phenyl ring, and those about the nitroxide nitrogen–phenyl ring bond.

L. J. Berliner: The reason I brought up the point for directly inserting the label stems from membrane studies, in which it is felt that a linear labeled molecule would be better in that it could associate more closely with neighboring chains and thereby reflect motions of the whole assembly. Whereas, if the label were pendant it would probably only reflect very local motions.

W. G. Miller: Such a consideration would probably only apply when discussing the motion in highly crystalline polymers. In most cases we are dealing with quite amorphous systems and this point of direct insertion is not as critical.

P. L. Kumler (State University of New York, Fredonia, New York): Are you suggesting, using a nitroxide, that one incorporate the nitrogen into the backbone of the polymer?

L. J. Berliner: Right, essentially I would envision the case of having di-*t*-butyl nitroxide, where instead of methyls you just continue on in the chain as a case of minimum perturbation.

P. L. Kumler: Has anyone made any systems like that?

L. J. Berliner: The azethoxyls come close but they have that bridge on them. I think there's been a few analogs of di-*t*-butyl nitroxide made but so far they are all fairly short chain molecules.

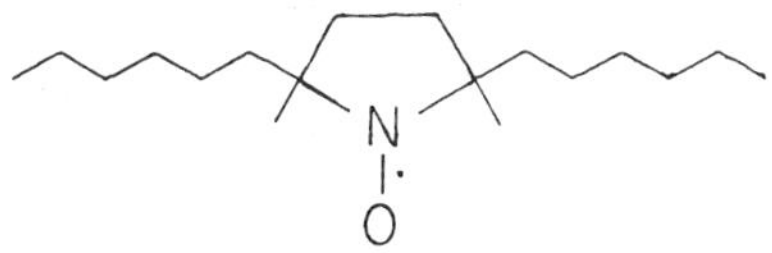

N
|·
O

G. G. Cameron (University of Aberdeen, Old Aberdeen, Scotland): There has been one synthetic polymer labeled in a manner similar to what you describe although it was done probably more by accident than by design. On heating amine-terminated nylon 6 in air secondary amine groups are formed, possibly by deamination between two amine ends, and on oxidation these lead to nitroxide groups within the chain. There is no bridge in this case.[2]

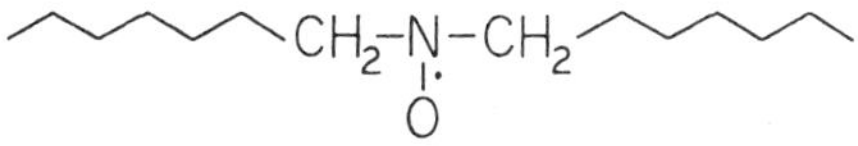

CH_2-N-CH_2
|·
O

L. J. Berliner: That can in fact be done with several polyamides. They're not particularly stable but in aprotic solvents will last long enough to be able to do simple experiments on them.

W. G. Miller: The polymer of polyethylene which Bullock, *et al.*[3] synthesized with carbon monoxide is the best literature example for coupling a label right into the chain. But to say that this will exactly represent polyethylene motion is like saying polyethylene and poly(ethylene oxide) have the same chain dynamics, inasmuch as the ESR spectrum is a reflection of nitroxide motion.

P. Törmälä (Tampere University of Technology, Tampere, Finland): I would like to divide this question into two parts, studies in solution and studies of the solid state, in these I see principal differences. We haven't worked on spin labeled solution studies but our data on labeled solids and polymer melts seems to be following solution study data. In the case of melts or solids there is no real difference between probes and labels, one reason may be the tight structure of the polymer solid such that these bulky nitroxides cannot move without the cooperative motion of the surrounding polymer segments. Thus, in solids it is best to use probes since we require cooperative segmental motion anyway, we will not be introducing perturbation into the chain. It is necessary to use different probes to eliminate the effects of radical structure and size from the data.

I would like to emphasize that the points on a frequency map from ESR data are very accurate in comparison with other measurement techniques, e.g., dielectric or mechanical techniques. We have utilized a broad range of probes and labels in studying polyethylene and have found that the data for quite different types of probes are all about the same, suggesting little practical difference between various types of labels.

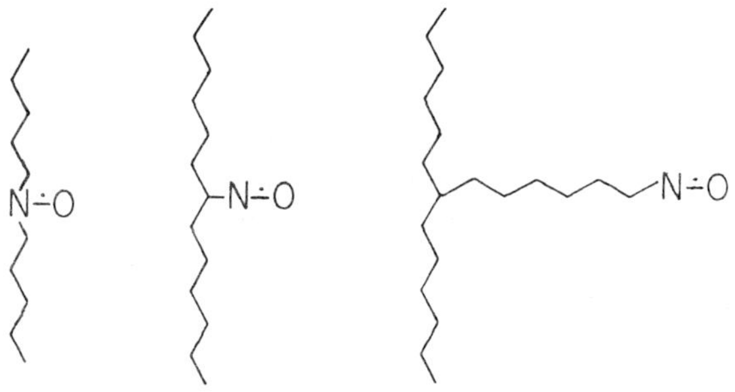

B. Rånby (Royal Institute of Technology, Stockholm, Sweden): Up to this point, I have not seen an example of a label being small enough to be incorporated into a polymer crystal lattice, without either disturbing the lattice or being excluded from it. There is of course great interest in being able to incorporate such a small label that could crystallize into the lattice and report on the mobility present there. There should be serious attempts made to make such a labeled system; even if we have to work on polyamide systems it would still be of interest. We would most probably need the nitrogen in the chain and the oxygen outside as nitrogen has the available valences.

P. L. Kumler: We would like to make a system with the nitroxide nitrogen in the chain

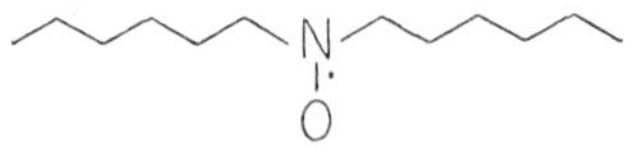

Perhaps the best way to do this is by a spin trapping technique. The basis behind a spin trap experiment is that a radical having a relatively short lifetime is reacted with an appropriate spin trapping molecule to form a substantially more stable radical.[4]

The most common kinds of spin traps used are either the nitroso compounds or the nitrones. Although the chemistry involved is similar, there are some importance differences.

$$R\cdot + O{=}\ddot{N}{-}R' \longrightarrow R{-}N(O\cdot){-}R' \quad \text{(nitroso spin trapping)}$$

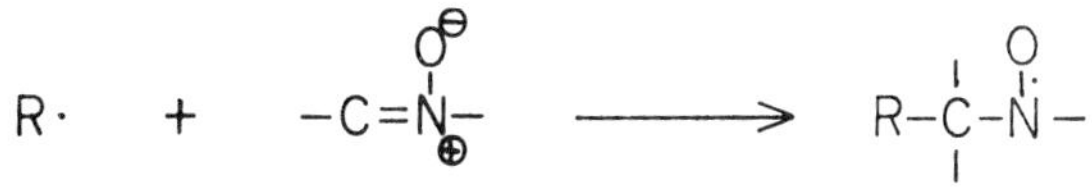

$$R\cdot + {-}C{=}N^{\oplus}(O^{\ominus}){-} \longrightarrow R{-}C{-}N(O\cdot){-} \quad \text{(nitrone spin trapping)}$$

If you could react a polymer end labeled with a nitroso group with a growing polymer radical you would get a polymer having the nitroxide label in the middle of the chain. Simply reacting a growing polymer radical with nitroso-*t*-butane would result in an end labeled polymer. Nitrones can be used similarly, in this case the free radical addition occurs at carbon rather than at nitrogen as in the nitroso compounds.

Some common nitroso and nitrone compounds are shown here.

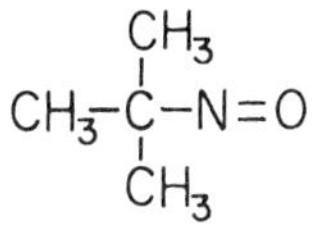

$(CH_3)_3C{-}N{=}O$ — *nitroso-t-butane*

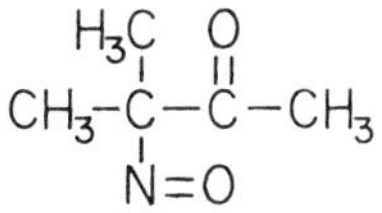

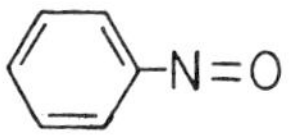

$CH_3{-}C(CH_3)(N{=}O){-}C({=}O){-}CH_3$

$C_6H_5{-}N{=}O$ — *nitroso benzene*

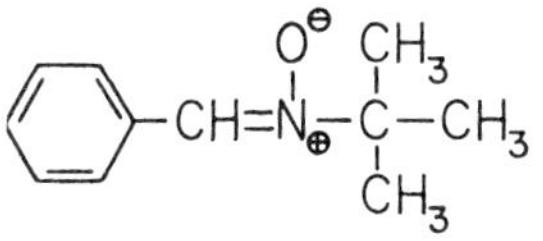

$C_6H_5{-}CH{=}N^{\oplus}(O^{\ominus}){-}C(CH_3)_3$ — *phenyl-t-butyl nitrone*

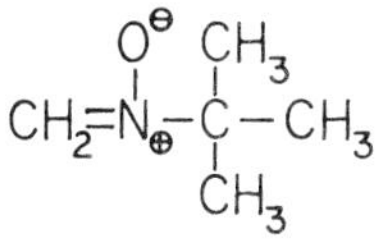

$CH_2{=}N^{\oplus}(O^{\ominus}){-}C(CH_3)_3$ — *t-butyl nitrone*

There are a few distinct differences in the spin adducts formed from these two spin trapping systems. In the nitrone spin adduct the trapped radical is attached to carbon and thereby removed one atom from the paramagnetic center. In the nitroso spin adduct the trapped radical is added onto the nitrogen directly.

The two adducts give different spectroscopic information because of their different chemical structures. The nitroso adduct gives better structural information about R because it is directly attached to nitrogen. For example, the hyperfine splitting pattern will indicate directly how many protons are attached to the carbon atom adjacent to nitrogen. In the nitrone experiment, you're one more atom removed and you don't get this kind of information.

One can trap most any kind of available radical, not necessarily just carbon radicals, and if it possesses a net nuclear spin an additional hyperfine splitting would be observed. With carbon radicals you can get information from the magnitude of the nitrogen hyperfine as to what kinds of substituents are attached to carbon, whether they're electron donating or electron withdrawing. Typical coupling constants are on the order of 13–16 gauss for carbon radicals; the specific value depends on the substituents attached. A trapped alkoxy radical gives a nitrogen hyperfine splitting on the order of 27–28 gauss. Sulfur radicals will form spin adducts with typical coupling constants of 17–19 gauss. A complication in using nitroso compounds is their tendency to form dimers. Some nitroxide spin adducts may not be particularly stable, e.g., the acyl radical spin adduct; most carbon based adducts are generally quite stable though. A possible side reaction involves the homolysis of the nitroso compound either photochemically or thermally to give a carbon radical and N—O; the carbon radical readily reacts with another introso molecule to give a very stable adduct.

G. G. Cameron: The idea of using a polymer with a nitroso end group is a good one if you want to place the nitroxide in the middle of the chain, but the synthetic route proposed, i.e., via a growing polymer radical, is not advisable as many such radicals will die by natural termination processes rather than react with the nitroso groups. A better way to do this is through a carbanion or "living" polymer system. Nitroso compounds will also react with carbanions as we have a positive center in the nitroso molecule.

$$R^{\ominus} + R'\!-\!N\!=\!O \longrightarrow R\!-\!\underset{R'}{\underset{|}{N}}\!-\!O^{\ominus} \xrightarrow[\text{oxidation}]{\text{hydrolysis}} R\!-\!\underset{R'}{\underset{|}{N}}\!\dot{-}\!O$$

The product from the first step is a salt of a hydroxylamine (the counterion has been omitted) which is readily hydrolyzed and oxidized to the nitroxide. If R and R′ are both polymeric then we have a nitroxide radical as part of the backbone.

P. L. Kumler: The dilemma remains however, if we could make a labeled polyethylene this way, we really don't have polyethylene. I would think this would be a very severe perturbation, even more so than by using spin probes.

J. Sohma (Hokkaido University, Sapporo, Japan): One method to induce mechanical damage in polymers is by ultrasonic irradiation in solution causing chain scission. The radical chain ends can be trapped using spin trapping agents such as the nitroso compounds Kumler just mentioned; my favorite spin traps include

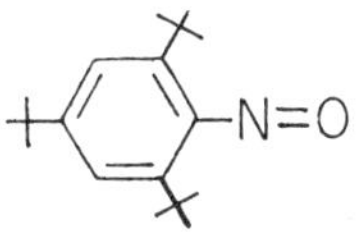

(BNB)

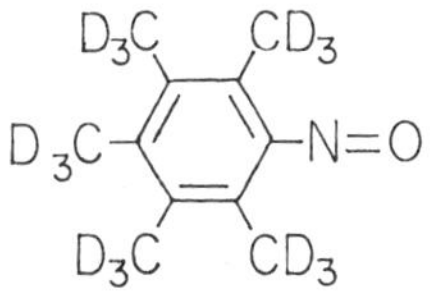

After chain scission in polypropylene two different spin adducts are formed. One spin adduct

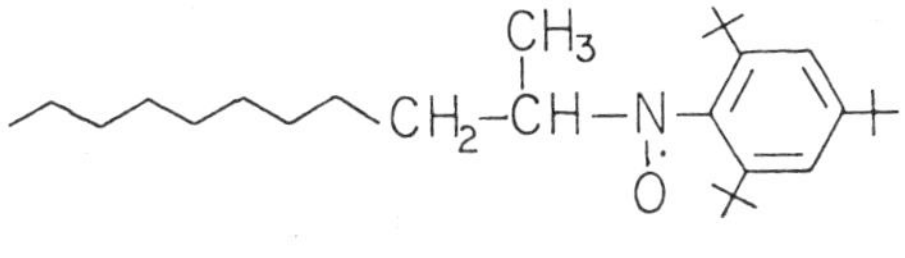

gives a double triplet spectrum since the nitrogen hyperfine is split by a single adjacent hydrogen atom. The other spin adduct

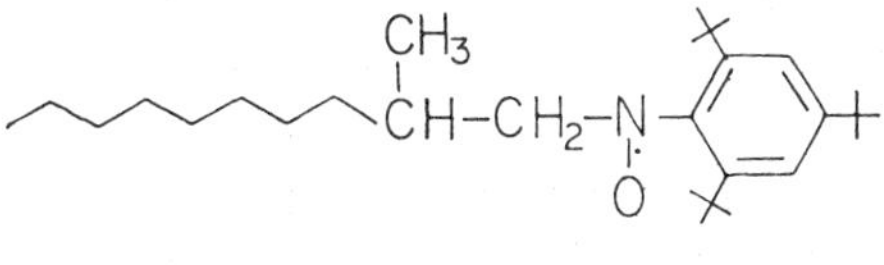

gives a triple triplet spectrum; here two adjacent hydrogen atoms contribute to the hyperfine structure. We have found this to be a very easy way to end label polypropylene.

A. T. Bullock: While this is an interesting example of spin labeling, it is probably of more use in telling us something about ultrasonic degradation rather than being a useful spin labeling technique. I would guess that ultrasonic degradation would give a whole series of chain lengths. If you're interested in obtaining correlation times, you will run into problems when you get down below 100 units, as there is a chain length dependence on correlation time below that point. I suspect what you'll observe is a distribution or envelope of correlation times.

J. Sohma: This may be possible but it is not very probable because we maintain the same molecular weight distribution before and after ultrasonic irradiation.

G. G. Cameron: I would have to disagree. You could end up with the same distribution only if you began with the most probable distribution and chain scission was purely random. It is well known, however, that ultrasonic degradation is not a random process; long chains break in preference to short ones.

J. Sohma: We have checked our samples using gel permeation chromatography and the molecular weight distribution seems to remain unchanged.

N. Kusumoto (Kumamoto University, Kumamoto, Japan): We have done experiments on spin labeling polyethylene using spin traps and γ irradiation at low temperatures. We prepare a solvent swelled sample, add a nitrone spin trapping agent, cool it down to liquid nitrogen temperature, and γ irradiate it. The ESR spectrum at this temperature shows a sextet arising from the chain radical specie $\sim\!\sim CH_2-\dot{C}H-CH_2 \sim\!\sim$
As the sample is heated up to room temperature the spectrum becomes the typical three-line asymmetric spectra of a nitroxide. The temperature at which spin trapping starts to occur is around $-50°C$, the point at which increased molecular motion begins.

R. F. Boyer: Getting back to Miller's experiment on spin labeled methacrylate copolymerized with styrene, in the bulk you probably don't have just a single styrene unit rotating but perhaps 3, 4, 5, or 6 units—some statistical segment length. I would think this might be the situation near T_g, but certainly at higher temperatures in the melt or in solution localized crankshaft motion would be possible.

W. G. Miller: In my view, in the bulk significantly below T_g there is no significant motion about the backbone, the label is simply rotating about its tether bond. As T_g is approached backbone motions begin to come in and characterization of the motion becomes impossible.

L. J. Berliner: How many neighboring units are actually influencing the motion of the label in solution and in the bulk above and below the glass transition?

W. G. Miller: In solution, the influence of one or so monomer units on either side of the labeled unit may be reflected. In the bulk it depends on what the label is and below the T_g on what kind of motion is present. We haven't seen any real backbone motion below T_g with our labeled systems and as we approach T_g the motion becomes complex and very difficult to characterize.

A. T. Bullock: Spin labeling and ^{13}C measurements in solution show that for polystyrene certainly and for one or two other polymers the motion is isotropic. In opposition to Miller's statement, no way can you get isotropic motion about the label just by the rotation of one unit on either side, I would guess that you would need three or four units on either side before getting anything like isotropic motion.

L. J. Berliner: There is a way to verify this question using these cyclic nitroxides—the doxyls, proxyls, and azethoxyls—by incorporating them with their different preferred axes into the chain. If the chain motion were of an isotropic nature, it would show up differently for each label.

A. T. Bullock: Another way to estimate how many units are involved before isotropic motion can occur would be by looking at an ABA triblock copolymer, where the A blocks are very rigid and the B block bears the spin label. If you could vary the length of B down to a few monomer units and look at the variation of correlation time you should be able to get at least a semiquantitative answer. Needless to say the practical chemistry to do this has not been worked out yet.

G. W. Eastland (Saginaw Valley State College, University Center, Michigan): Kumler mentioned their T_{50G}–T_g correlation curve constructed from 19 different polymers using BzONO. However, they've found that a few polymers don't seem to fit the correlation and one in particular is poly(methyl methacrylate). The T_{50G} consistently came out too high and there wasn't a good explanation for why. The PMMA sample originally used was probably prepared by solution polymerization with a calcium acetate suspending agent. The idea we now have is that trace amounts of calcium were acting as a catalyst to transesterify our spin probe onto the polymer. We started out with a spin probe in PMMA and by accident may have spin labeled the polymer.

We've looked at a series of PMMA's prepared by different means with two probes, TEMPONE and BzONO. In the samples free of transition metal impurities we get the proper T_{50G} and corresponding T_g for PMMA. With TEMPONE, which is incapable of transesterification, our data seem to suggest no difference in the T_{50G} for any of the various PMMA samples.

R. F. Boyer: Sohma's paper on spin labeled PMMA[5] in which his label was essentially the nitroxide portion of our BzONO probe was the clue to this observation. He reports a T_{50G} for labeled PMMA of 146°C which is just what we were getting with supposedly, our probe. We have a comparatively infinite amount of PMMA and trace amounts of transition metal impurities and BzONO so we may expect the labeling reaction to be driven essentially to completion. This could be a generally useful method to label polyacrylates, polymethacrylates, and copolymers.

L. J. Berliner: A topic which hasn't been mentioned yet is the possibility of taking advantage of spin-spin interactions in multiply labeled polymers to get a measure of intramolecular distances in a polymer. For example, a system undergoing mechanical deformations like stretching or bending, with say the spin labels spaced 10–20 Å apart along the chain.

A. T. Bullock: Luckhurst[6] has done some work along those lines, not with polymers but with dinitroxides having several units between them. Problems arose in determining the form of the exchange interaction as a function of the relative orientation of the two nitroxide groups.

L. J. Berliner: In a solid you would generally have immobilized nitroxides with very strong dipole-dipole interactions resulting in dipolar splittings as large as 400–500 gauss. If you have a separation of something like 6 Å you will get extrema out to about 200 gauss, when the separation is 4 Å the separation increases to 400 to 500 gauss.[7] The idea would be to use this information as a molecular strain gage;[8] even if you didn't know the exchange interaction exactly you might still be able to obtain some semi-quantitative measures.

A. T. Bullock: Kumler and I have briefly speculated on the following suggestion for an alternative labeling method which would give an iminoxyl label rather than a nitroxide label. Using a strong base one could generate a carbanion which would autoeliminate to give a carbene, reacting this then with N—O to obtain the iminoxyl radical.

$$\sim\!\mathrm{C(H)(Cl)}\!\sim \xrightarrow{B^{\ominus}} \sim\!\mathrm{C^{\ominus}(Cl)}\!\sim \longrightarrow \sim\!\ddot{\mathrm{C}}\!\sim \xrightarrow{NO} \sim\!\mathrm{C(=N\dot{-}O)}\!\sim$$

There are a couple advantages to such a label. First, we don't have free rotation of the label so we would be looking at chain motion exclusively. A second advantage is that the anisotropic coupling tensor components are considerably larger than those for nitroxides. The correlation time at, say, T_{50G} for the iminoxyl label is about half the value for the nitroxides, so the data points would appear at another frequency in the relaxation map.

L. J. Berliner: Getting back to the ABA system Bullock mentioned, it might be of value to incorporate several azethoxyl labels into the B block and study them by taking advantage of spin-spin interactions.

G. G. Cameron: The problem we encounter in these systems is the need for a pair of very hard A blocks and a very soft B block, whereas the hard blocks are usually easily labeled the soft block is generally difficult to label. It would be very easy to make an ABA block copolymer with a hard, labeled B block flanked by soft A blocks.

As a comment to the iminoxyl label Bullock just presented, if you attack poly(vinyl chloride) as suggested with base you will almost certainly have elimination of HCl. A better system to use would be an α-Cl-acrylate, which by base decomposition might generate the carbene you want.

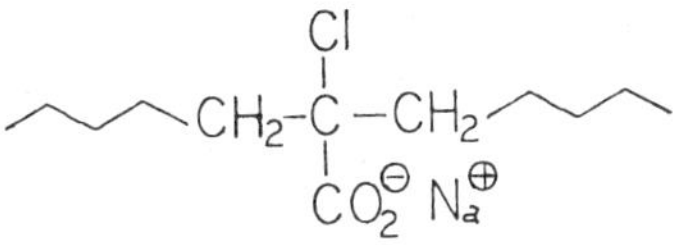

$$\sim\sim CH_2-\overset{Cl}{\underset{CO_2^{\ominus}\ Na^{\oplus}}{C}}-CH_2\sim\sim$$

L. J. Berliner: In biochemical studies we've used ^{19}F as a probe and found it quite valuable as the perturbation involved in substituting a fluorine atom for a proton is quite small.[9] A particular advantage in substituting a trifluoromethyl group for a methyl group is that you get a single line with three times the intensity.

A. T. Bullock: Fluorine may be a good structural probe but in dynamic studies my feeling is that deuterium would be an even better probe. I'd be a bit surprised if fluorine doesn't perturb the system as you're dealing with a very polar atom, whereas deuterium would have no polar effects. Also, ^{19}F relaxations are a complicated mixture of dipole-dipole interactions and chemical shift interactions. Deuterium is almost invariably determined by the quadrupole relaxation which is well defined with respect to the C—D bond.

L. J. Berliner: Anyone interested in deuterium resonance should look at the work of M. Bloom, I. C. P. Smith, and J. Seelig[10] who have done quite a bit of work recently in the area of fatty acids and liquid crystals.

One last topic we should discuss briefly is the potential of ESR as a routine, quality control instrument in industrial applications. Would it be practicable in an industrial setting for someone to take a sample and do a quick spin assay of it?

A. Yelon (Ecole Polytechnique, Montreal, Canada): General Electric is presently using ESR as a quality control technique on their ceramics production line to check essentially the oxidation state of the iron in the ceramic. Obviously, if the test required is best done by ESR and you want to do it badly enough, then it will be feasible to do even in an industrial setting.

G. W. Eastland: A basic problem in speaking about quality control is increasingly the demand to find out how much rather than what's there, and getting good quantitative measurements with ESR are fraught with problems. With very low concentrations atomic absorption or a similar technique will most of the time be just as sensitive and will give a better quantitative measure than ESR can.

R. F. Boyer: About 10 years ago Dow Chemical made a very serious attempt to apply ESR to various polymer problems. Rabold[11] tried probes in all of the thermoplastics and also in various latex systems and nowhere was he able to connect. So, Dow Chemical gave away not one but two ESR instruments to local colleges. But this has also happened with an ultracentrifuge, dynamic mechanical testing equipment with torsion pendulum, and dielectric equipment, all sorts of equipment normally considered as standard characterization tools. In all of these cases management concluded that the instruments were not showing profits, today.

B. Rånby: You cannot justify an ESR to study polymerization reactions because we know the end results can be obtained by other means less costly. However, it may be of use in evaluating material purity, e.g., in studying radical products following radiation sterilization or to insure the absence of radicals in pharmaceutical products.

P. Törmälä: We have been studying the degree of milling of cellulose fibers recently. During milling, fiber surfaces are broken and large numbers of active hydrogen groups are present which form hydrogen bonds between different fibers. The milling process must be stopped at a certain point before too many fibers are broken resulting in poor mechanical properties of the paper. Using the TEMPO probe as a surface agent to report on the state of the fibers, we have found that the correlation time changes as a function of milling time. We hope to develop this into a routine laboratory method for the paper industry.

We have also studied the plasticization of poly(vinyl chloride) and found that the correlation time changed as a function of plasticizer content. The PVC studies would seem to be important as the commercial products of this polymer are generally plasticized.

J. Sohma: I know of two industrial applications of ESR in Japan. One is to study a polymerization scheme combined with a characterization of the catalysis mechanism by a spin coupling method, and the other is a test of the peroxy radical content in dairy products.

A. M. Bobst (University of Cincinnati, Cincinnati, Ohio): The reason spin assaying was a marketable failure was not so much that it was bad but that a radio assay was developed and introduced concurrently costing substantially less. To compare costs, you can buy a scintillation counter for $10,000 to do radio labeling, whereas an ESR will cost $100,000 and upwards. The only point to contest the cost is that radio labeling is pollution whereas spin labeling is not. Up until last year Bruker sold an ER-10 for $35,000 and Varian used to sell a cheap ESR for $8000, but these instruments were unsatisfactory. If you want a good ESR you've got to pay for it and unless this price barrier can be overcome ESR will never become a viable quality control tool.

R. F. Boyer: Kumler and I have discussed the fact that ESR seems to measure number average molecular weight. Granted you have to calibrate the system, but anyone who's worked with osmometers or vapor pressure lowering or freezing point lowering of low molecular weight compounds may want to consider using an ESR if they have access to one.

A. T. Bullock: I recently refereed a research proposal where ESR might possibly be used in quality control. A certain powder that was used on surgeon's gloves had been implicated in post-operative complications. The powder was sterilized by radiation and the proposal was to look at whether radicals were formed during the sterilization procedure.

L. J. Berliner: To close this last session of the Symposium I'd like to express my thanks to Boyer for organizing this symposium and bringing the various experts together to discuss their work.

R. F. Boyer: I'd like to thank each participant for making this Symposium a truly worthy endeavor and for the very fine discussions we've had on the various applications of ESR for studying molecular motion in polymeric systems. Thank you.

REFERENCES

1. Z. Veksli and W.G. Miller, *Macromolecules*, **10**, 686 (1977).
2. T.C. Chiang and J.P. Sibilia, *J. Polym. Sci., A-1*, **10**, 605 (1972).
3. A.T. Bullock, G.G. Cameron, and P.M. Smith, *Europ. Polym. J.*, **11**, 617 (1975).
4. E.G. Janzen, *Accounts Chem. Res.*, **4**, 31 (1971).
5. M. Shiotani and J. Sohma, *Polymer J.*, **9**, 283 (1977).
6. G.R. Luckhurst, *Mol. Phys.*, **10**, 543 (1966).
7. Z. Ciecierska-Tworek, S.P. Van, and O.H. Griffith, *J. Mol. Struct.*, **16**, 139 (1973).
8. G.R. Luckhurst in "Spin Labeling: Theory and Applications", (L.J. Berliner, ed.), Academic Press, New York, p. 133 (1976).
9. L.J. Berliner and B.H. Landis in "11th Jerusalem Symposium on Quantum Chemistry and Biochemistry", Reidel-Dordrecht, Holland, in press (1978); also J.T. Gerig in "Biological

Magnetic Resonance", (L.J. Berliner and J. Reuben, eds.), Vol. I, Plenum Publ., New York, p. 139 (1978).
10. These authors have current reviews in preparation or press, the reader should check the most recent literature; J. Seelig in *Biochem. biophys. Acta, Biomembrane Reviews*, in press (1978).
11. G.P. Rabold, *J. Polym. Sci., A-1*, 7, 1203 (1969).

Author Index

Numbers preceding those given in parentheses refer to the pages on which complete references are listed; reference numbers on these pages are given in parentheses.

Subject Index